Edited by
Jürgen Popp, Valery V. Tuchin,
Arthur Chiou, and
Stefan H. Heinemann

Handbook of Biophotonics

Related Titles

Popp, J., Tuchin, V. V., Chiou, A., Heinemann, S. H. (eds.)

Handbook of Biophotonics

3 Volume Set

2012

ISBN: 978-3-527-40728-6

Popp, J., Tuchin, V. V., Chiou, A., Heinemann, S. H. (eds.)

Handbook of Biophotonics

Volume 1: Basics and Techniques

2011

ISBN: 978-3-527-41047-7

Popp, J., Tuchin, V. V., Chiou, A., Heinemann, S. H. (eds.)

Handbook of Biophotonics

Volume 2: Photonics for Health Care

2011

ISBN: 978-3-527-41048-4

Popp, J., Strehle, M. (eds.)

Biophotonics

Visions for Better Health Care

2006

ISBN: 978-3-527-40622-7

Edited by
Jürgen Popp, Valery V. Tuchin, Arthur Chiou, and Stefan H. Heinemann

Handbook of Biophotonics

Vol. 3: Photonics in Pharmaceutics, Bioanalysis and Environmental Research

WILEY-VCH Verlag GmbH & Co. KGaA

The Editors

Prof. Jürgen Popp
Institute of Physical Chemistry
Friedrich Schiller University of Jena
Jena, Germany

and

Institute of Photonic Technology e.V.
Jena, Germany

Prof. Valery V. Tuchin
Department of Optics & Biophotonics
Saratov State University
Saratov, Russia

Prof. Arthur Chiou
Department of Biophotonics Engineering
National Yang Ming University
Taipei, Taiwan, China

Prof. Stefan H. Heinemann
Institute of Biochemistry and Biophysics,
Friedrich Schiller University of Jena
Jena, Germany

Editorial Assistant:

Dr. Marion Strehle
Institute of Physical Chemistry
Friedrich Schiller University of Jena
Jena, Germany

Dr. Thomas Mayerhöfer
Institute of Photonic Technology e.V.
Jena, Germany

Cover
Background: Frozwn sections of mammacarcinoma (Tissue: Brigitte Mack, LMU; Immunohistochemistry: Sabine Sandner, LMU; TKTLI-antibody: R-Biopharm, Darmstadt)

Small image: Electronic DNA chip, Institute of Photonic Technologies (IPHT), Jena.

Library of Congress Card No.: applied for

British Library Cataloguing-in-Publication Data
A catalogue record for this book is available from the British Library.

Bibliographic information published by the Deutsche Nationalbibliothek
The Deutsche Nationalbibliothek lists this publication in the Deutsche Nationalbibliografie; detailed bibliographic data are available on the Internet at http://dnb.d-nb.de.

Composition Thomson Digital, Noida, India
Printing and Binding Strauss GmbH, Mörlenbach
Cover Design Adam Design, Weinheim

Printed in the Federal Republic of Germany
Printed on acid-free paper

Print ISBN: 978-3-527-41049-1

Contents

List of Contributors

José M. Amigo
University of Copenhagen
Faculty of Life Sciences
Department of Food Science Quality and Technology
Rolighedsvej 30
1958 Frederiksberg C
Denmark

Maxim E. Darvin
Charité – Universitätsmedizin Berlin
Klinik für Dermatologie, Venerologie und Allergologie
Center of Experimental and Applied Cutaneous Physiology (CCP)
Campus Charité Mitte
Chariteplatz 1
10117 Berlin
Germany

Colette C. Fagan
Biosystems Engineering, Bioresources Research Centre
UCD School of Agriculture, Food Science and Veterinary Medicine
University College Dublin
Rm 110
Agriculture and Food Science Centre
Belfield, Dublin 4
Ireland

Joachim W. Fluhr
Charité – Universitätsmedizin Berlin
Klinik für Dermatologie, Venerologie und Allergologie
Center of Experimental and Applied Cutaneous Physiology (CCP)
Campus Charité Mitte
Chariteplatz 1
10117 Berlin
Germany

Günter Gauglitz
Eberhard Karls Universität Tübingen
Mathematisch-Naturwissenschaftliche Fakultät
Faculty Division Department
Institut für Physikalische und Theoretische Chemie
Auf der Morgenstelle 18
72076 Tübingen
Germany

Aoife A. Gowen
Biosystems Engineering, Bioresources Research Centre
UCD School of Agriculture, Food Science and Veterinary Medicine
University College Dublin
Rm 110
Agriculture and Food Science Centre
Belfield, Dublin 4
Ireland

Kurt Hoffmann
RWTH-Aachen University
Institute for Molecular Biotechnology
Forckenbeckstraße 6
52074 Aachen
Germany

Christoph Janzen
Fraunhofer Institute for Laser
Technology
Bioanalytics
Steinbachstraße 15
52074 Aachen
Germany

Andreas Kandelbauer
Reutlingen University
School of Applied Chemistry
Alteburgstraße 150
72762 Reutlingen
Germany

Rudolf W. Kessler
Reutlingen University
School of Applied Chemistry
Alteburgstraße 150
72762 Reutlingen
Germany

Juergen Lademann
Charité - Universitätsmedizin Berlin
Klinik für Dermatologie,
Venerologie und Allergologie
Bereich Hautphysiologie
Charitéplatz 1
10117 Berlin
Germany

Torsten May
Institute of Photonic Technology
Quantum Detection Department
Albert-Einstein-Straße 9
07745 Jena
Germany

Martina C. Meinke
Charité - Universitätsmedizin Berlin
Klinik für Dermatologie,
Venerologie und Allergologie
Bereich Hautphysiologie
Charitéplatz 1
10117 Berlin
Germany

Hans-Georg Meyer
Institute of Photonic Technology
Quantum Detection Department
Albert-Einstein-Straße 9
07745 Jena
Germany

Robert Möller
Institute of Photonic Technology
Spectroscopy / Imaging Department
Albert-Einstein Straße 9
07745 Jena
Germany

Mario Mordmüller
Clausthal University of Technology
Institute of Energy Research and
Physical Technologies (IEPT)
Energy Campus
Am Stollen 19
38640 Goslar
Germany

Rozalia Orghici
Fraunhofer Heinrich Hertz Institute
Fiber Optical Sensor Systems
Am Stollen 19
38640 Goslar
Germany

Günther Proll
Eberhard Karls Universität Tübingen
Mathematisch-Naturwissenschaftliche Fakultät
Institut für Physikalische und Theoretische Chemie
Auf der Morgenstelle 18
72076 Tübingen
Germany

Manfred Rahe
Sartorius AG
Weender Landstraße 94–108
37075 Göttingen
Germany

Wolfgang Schade
Fraunhofer Heinrich Hertz Institute
Fiber Optical Sensor Systems
Am Stollen 19
38640 Goslar
Germany

and

Clausthal University of Technology
Institute of Energy Research and Physical Technologies (IEPT)
Energy Campus
Am Stollen 19
38640 Goslar
Germany

Michael Schaefer
Universität Leipzig
Medizinische Fakultät
Rudolf-Boehm-Institut für Pharmakologie und Toxikologie
Härtelstraße 16–18
04107 Leipzig
Germany

Astrid Tannert
Universität Leipzig
Medizinische Fakultät
Rudolf-Boehm-Institut für Pharmakologie und Toxikologie
Härtelstraße 16–18
04107 Leipzig
Germany

Ulrike Willer
Clausthal University of Technology
Institute of Energy Research and Physical Technologies (IEPT)
Energy Campus
Am Stollen 19
38640 Goslar
Germany

Part One
Process Control and Quality Assurance

Handbook of Biophotonics. Vol.3: Photonics in Pharmaceutics, Bioanalysis and Environmental Research, First Edition.
Edited by Jürgen Popp, Valery V. Tuchin, Arthur Chiou, and Stefan Heinemann

1
Industrial Perspectives

Andreas Kandelbauer, Manfred Rahe, and Rudolf W. Kessler

1.1
Introduction and Definitions

1.1.1
Introduction

The concept of Quality by Design (QbD) is based on the evident fact that a high level of quality of intermediate or final products cannot be achieved by testing the products but needs to be implemented into the products by intelligently designing the whole manufacturing process. To be able to do so, a comprehensive understanding of the process on a causal basis is required. Only such an understanding ensures reliably defined and consistent quality levels by operating stable and robust processes and, for instance, in the case of pharmaceutical products, allows real-time release of the manufactured goods. The technology that enables translation of the concept QbD into industrial reality is process analytical technology (PAT). PAT is generally considered a science-based and risk-based approach for the analysis and improved control of production processes [1]. Although initially conceived by the FDA for application in the pharmaceutical sector, the PAT initiative continues to grow in importance also for related industries in the applied life sciences, for example, biotechnology, the food industry, as well as the chemical industry [2].

1.1.2
Historical Aspects

In the course of the past decades, the industrial landscape has undergone many changes which were mainly dominated by the shift from a supplier-dominated market to a customer-dominated market (Figure 1.1, [3]). Due to the rebuilding after the Second World War, in the 1950s, the overall product demand exceeded the general capacity supplied by industry. Hence, the quality of a product was mainly defined by the producer's view of what quality was (compare the partial analytical understanding of quality) [4]. This situation has changed dramatically since then. Today the industry faces a market situation that is often characterized by an intense

Handbook of Biophotonics. Vol.3: Photonics in Pharmaceutics, Bioanalysis and Environmental Research, First Edition.
Edited by Jürgen Popp, Valery V. Tuchin, Arthur Chiou, and Stefan Heinemann

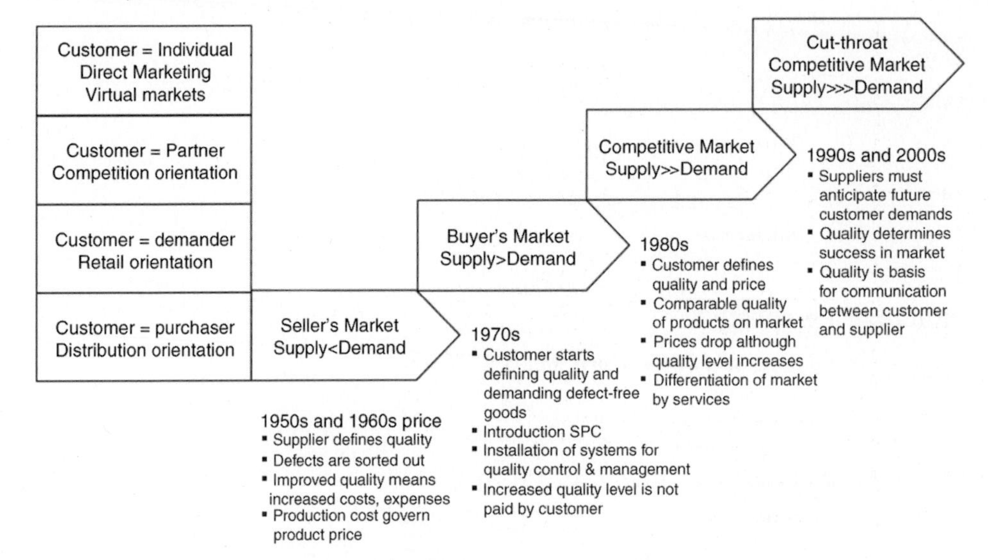

Figure 1.1 Changes in market situation during the past decades and concomitant adaptation of quality concepts (modified after [3]).

cut-throat competition in many branches, by ever increasing product complexity and diversity and by growing customer awareness of high product quality and functionality. In addition, more stringent legislative and regulatory measures imposed by society comprise an increasingly powerful driving force towards comprehensive public health and environmental compatibility of industrial processes. In contrast to the earlier years of industrialization, producing enterprises nowadays can no longer stratify their market position by increasing the mass-per-hour throughput of a certain product. Today, the sustainable success of an industrial company depends more critically than ever on the cost-effective realization of customer-tailored products of high quality that flexibly meet the rapidly changing end-customer's demands.

Reflecting this overall trend towards an increased focus on custom-made quality, increasingly sophisticated and holistic quality management systems have been developed over the years (Figure 1.2), ranging from simple inspection of the finished parts for defects and elimination of inferior ones, over implementing increasingly complex quality systems to avoid the production of any defective parts during manufacturing, to the modern views of process oriented, integrated and comprehensive total quality management systems. It is in this context that the modern concepts of PATs and QbD have to be reviewed. Before this, some general definitions and remarks on quality and process control are given.

1.1.3
Definition of Quality: Product Functionality

Quality may be best defined as product functionality [6]. Several levels of functionality can be identified (see Table 1.1) that are related to the various contexts a product and

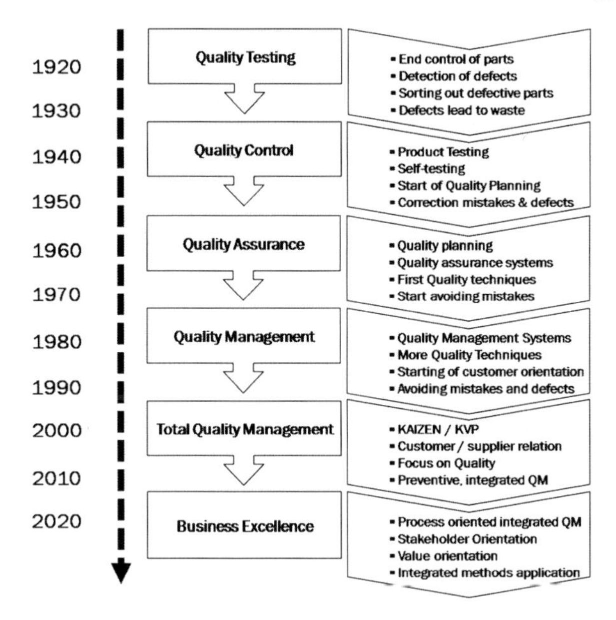

Figure 1.2 Historical development of important quality concepts and their major elements (modified after [5]).

Table 1.1 The various levels of functionality [6].

Functionality level	Description	Examples
Fundamental functionality	Basic properties based on the chemical composition and morphological constitution of the material	Is the material as it should be? Content of ingredients, particle size distribution, distribution of active compounds/traces/ dopants within material, and so on
Technical functionality	Behavior during the production process	Is the product processable? Flow behavior, mixing properties, purification and down-streaming properties, and so on
Technological functionality	Required performance profile for the intended use	Is the product usable? Hardness, strength, efficacy, durability,
Value-oriented functionality	Cost : benefit ratio	Is the tailored quality level appropriate? Displayed product features versus price
Sensory functionality	Appearance and design	Is the product appealing? Haptic behavior, product smell, visual appearance, and so on

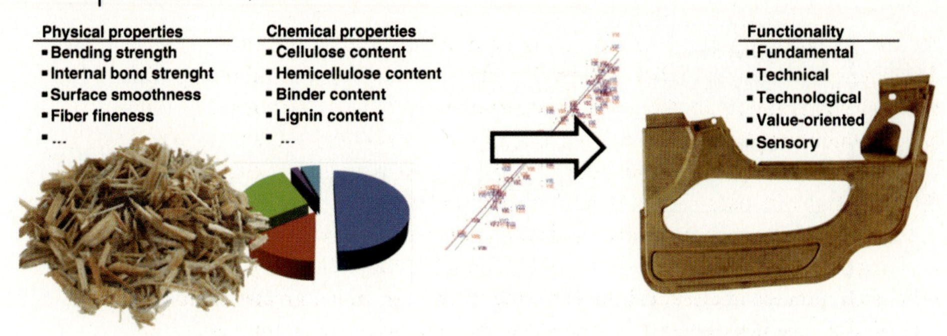

Figure 1.3 Relationship between measurement and product functionality.

its intermediates experience throughout the whole life cycle from manufacturing to the end-use and its disposal. Besides its fundamental functionality, which relates to the basic physical and chemical properties, such as composition and components distribution, surface properties or morphology of a product, the product must also be processable during manufacturing (technical functionality) and must fulfill the customer's requirements during the end-use (technological functionality). Moreover, the extent to which certain quality levels are realized in the product (value-oriented functionality) and its design-related properties (sensory functionality) are also important aspects. In the ideal case, single measured values from various methods are combined and mathematically related to all these different functionality levels (Figure 1.3).

According to the Kano model [7], a product complies with the customer's requirements and expectations when it displays basic, performance and excitement functionalities [8]. All performance characteristics of a group of objects or services intended to successfully populate a certain product niche will steadily have to be improved, expanded and further developed with time, since the customer will get used to the features and functions. This provides the driving force for continuous product development and product innovation. Due to cost issues and a desirably short time-to-market of novel products, only companies that are able to handle very short innovation cycles will sustain economic success. Knowledge-based production, based on a QbD approach and realized by process analytics, is the key element in achieving such short innovation cycles. Furthermore, it allows flexible response to sudden changes in customer's expectations, since the processes to translate the quality characteristics into product features are causally understood.

In this context, two main aspects need to be considered when introducing process analytical tools, for example, on-line spectroscopy, into manufacturing: *quality monitoring* and *product functionality design* [6]. Usually, the identification and quantification of a direct relation between, for instance, the measured spectral information and a target compound like a pharmaceutical ingredient is attempted. In many cases, univariate target responses, such as concentration, purity, or extent of conversion, and so on are determined and compared with standard values. Thereby, deviations of characteristic process parameters may be determined in real-time and the quality of the manufacturing process may be monitored and controlled. Due to the recent developments of stable on-line and in-line instrumentation, in combination with complex chemometric toolboxes, a robust calibration of such relations is possible and, therefore, applications of process analytics in industry are numerous. In most cases, off-line measurements can be directly substituted by on-line spectroscopy, with the advantage of a possible 100% quality control of a specific response.

However, while process analytics is already widely recognized as a powerful tool to perform such *quality monitoring*, the full potential of this technology is by far not exhausted; process analysis can be exploited to an even much greater extent with economical advantage when it is embedded in a philosophy of continuous process improvement and *product functionality design*. Currently, process analytical data obtained from measurements on intermediate product stages during the running manufacturing process are only very rarely related to the performance of the final product or to the final application properties of the product. However, it is possible to relate process analytical information even to product functionality when a consequent transition from considering univariate data (single parameters and responses) to multivariate data analysis (multiple parameters and responses) is performed. Product functionality may be defined as the fundamental chemical and morphological properties of the material, by its performance through manufacturing, by the technical properties for the final application, and certainly also by its cost/performance ratio. If objectively classifying data for these definitions exist, a direct correlation to, for example, the spectral information in the case of on-line spectroscopy is possible. The exact nature of the individual signature of a spectrum (the "*spectral fingerprint*") is always dominated by the morphology and chemistry of the substrate, due to its substance specific absorbance and scattering behavior. The relative contributions of these two components to the measured spectrum depend on the wavelength of the interaction, on the angle of illumination of the substrate, on the angle of detection, on the difference in refractive indices, and on the particle size and particle distribution. In Figure 1.4 this is illustrated using the example of tablet spectroscopy [9].

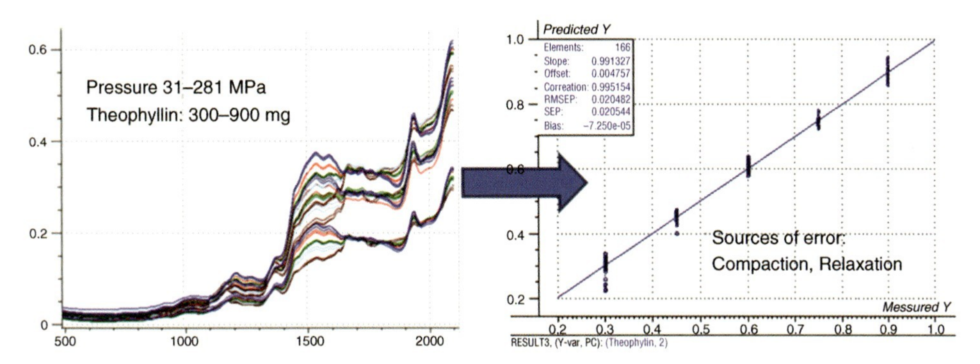

Figure 1.4 The information contained in wavelength-dependent scattering and absorption spectra of tablet samples compacted at different pressures (ranging from 31 to 281 MPa) and containing different amounts of theophyllin can be used to model the theophyllin content [9].

This multivariate information can be used not only to calculate the dependence of a single target (*quality monitoring*), but also allows full and overall classification of the sample quality (*functionality design*). This is especially true for the characterization of solids and surfaces by means of diffuse reflectance spectroscopy. Within the concept of QbD/PAT, a knowledge-based manufacturing is attempted which relies heavily on the combination of various sources of information using chemometric methods like principal component analysis, PLS, multivariate curve resolution, or other multivariate calibration methods.

1.1.4
Quality Control

The term *quality control* may generally be defined as a system that maintains a desired level of quality [10]. In general, this may be accomplished by comparing a specific quality characteristic of some product or service with a reference, and if deviations from the desired state are detected, taking remedial action to reinstate the targeted quality level. Similarly, *process control* may be defined as appropriate measures to readjust the state of a process upon some observed undesired deviation. *Process analytics* provides the required information on the state of the process. While *process analytics* deals with the actual determination of specific data using process analytical devices, like HPLC, optical spectroscopy or other sensors, *process analytical technology* is a system for designing, analyzing and controlling manufacturing by timely measurements [1] (see below), already with quality determination in mind. *Process analysis* is the comprehensive analysis of the industrial process including every single activity involved in the manufacturing of the product. Thereby, all material and virtual flows are considered. Historically, the *PAT initiative* roots in a comprehensive approach to realizing *process analysis* on an instrumental basis. PAT is the essential tool to realize the concept of *Q bD*, which has recently been further developed to the even more comprehensive approach of product quality life-cycle implementation (PQLI) [11, 12].

Essentially, quality control is accomplished by *off-line quality control* procedures, *statistical process control* and, to a lesser degree, by *acceptance sampling plans*. *Off-line quality control* involves selecting and defining controllable product and process parameters in such a way that deviations between process output and a standard will be minimized [10]. A typical tool for such a product or process design is the statistical experimental design approach or design of experiment (DoE). Quality is here basically defined "off-line" before the process has actually been implemented or started. *Statistical process control* (SPC) in contrast compares the results or output of a process with the designated reference states and measures are taken when deviations from a desired condition of a process are statistically significant. When the process is poorly designed (by inappropriate off-line quality control measures, that is, unsuitable or sub-optimal processes) these deviations may be large and cannot be compensated for by statistical process control. Hence, it is obvious that off-line quality control by well-designed processes which are based on a thorough understanding of the effects of the involved process factors on the critical quality features of the product will govern the achievable product performance, or in other words: quality cannot be tested into products afterwards.

1.1.5
Quality Assurance

Quality assurance relates to a formal system that ensures that all procedures that have been designed and planned to produce quality of a certain level are appropriately followed. Hence, quality assurance acts on a meta-level and continually surveys the effectiveness of the quality philosophy of a company. Internal and external audits, standardized procedures and comprehensive documentation systems (traceability) are important tools to accomplish this "watchdog" function within the company. Strict process descriptions determining every single step required during manufacturing a product, including the required appraisal procedures, may be defined, and deviations from these fixed procedures may be indicative of potential deteriorations in quality; instruments like the Good Manufacturing Practice approach or ISO certifications are typical for quality assurance on a highly sophisticated level. However, defined procedures and certification alone do not necessarily lead to improved performance or functionality of a product; obeying agreed-on procedures merely guarantees conformance within a specifically designed process. Pre-defined and fixed processes that are certified and commissioned by the regulatory authorities, like for instance the manufacturing process of a specific drug by a pharmaceutical company which is accepted and granted by authorities like the Food and Drug Administration (FDA) may even prove to be inflexible, sub-optimal and difficult to develop further. Since every small deviation from the standard routine processing is considered a potential quality risk and, especially in the case of pharmaceuticals or biologicals, may comprise a potential health hazard, all such deviations are required to be communicated to the authorities. Process improvements or further adaptations that usually require significant redefinition of process parameter values need renewed approval by the authorities, which in most cases is time and cost intensive. Thus, in a sense, quality assurance may even be counterproductive to process improvement and impede the establishment of higher quality levels in products. To overcome these limitations of current quality assurance policies, during the past years, the FDA has promoted the PAT initiative which, in a similar form, is also supported by the European Medicine Agency (EMA).

1.2
Management and Strategy

1.2.1
PAT Initiative

The major incentive behind the *PAT initiative* of the FDA is defined in the FDA-Guidance "PAT – a Framework for Innovative Pharmaceutical Development,

Manufacturing and Quality Assurance" [1]: "PAT is a system for designing, analyzing, and controlling manufacturing through timely measurements (i.e., during processing) or critical quality and performance attributes of raw and in-process materials and processes, with the goal of ensuring final product quality. It is important to note that the term *analytical* in PAT is viewed broadly to include chemical, physical, microbiological, mathematical, and risk analysis conducted in an integrated manner." Within the PAT initiative, the manufacturers of pharmaceutical compounds are motivated to undergo a transition from the currently used strategy of testing random samples at the end of the pipeline for their compliance and otherwise strictly sticking to approved routine procedures, towards a causal understanding of the process by means of process-accompanying and process-controlling measurements and tests. This PAT recommendation is valid also for other branches of industry, such as the food industry or biotechnology. By using powerful process analysis in the sense of PAT, manufacturing processes may be controlled and directed towards the desired levels of quality; moreover, PAT also contributes to the resource-efficiency of the production process by, for instance, minimizing the emission of carbon dioxide or reducing the energy consumption. Ideally, a 100% control of the manufactured goods is accomplished by using on-line and in-line sensors. It is anticipated that, by an integrative and system-oriented approach based on process analysis and process control, the industry will experience significant competitive advantages in the manufacturing of high-quality, customized products. Explicitly, the following goals are pursued with employment of process analysis and process control tools (PAT):

- Increase in productivity and product yield
- Minimization of energy and resources consumption
- Minimization of expenses for safety issues in the production facility
- Decreased number of customer complaints
- Increased operational flexibility
- Anticipating maintenance and process-integrated self-diagnosis
- 100% constant and certified quality

PAT will increase the production efficiency by

- Deep understanding of the production process
- Integration of quality into process steps
- Reduction of quality overhead costs
- Higher production quality
- Lower production costs
- Self-adjusting production processes

With the implementation of PAT it will be possible to pursue product-functionality design or a quality by design approach from the very beginning of the product conception. PAT targets a comprehensive feed-forward control approach and adaptive process analysis systems. Implementing PAT consequently requires application of the following modules:

1) Risk analysis of the production process (e.g., by failure mode and effects analysis, FMEA)
2) Process analytics (sensors, spectrometers, etc.)
3) Process control systems (SPC, MSPC)
4) Statistical experimental design (DoE)
5) Multivariate data analysis

This PAT toolbox and its interplay is depicted schematically in Figure 1.5.

1.2.2
PAT Toolbox

Hence, in contrast to a widely anticipated false view, PAT is not restricted to single devices for process analysis; PAT covers numerous tools included multivariate statistical methods for data design, data gathering and data analysis, process analytical sensors, control systems in the manufacturing, testing and admitting of products, as well as measures for continuous process improvement and methods of knowledge management. One of the most important groups of on-line sensors is the *spectroscopic sensors*, using the interaction between electromagnetic radiation and matter for material characterization [6]. Another important group of PAT sensors are based on *chromatographic methods* which employ various types of physical-chemical separation principles to physically de-convolute complex reaction mixtures into single components which may subsequently be identified and quantitatively determined [13]. Yet another and completely different group of sensors is the so-called *soft sensors* [14]. The basic principle behind soft sensors is that some material properties

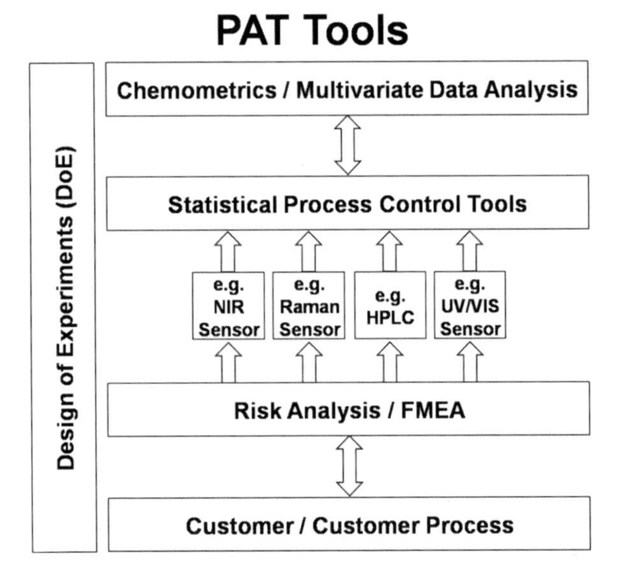

Figure 1.5 The PAT toolbox.

may not be measured directly as physical or chemical material parameters but can only be deduced indirectly by analyzing secondary variables, which in turn can be related to the target property by mathematical models. A very interesting recent account of soft sensors in the PAT context is given in Ref. [14]. With soft sensors, basically two main groups can be distinguished. On the one hand there are purely data driven models which involve no *a priori* knowledge of biological, physical or chemical interrelationships between the various categories of variables. Such models are called *black box models* and have the advantages of requiring no deep process understanding and relative ease of implementation. However, they may over-fit the observed data and be restricted to pure descriptions of the data without yielding true causal relationships which are required for a knowledge-based quality design. Tools often used are artificial neuronal networks, evolutionary algorithms, chemometric models like partial-least squares (PLS), principal component analysis (PCA), principle component regression (PLR), or support vector machines (SVR). *White box models,* in contrast, are based on known physical or chemical relationships based on kinetic or thermodynamic equations. If *a priori* information is integrated into data driven models, so-called *gray box models* are employed [14]. It is evident that numerous mathematical methods and algorithms are also included in PAT and, hence, PAT is not restricted to specific sensors that record specific physical signals. There are numerous requirements from the industrial user for process analytical technologies. The most important ones are summarized in Figure 1.6 [15].

1.2.3 The Concept of Quality by Design (QbD)

The basic concept behind PAT and QbD in the context of regulatory authorities of the pharmaceutical industry has only recently been summarized very concisely by

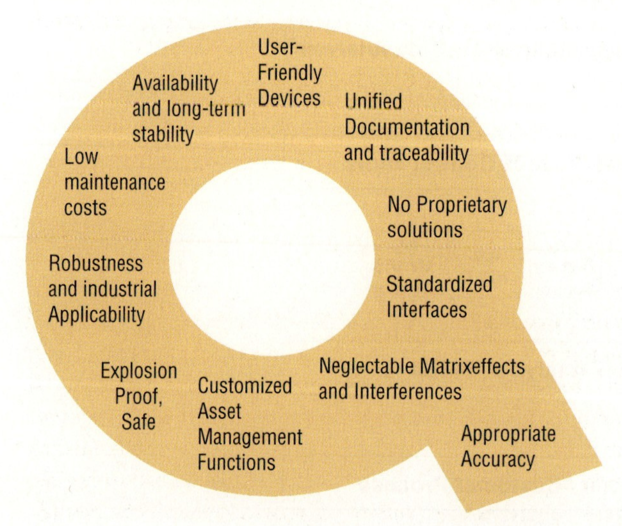

Figure 1.6 User requirements for PAT tools Source: [15].

Schmidt-Bader [16] and in the following paragraphs some of the most relevant aspects from this article will be adapted and summarized. The key aspect with a QbD approach is that quality should not simply be tested at the end after the manufacturing process has finished but should be considered already in the early phase during the conceptual design of the product, and later at all stages of its manufacture. "Quality cannot be tested into products; quality should be built in by design" [1]. In consequence, quality becomes already a matter of product development and, hence, is also strongly dependent on prior research activities into how the desired product features may be realized by industrial processes. The whole processing cycle, ranging from the early developmental stage when the product and its quality features are designed and planned, based on the input from the customers, over the product realization and production phase to its final distribution and end-use is included in such a perspective, and the manufacturer is now in the situation that he needs to demonstrate a causal process understanding throughout the whole cycle, starting from the early phases of product development to the routine production which allows guaranteed compliance with the required critical quality attributes (CQAs) during all steps. This can only be brought about by employing scientific methods. "Using this approach of building quality into products, this guidance highlights the necessity for process understanding and opportunities for improving manufacturing efficiencies through innovation and enhanced scientific communication between manufacturers and the Agency" [1].

All in all, with the PAT initiative the industry faces a shift in paradigm regarding the future of intelligent process and quality control from a *quality by testing* approach towards a *quality by design* approach. This paradigm change offers many opportunities for business excellence. Primary goals in the context of QbD are

- Assurance of a reproducibly high quality of the intermediate and final products
- Reduction of the manufacturing costs
- The promotion and advancement of novel, innovative technologies for quality assurance and process optimization
- The generation of in-depth, causal understanding of manufacturing processes

The transition from QbT to QbD is characterized by numerous significant changes in quality philosophy which are summarized in Figure 1.7 and Table 1.2.

1.2.4 ICH

The concept of QbD as developed by the FDA has been pushed towards realization during the past years mainly by the International Conference on Harmonization of Technical Requirements for Registration of Pharmaceuticals for Human Use (ICH) [18]. The motivation behind ICH was to stimulate world-wide a common and more flexible, science-based approach to the admission of pharmaceuticals. The main focus lies in an international consensus between regulatory authorities and industrial companies regarding the quality of pharmaceutical compounds: "Develop a harmonized pharmaceutical quality system applicable across the life

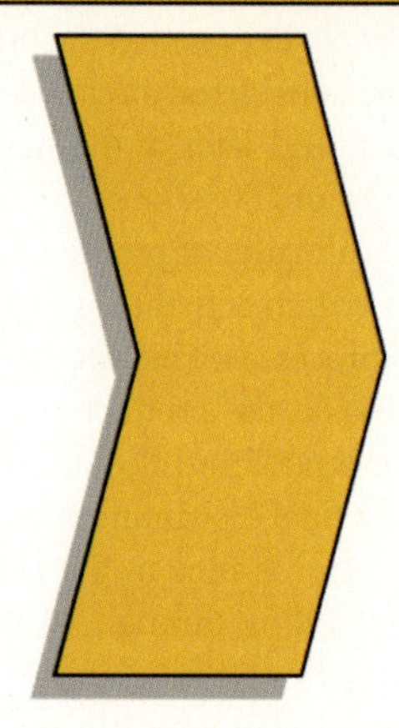

Figure 1.7 Paradigm change by QbD (Courtesy of JH & Partner CATADIA Consulting GmbH, Germany).

cycle of the product emphasizing an integrated approach to risk management and science".

In the ICH Q8 document [19] quality in the context of the QbD concept is defined as follows: "Quality is the suitability of either a drug substance or drug product for its intended use. This term includes such attributes as the identity, strength, and

Table 1.2 Major differences between QbT and QbD (modified after [17]).

	Quality by Testing (QbT)	Quality by Design (QbD)
Development	• Empirical approach • Importance of random findings • Focus on optimization	• Systematic approach • Multivariate strategies • Focus on robustness
Manufacturing	• Fixed • Based on defined specifications	• Variable within design space • Based on knowledge • Supported by robust processes
Process control	• Retrospective analysis • Based on in-process control quality is determined • Data variability is not completely understood • Focus on reproducibility	• Prospective analysis • PAT tools control critical parameters, quality is predicted • Data variability has been subject of research and is completely (causally) understood • Focus on PAT and QbD concepts
Control strategy	• Feed-back control • Control by testing and inspection	• Feed-forward control • Knowledge- and risk-based quality assurance
Product specification	• Acceptance criteria depend on data of specific product charge	• Acceptance criteria depend on end-user benefit

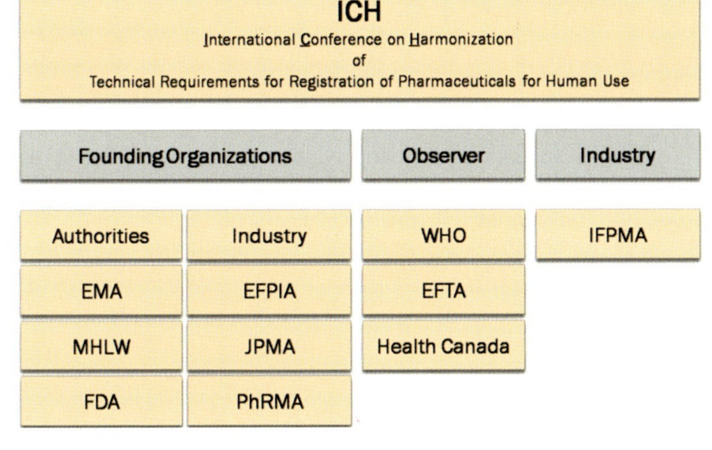

Figure 1.8 The organizational structure of the International Conference of Harmonization of Technical Requirements for Registration of Pharmaceuticals for Human Use (ICH) [18].

purity." [19] Besides considering the traditional definition of drug quality in the sense of the law, the ICH definition also includes all processes and parameters that might have an impact on the quality of the drug and, as documented in the ICH Q6 publication [20], ICH clearly assigns full responsibility to the industrial manufacturers to provide the required level of quality of a drug by appropriate measures: "Specifications are critical quality standards that are proposed and justified by the manufacturer and approved by the authorities" [20].

ICH was founded in 1990 by six independent organizations representing the regulatory authorities as well as the industry involved in pharmaceutical research in USA, Europe and Japan (see Figure 1.8).

The organizations directly involved are

- European Medicines Agency (EMEA)
- European Federation of Pharmaceutical Industries and Associations (EFPIA)
- Food and Drug Administration (FDA)
- Pharmaceutical Research and Manufacturers of America (PhRMA)
- Ministry of Health, Labor and Welfare (MHLW)
- Japan Pharmaceutical Manufacturers Association (JPMA)
- International Federation of Pharmaceutical Manufacturers and Associations (IFPMA)

The European regulatory authorities are represented by the Committee for Medicinal Products for Human Use (CHMP) as a subsection of the EMA. The European pharmaceutical market is represented by EFPIA which represents 29 national pharmaceutical associations in Europe and 45 of the most important industrial companies. For the American authorities, the FDA is involved with the Center for Drug Evaluation and Research (CDER) and the Center for Biologics Evaluation and Research (CBER). The researching American pharmaceutical industry is involved with the PhRMA which represents 67 industrial enterprises and 24 research organizations from the US. The Japanese authorities are represented by the Pharmaceuticals and Medical Devices Agency (PDMA) and the National Institute of Health Sciences (NIHS) while the Japanese industry is represented by the JPMA which heads 75 Japanese pharmaceutical companies and 14 national committees. Pharmaceutical companies and associations situated in other countries all over the world, including threshold and developing countries, are represented by the IFPMA, which is a non-profit, non-governmental organization comprising numerous companies involved in pharmaceutical research, biotechnology and vaccine manufacture.

Additionally, three international organizations of observing status are involved in ICH, whose most important role is to mediate between ICH and non-ICH member countries. These three organizations are

- World Health Organization (WHO)
- European Free Trade Association (EFTA)
- Health Canada

1.2.5 The Concept of a Design Space

For an understanding of the strategy behind PAT/QbD, the concept of a design space is of special importance. The ICH Q8 document [19] defines a design space as a multidimensional correlation, that is, the combination and interaction of numerous production factors governing the built-in quality of a (pharmaceutical) product. Within such a design space, the complex interplay between input variables, like properties of raw materials, process parameters, machine properties or user-effects are completely understood on a causal level (see Figure 1.9).

Causal understanding is achieved by identifying and quantifying the effects of critical factors on product quality at any stage of the process by multivariate mathematical models. A design space can be obtained by applying the design of experiments (DoE) approach which has been established as an important tool in quality management since the 1980s [21, 22], for example, in the Six Sigma concept [10, 23].

As pointed out in the ICH Q8 document [19], from a regulatory point of view, a process would be considered as conforming as long as it is carried out within the pre-defined design space. Since the exact trace within the design space is of no importance, manufacturing a pharmaceutical or any other product becomes more flexible. Currently, even small deviations from pre-defined values of process parameters need to be addressed with the authorities in a time-consuming and cost-intensive procedure. Such deviations from a specific path within the design space would no longer matter as long as a defined level of final quality can still be guaranteed.

As an example to illustrate this shift in process philosophy in the pharmaceutical industry, an industrial process will be discussed schematically. Consider the chemical synthesis of a given compound X. According to conventional philosophy, the

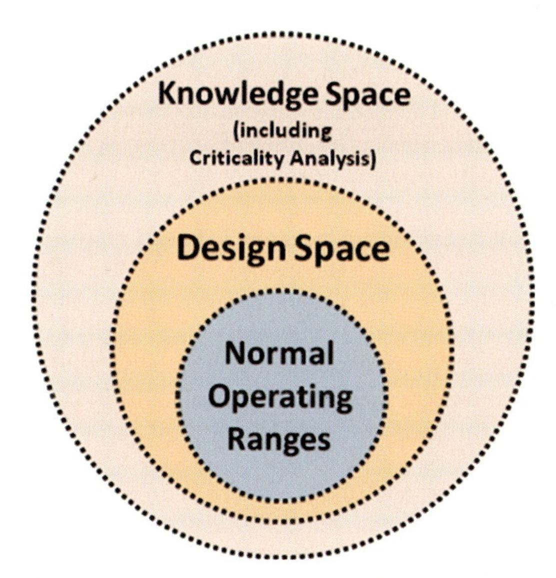

Figure 1.9 Schematic representation of the design space which is based on a knowledge space. The control strategy with a QbD approach is to maintain a process within the design space.

preparation process would have been required to be defined very specifically in terms of production conditions, such as specified reaction temperature, reaction time and, say, catalyst concentration. To conform with the certified procedure all efforts are focused on keeping the process within defined limits. Process analytical technology would be employed to control the parameter settings for temperature and catalyst concentration within agreed narrow boundaries to carry out the process in the predefined way. The target in the traditional philosophy of QbT would be to maintain a process "180 min at 120 °C with 5% of catalyst", all significant deviations therefrom would have to be documented and announced. In a QbD approach, in contrast, the target would be "95% conversion plus defined product specifications" and all feasible combinations of temperature, time and catalyst concentration (= design space) leading to this end would be allowed. Thereby, a quality (and risk) based approach to production is realized and much regulatory effort and certification cost can be avoided.

This offers process analytical technology a decisive role in manufacturing. Process analytical tools would not be used to "measure and control the quality of the product" or intermediate stages thereof, but instead would be employed to determine the current "position of the process in the design space" allowing prediction of the expected degree of quality and adjusting accordingly subsequent process steps to guarantee an agreed final level of quality.

Accordingly, validation of the manufacturing procedure would be focused on the process analytical methods that allow complete control of the design space. Robust process analytical technologies are the key requirement for monitoring and controlling the multivariate design space [16]. Analytically robust methods yield precise and accurate results under modified working conditions. Analytical ruggedness is of importance, too. This means that analytical methods are reproducible under varying analytical circumstances, such as different laboratories, different laboratory personal or different measurement devices. Robustness and reliability of analytical methods as well as their continuous improvement will be of increasing importance for assuring the quality and safety of products. In consequence, industrial processes will become more robust. The robustness of an industrial process may be defined as its tolerance against variability introduced by fluctuations in raw material quality, variations in process and environmental conditions, and variations introduced by equipment (e.g., deterioration with time) and human resources (e.g., habits, moods). With robust processes, companies may be allowed to produce their goods with a higher level of independence at lower cost and still improved guaranteed quality levels.

Currently, manufacturing processes are defined and certified at an early stage when only incomplete information is available on the influence of variations and fluctuations in raw material quality and process parameters. Improvements based on later experience are difficult to implement. Knowledge-based manufacturing allows rapid and continuous adaptation of the process to varying starting conditions and allows cost-effective further development and quality improvement.

A major element within the philosophy of QbD is the exploitation of PAT to accomplish this transition towards a knowledge-based production.

1.2.6 Implications for Other Branches of the Life Sciences

1.2.6.1 General Remarks

Although the basic impetus for pursuing PAT/QbD approaches arises from restrictions imposed by the rigorous legislative and regulatory framework in the context of the good manufacturing policy (GMP) in the pharmaceutical industry, which lead to complicated and cost-intensive approval processes caused by already moderate modifications or even improvements in the production process, the implications of this approach are certainly also of importance for other branches of the life sciences. For example, in biotechnology and the food industry, the total control over the design space in the growth of microorganisms and the harvesting of compounds produced by them has also been targeted recently [24]. Compared to the pharmaceutical industry, however, there are several peculiarities that have to be overcome by these industries, many of which deal with process analytical research questions. Although many of the chemical processes employed in the pharmaceutical industry which are desired to result in well defined compounds of high purity with a very specific biological activity are complex systems governed by numerous factors like concentration, composition, temperature, pH, and so on, in comparison to biotechnological or biosynthetic processes they seem rather simple and well-defined with respect to the possible synthetic pathways encountered.

1.2.6.2 **Biotechnology**

In biotechnology, due to the introduction of living cells or even mixed cultures, another level of complexity is introduced which usually renders classical chemical reaction engineering insufficient to develop a full understanding of the relevant design space. Considering, for instance, recombinant protein expression by genetically modified microorganisms, this typically involves numerous inter- and intracellular interactions, growth and diffusion phenomena, and very complex chemical reaction cascades within the living cells of the microorganisms. Expression of the foreign target compound by the organism has to be balanced with the energy and material demands of the growing microorganism. Intracellular transport phenomena and metabolic processes have to be included in the quantitatively and causally determined design space, as well as external factors like organism selection, fermentation medium development, process parameters, and scale-up effects. Hence, like in conventional chemical or pharmaceutical synthesis, the reactor system may not be treated as a black box; in the case of biotechnology the living microorganism systems need also to be scientifically understood. This imposes great challenges on process analytical technologies and is reflected by numerous recent attempts to monitor in real-time fermentation processes and the cultivation of microorganisms. Although subsequent process steps in biotechnology, such as down-streaming, that is, the purification and enrichment of the desired compounds from the fermentation broth, are also important, the major focus in the application of PAT/QbD concepts in biotechnology clearly lies in the fermentation step. Quality improvement in the course of the down-streaming may be achieved only to a certain degree, and is usually achieved only via a sequence of several process steps and, hence, is rather time-consuming and cost-intensive. Therefore, the development and adaptation of suitable in-line measurement technologies that lead to an improved understanding of microbial fermentation is the most effective and promising way. Again, in comparison to the pharmaceutical industry which was our starting point, specific problem solutions of process analytical tools for biotechnology must, among others, consider that the sterility of the reaction system must be ensured, that many of the targeted analyte species are present only in very small amounts, and that there is usually a large influence of the surrounding medium, which in general is rather complex and not constant with time.

1.2.6.3 **Food Industry**

Another level of complexity is typically added when the food manufacturing industry is considered from a holistic point of view. Here again, purely physical and chemical processes similar to classical chemical and pharmaceutical industries or fermentation processes as in biotechnology may be involved in the manufacturing of food. However, the growth, harvesting and processing of multi-cellular organisms and complex objects has to be taken into consideration, and should actually be included in the causal analysis of the variation pattern observed in food manufacturing. Again, powerful process analytical technologies are required that include measurement and processing of reliable data. However, in the context of establishing PAT/QbD approaches in the manufacturing industry, the food industry plays a special role, not only because of the complexity of the involved processes but also because it displays a disproportionately low level of automation in comparison to the chemical and pharmaceutical industries or other branches. Hence, the demand for process analytical technologies is especially high. It can be safely assumed that due to the ever increasing cost pressure, to globalization, and to the increasing requirements regarding quality assurance and food safety the food industry will develop to an important emerging market for the automation industry in the coming years. Besides activities directed at the rationalization of not-yet automated processes, two main fields will be of major importance in the food industry, (i) pro-active process and quality management, involving the integration of quality assurance in the production and an improvement in equipment availability; and (ii) the tracking and tracing of goods, that is mainly targeted at an increase in food safety and a reduction in food deterioration caused by microbial decay processes. This implies, in turn, an increasing demand for the engineering and development of appropriate process analytical technologies, suchas, among others, in-line sensors and data extraction tools [25].

1.2.6.4 **Summary and Outlook**

In any industry, process analytical technologies will, hence, gain in importance in the near future. Process analysis as one of the major tools within PAT is concerned with chemical, physical, biological and mathematical techniques and methods for the prompt or real-time acquisition of critical parameters of chemical, physical, biological or environmental processes. The aim of process analysis is to make available relevant information and data required for the optimization and automation of processes (process control) to assure product quality in safe, environmentally compatible and cost-efficient processes [12].

Not only does time-resolved information need to be retrieved and used as a controlling input, but, as becomes most evident from the outlined increase in complexity of the subjects that are dealt with in the life sciences, the retrieved information should also be space-resolved (chemical imaging). While in rather “simple” aqueous reaction systems with quite rapid adjustment of (dynamic) equilibria during chemical synthesis a spatial resolution may not necessarily be required, it is obvious that the distribution of a drug in a pharmaceutical formulation, or the spatial distribution of pathogens on food crops may be of critical importance to the overall quality of the manufactured goods.

The major improvements for industrial processes that are brought about by pursuing a QbD approach with the concomitant rigorous application of PAT tools may be summarized as follows:

- Assurance of reproducibly high quality of intermediates and final products
- Reduction of the manufacturing costs
- Continuous process optimization with respect to an improved exploitation of the employed material and energy resources
- Improved yields of high-quality end-product

- Improved safety and environmental compatibility of the industrial processes
- Stimulation of novel technologies for quality assurance and process optimization
- Generation of a causal understanding of the manufacturing process

1.3 Toolboxes for Process Control and Understanding

1.3.1 Introduction: Causality

Process control and understanding is an important feature for a future knowledge-based manufacturing. Although on-line process control has been well known for the last 20 years, the aspect of PAT in the pharmaceutical industry has become a driving force for recent activities [1]. However, "quality" has different meanings to different companies. For instance, for large companies that produce a standardized material, the major target associated with quality is ensuring close adherence to the defined product specification. In contrast, for smaller companies, "quality" means, in many cases, guaranteeing the flexibility to fit the end-product requirements in relation to rapidly changing market needs. Moreover, nowadays the concept of quality often goes far beyond a specific product but embraces the concepts of plant quality and total quality management (TQM). However, what all views of quality have in common is their dependence on information about the intrinsic properties of a product and knowledge of the relationship between plant parameters and product functionalities.

On-line and inline quality control and process optimization will only be successful when based on appropriate process analysis and process understanding, that is, the analysis of the connection (cause and effect) between process parameters and the quality characteristics of the final product with its specifications. Process analytics in this sense means, therefore, understanding the causal relation between measurement and response. By definition, causality is the strict relationship between an event (the cause) and a second event (the effect), where the second event is a consequence of the first [26]. Very often in chemometrics only descriptive or statistical knowledge is produced which fits the special data set but cannot be used as a general model. Figure 1.10 visualizes the different levels of knowledge for process understanding.

A straightforward "cooking recipe" for knowledge-based production integrates the following procedures (see also Section 1.2.2, Figure 1.5):

Step 1: Detailed analysis of the process and risk assessment
Step 2: Selection of the process analytical toolboxes
Step 3: Define the design space and design the necessary experiments
Step 4: Multivariate data analysis of the data
Step 5: Define the control system and integrate system into production

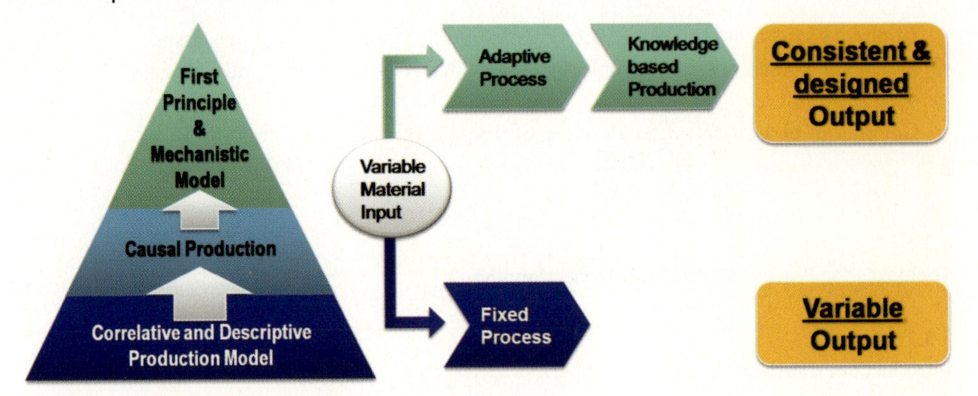

Figure 1.10 Road map for process understanding.

1.3.2
Sampling

In homogeneous systems any sample which is taken will be representative of the whole system. In heterogeneous systems it is difficult to find a way to extract a sample (= fragment) which represents the average of the material. In practice, no sample will be strictly identical to the material as it is a matter of scale [27].

There are four general sampling strategies that are used for process analysis[1]:

- **Withdrawal**: A portion of the process stream is manually withdrawn and transferred into a suitable container for transport to the analyzer.
- **Extractive sampling**: A portion of the process stream is automatically taken to the analyzer. This may take place either on a continuous basis or at frequent intervals.
- ***In situ* probing**: A probe is inserted into the process stream or vessel and brought into contact with the sample.
- **Non-invasive testing**: Either a “window” into the process stream or another mode of non-contact measurement is used in order to account for interaction of the analysis system with the process material

All four approaches have certain advantages and drawbacks and there is hardly ever a clearly right or wrong approach to sampling.

In *process analytics*, one can distinguish between off-line, at-line, on-line and in-line measurement methods [6]. In the case of *off-line* measurements, samples are withdrawn from the process and analyzed in a laboratory environment which is spatially clearly separated from the industrial equipment. Thus off-line analysis always exerts significant lag times between recognizing and counteracting irregularities. With *at-line* measurements, the sample is withdrawn from the process flow

1) D. Littlejohn, personal communication.

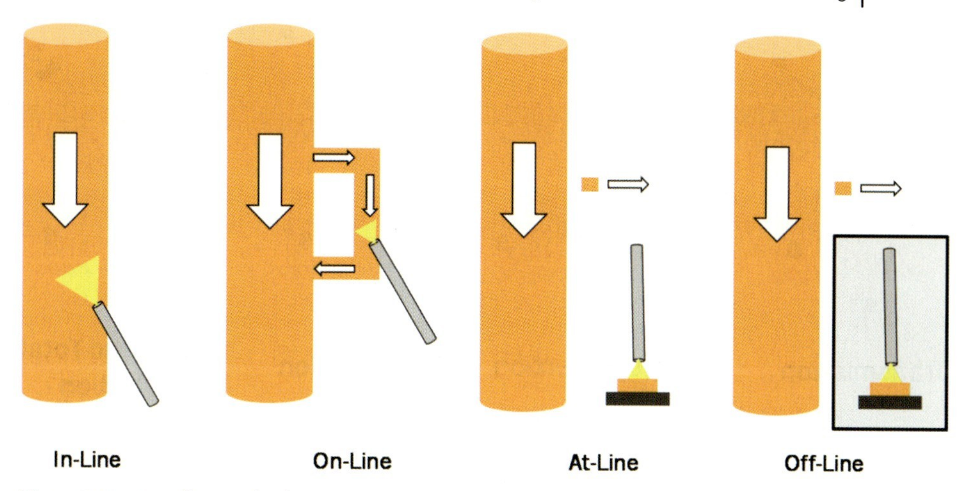

Figure 1.11 Sampling methods.

and analyzed with analytical equipment that is located in the immediate environment of the industrial equipment. Hence, the reaction time for countermeasures is already significantly reduced. Due to the industrial proximity it is often observed that at-line analytical equipment is more robust and insensitive towards process environment but less sensitive or precise than laboratory-only devices. In the case of *on-line* measurements, samples are not completely removed from the process flow but temporarily separated, for example, via a by-pass system which transports the sample directly through the on-line measurement device where the sample is analyzed in immediate proximity to the industrial machining and is afterwards reunited with the process stream. When *in-line* devices are used, the sensor is directly immersed into the process flow and remains in direct contact with the unmodified material flow. Figure 1.11 illustrates various sampling modes realized with in-line, on-line, at-line and off-line sampling. Typical probes for interaction of electromagnetic irradiation with samples are shown schematically in Figure 1.12 for various spectroscopic techniques (transmission, diffuse reflectance, transflection and attenuated total reflectance spectroscopy).

In the ideal case of process control, 100% of the processed elements or products are covered by analytical methods at any time and complete knowledge about the quality of the manufactured goods is obtained throughout the whole process. However, in most cases (when only off-line or at-line devices are used) this is not possible and then *accepted sampling plans* are required which define which and how many samples are inspected at certain intervals [6].

Sampling is always an issue no matter which measurement approach is used, but the nature of the challenge also varies with the approach. For instance, since with on-line analysis a sample may be spatially separated from the main process stream by means of a bypass system, on-line analysis has the advantage over *in situ* or in-line analysis in that the sample can be pre-conditioned (filtrated, extracted, constant

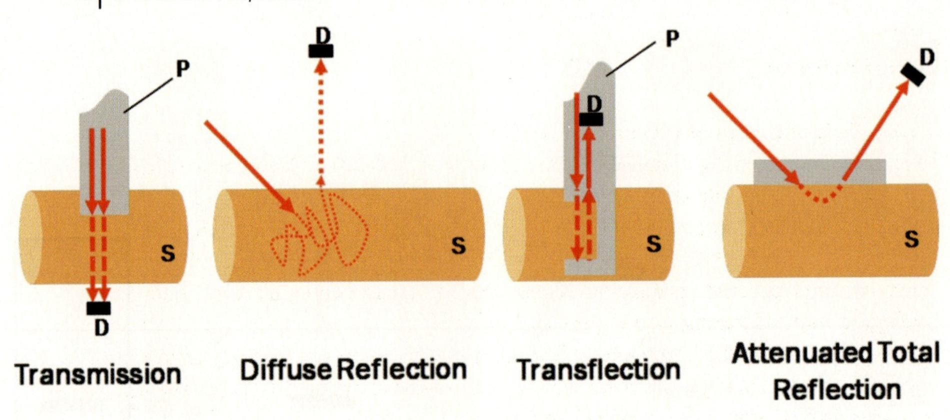

Figure 1.12 Schematic representation of typical spectroscopic probes (D, detector; P, probe; S, sample).

temperature etc.) prior to the analytical procedure. The main concern of sampling is to get access to a representative sample. Besides being representative of the process stream, some additional features need to be considered when selecting a certain sampling approach. Sampling systems must ensure that the sample is[1]:

- obtained and delivered safely,
- conditioned and presented reliably to the analyzer,
- compatible with the analyzer regarding the measurement conditions, such as temperature or pressure,
- obtained in a cost effective way, and
- representative also of the process stage, that is, the lag time from the process to the analyzer must be within an acceptable range

It is generally believed that about 75% of analyzer reliability problems are associated with the sampling system.

Sampling of solids presents the most significant problems. It is clear that finding the key component at low concentrations of a targeted analyte within a complex particulate system with high precision is a real challenge. As a rule of thumb, around 3000 to 10 000 particles are needed to obtain representative values. This can easily be realized when only small particles are present in the medium; however, it is almost impossible to calibrate a system with large particles. In this case, on-line calibration may be the best way to receive representative information over time.

Liquids are considerably easier to sample, unless there are two phases present or the liquids carry high levels of solid compounds.

Gases in general present the least problems, but can still be tricky where there are components close to their dew point. The NESSI consortium has focused on using modular sample system components in building block form that can be linked together to make complete conditioning systems which include pressure regulators, valves, pressure gauges and filters [28].

1.3.3 Process Validation

1.3.3.1 Role of Design of Experiments (DoE)

In January 2011 a new "Guidance for Industry: Process Validation: General Principles and Practices" was introduced by the FDA and will be outlined here [29]. This guidance describes process validation activities in three stages through the whole lifecycle of the product and process, (i) process design, (ii) process qualification, and (iii) continued process verification. These are the basic principles for smart or intelligent manufacturing (see Figure 1.13).

The process should be understood from first principles and from a scientific and engineering point of view. In most cases, the sources of variation during production may already be known by experience; however, only rarely are they really understood or quantified. Thus an important objective of process validation is to attribute all variations in the process and raw materials directly to the product variability. The perfect way to relate product variability to process and raw material changes is to use a DoE strategy [30]. However, it is sometimes difficult to select the appropriate parameters and parameter settings for the design. A parameter is a measurable value which can describe the characteristics of a system, for example, temperature, pressure, and so on. Very often, these parameters may be the factors (= independent variables) which predominantly influence the process and product quality (critical process parameter or factor (CPP)). DoE allows one to identify the relevant factors and to quantify their relative importance. It is logical that the factors which are of importance should be controlled. However, it is sometimes a difficult task to find the correct CPP. Multivariate data analysis (e.g., PCA) of historical data helps to select

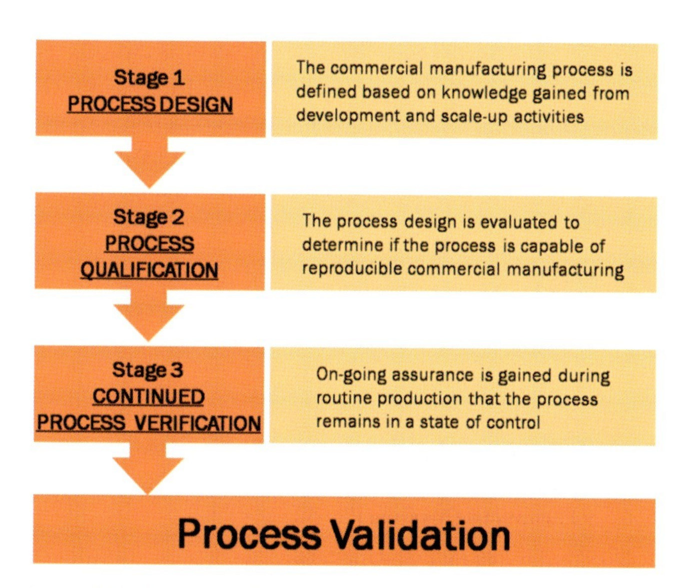

Figure 1.13 Basic principles of smart manufacturing as defined by the FDA [29].

the most important orthogonal parameters [6]. Orthogonality is an important prerequisite in parameter selection for DoE since it is desirable to vary CPPs independently of each other in order to influence the response properties of the product.

Outliers are usually a much better source of information than historical data based on "standard" production, as standard production data only show the variation within the regular production and, therefore, no systematic and significant variation of the factors can be deduced. It is also important to emphasize that a reliable design of the experiments should always include the possibility to evaluate the interaction terms between the factors. When the interaction is more important than the main factors this usually indicates that parameters are used which are not factors and, hence, the DoE has to be modified.

1.3.3.2 Role of Failure Mode and Effects Analysis (FMEA)

Another tool to extract the most important factors is risk assessment such as failure mode and effect analysis (FMEA) or cause and effect matrices [10, 31, 32]. This assists in effectively defining the design space, increases the awareness of the process risks, and yields a better understanding of the relationships between functionality and quality parameters. The critical quality attributes (CQA) are the data that best describe the characteristics that need to be controlled to ensure product quality. Figure 1.14 shows the principal steps involved in FMEA and Figure 1.15 illustrates how FMEA is situated within the production site of a product.

1.3.4 Measurement Technologies (How to Measure)

1.3.4.1 Selection of the Appropriate Technique

One of the key elements for process control is the selection of the best possible technology. Common techniques used in industry measure physical attributes such as conductivity or refractive index. They may be addressed as univariate sensors. Process chromatography can be used to separate the components of complex mixtures, but chromatographic methods are difficult to integrate in *in situ* and

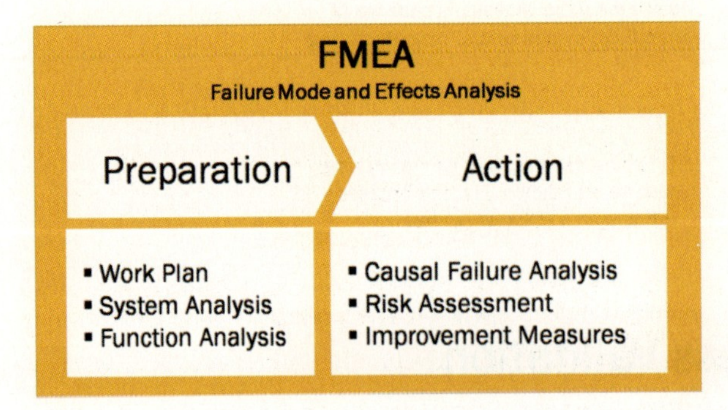

Figure 1.14 Basic approach behind failure modes and effects analysis (FMEA).

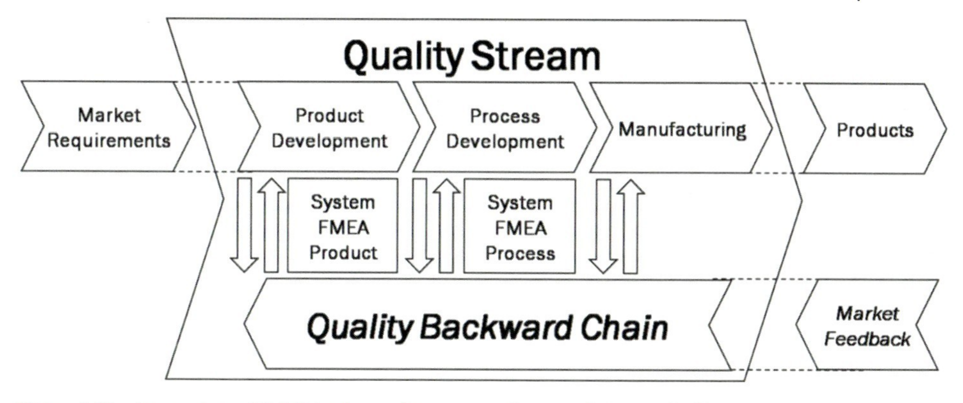

Figure 1.15 Integration of FMEA in the quality stream of a manufacturing facility.

in-line process control set-ups. Unlike optical spectroscopy, new technologies like on-line-NMR or terahertz spectroscopy, as well as mass spectroscopy are not yet standard equipment in in-line process analytics. A detailed overview of the different techniques is given in [6, 33]. Since optical spectroscopy is currently among the most prominent methods used in inline-process analytics, it is discussed in more detail in the following paragraphs.

Optical spectroscopy has developed into a widely used technique in process analytics. Depending on the measurement problem, a broad range of useful wavelength ranges and modes of interaction between electromagnetic radiation and the sample can be used (see Figure 1.16).

Key issues from a practical point of view, besides cost/performance, are the need for high sensitivity and selectivity, as well as the simplicity of application. Although

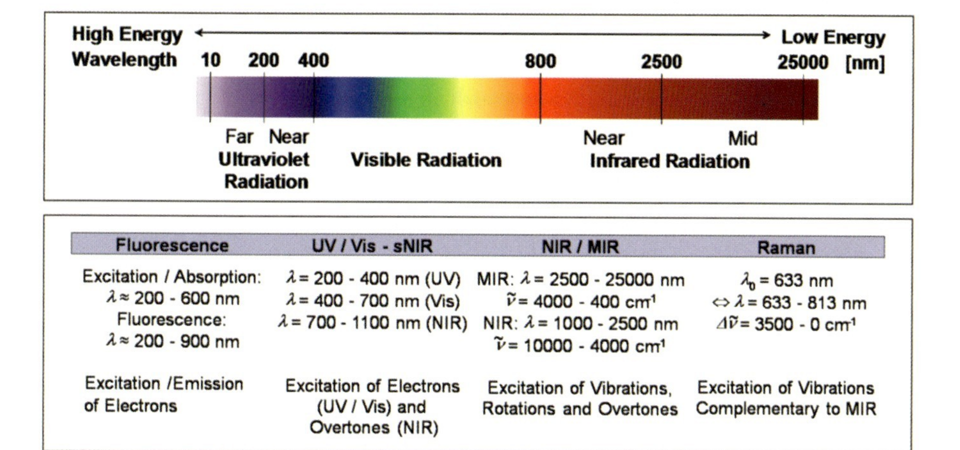

Figure 1.16 Selection of appropriate wavelength regions depends on the spectroscopic method used and the intended analytical application.

Table 1.3 Selection of the best possible inline technology in optical spectroscopy [6].

	UV/VIS/s-NIR	NIR	MIR	Fluorescence	Raman
Selectivity	+	++	+++	++	+++
Sensitivity	+++	+(+)	+++	+++ (+)	+(++)
Sampling	+++	+++	+	++	+++
Working in aqueous media	+++	+	+	++	+++
Applicability	+++	++	+	+	+
Process analytical tool	+++	+++	+	+	+++
Light guide	+++	+++	(+)	+++	+++
Signal	Absorption	Absorption	Absorption	Emission	Scattering
Sampling on-line/in-line	s, l, g	s, l	s, l, g	s, l (g)	s, l, (g)
Techniques	Transmission Reflectance ATR	Transmission Reflectance ATR	ATR (Transmission)	Reflectance Transmission	Reflectance
Relative costs	1	3–5	6–10	4–6	8–12

the basic layout of spectroscopic tools is always very similar (light sensors–sample contact area–detector), the various optical spectroscopic techniques are based on numerous different measurement principles. Ultraviolet- and visible (UV/Vis) spectroscopy is a highly sensitive technique for electronic transitions while mid-infrared (MIR) spectroscopy is specific for vibrational transitions. Since energy transitions between vibrational states of a molecule are highly substance-specific, peaks measured in the MIR region can be directly attributed selectively to fundamental moieties in a molecule. Near-infrared (NIR) spectroscopy is less sensitive due to lower yields of the higher order vibrational transition probabilities. However, although not easily directly interpretable, the major advantage with NIR is that, even at higher concentrations, no sample preparation (e.g., dilution) is needed. It is important to emphasize that both NIR and MIR spectroscopy are highly sensitive to water absorption. Table 1.3 shows a qualitative comparison of the advantages and disadvantages of the different optical spectroscopic tools.

1.3.4.2 **Working in Aqueous Systems**

Figure 1.17 shows the spectra of water in the NIR and MIR wavelength ranges. Due to the different absorption cross sections of the fundamental vibrations (MIR), the combination bands, first second and third overtones in the NIR, different path lengths must be used. The measured MIR spectrum is measured with a diamond ATR system with a mean path length around 5 μm.

The strong water absorption limits a broad application of these techniques in aqueous systems, for example, the study of fermentation processes. Raman spectroscopy may be advantageous over NIR and MIR spectroscopy in aqueous systems. In recent years, Raman spectroscopy has developed into a highly sensitive and versatile technique and, therefore, has proven a very suitable method in biotech-

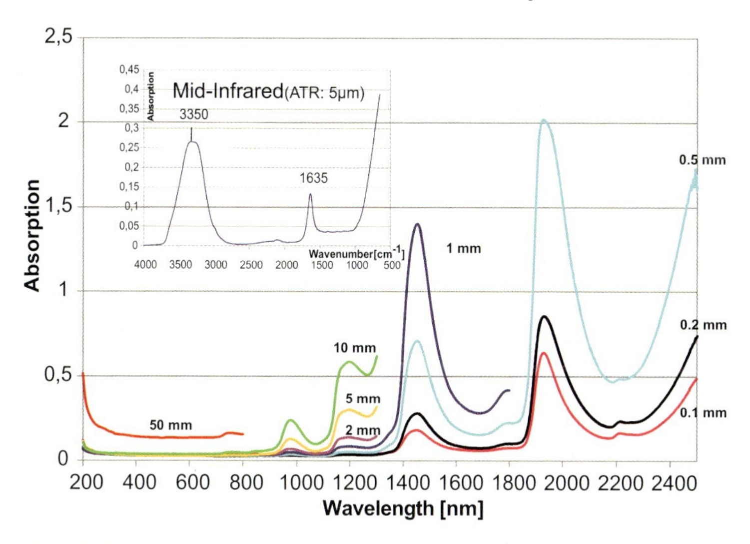

Figure 1.17 Absorption spectra of water in the NIR region with different path lengths and MIR (insert).

nology. For special applications fluorescence spectroscopy is certainly one of the most sensitive techniques in spectroscopic analysis.

1.3.4.3 **Trace Analysis**

Trace analysis is still a challenge in process analytics. Optical spectroscopy can cover a broad range of sensitivities and selectivities, as described before. One major advantage of NIR is that the absorption cross sections are generally low. Thus the technique can be used even at high concentrations. The typical detection limit for low concentration mixture components lies at around 1%; due to the high absorption coefficient for water as a trace component it is in the range down to 0.1% (or even 0.01%, depending on the system). In contrast, with MIR spectroscopy concentrations as low as 0.01% are easy to measure and standard detection limits can be even as low as 0.001%. Although Raman absorption cross sections lie typically around 10 orders of magnitude lower than in FTIR, due to recent development of extraordinarily sensitive detection systems Raman spectroscopy may approach the performance of FTIR spectrometers in the near future. UV/Vis and fluorescence spectroscopy are very sensitive techniques (in ppm, ppb and even lower), but lack selectivity.

As can be seen, the work horses in PAT are in many cases sufficiently sensitive. However, especially for applications in biotechnology when one is working in an aqueous environment at typically rather low metabolic concentrations, only chromatography in combination with mass spectroscopy may be a reasonable option [6, 33].

1.3.4.4 **Qualification of a Spectrometer**

Generally the quality of a spectroscopic inline control system can be described in terms of its spectral range, spatial resolution, non-linearity, S/N ratio (stray light), diffraction efficiency and stability. The parameters needed to characterize the systems are, for example:

- Spectral resolution
- Spectral linearity
- Absolute efficiency of the optics (throughput) and diffraction efficiency of the grating,
- Straylight (S/N), ghost line and ghost image properties.
- Wavelength stability

Spectral axis calibration is done with spectrally well known light sources for example, neon lamps, lasers, fluorescent systems. The background CCD signal (dark current) must be compensated for and – if possible – be minimized by cooling. The detector response to light varies from pixel to pixel and is also strongly wavelength-dependent. Moreover, the energy throughput of lenses and other optical elements also depends on the wavelength. These variations can be calibrated by measuring a white reference surface, storing this image, and then calculating the ratio between a measured sample image and this white image (after dark current subtractions). Light source color temperature drift and lighting spatial non-uniformity are also compensated for, as long as the texture of the reference and the target is similar in terms of specular and diffuse reflectance.

The spectral range defines the wavelength regions covered by the spectrometer. The spectral resolution is related to the monochromator system used and defines the power to resolve the narrowest spectral features in the electromagnetic spectrum. The bandwidth is defined as the full width at half maximum of the spectral line. It is important to notice that the optical resolution is different to the (digital) pixel resolution of, for example, a diode array spectrometer. The pixel resolution describes the number of digital points which are required to represent a peak in the spectrum. Usually the pixel resolution should be about 2 to 3 times higher than the optical resolution. The signal to noise ratio is the ratio of the radiance measured to the noise created by the instrument electronics and the detector.

For on-line process analysis some additional features like the frequency of maintenance and the frequency of recalibration are important and define, among other features, the cost of ownership. Details of the calibration procedures are defined in ASTM standards. The location of the analyzer must be compatible with the safety ratings of the end user area.

1.3.5 Data Analysis and Calibration (How to Process Data and How to Calibrate)

1.3.5.1 Introduction

The basic idea of multivariate data analysis is to extract useful information from data and to transfer this information into knowledge. Figure 1.18 visualizes the methodology of multivariate data analysis to extract useful information from multidimensional data sets.

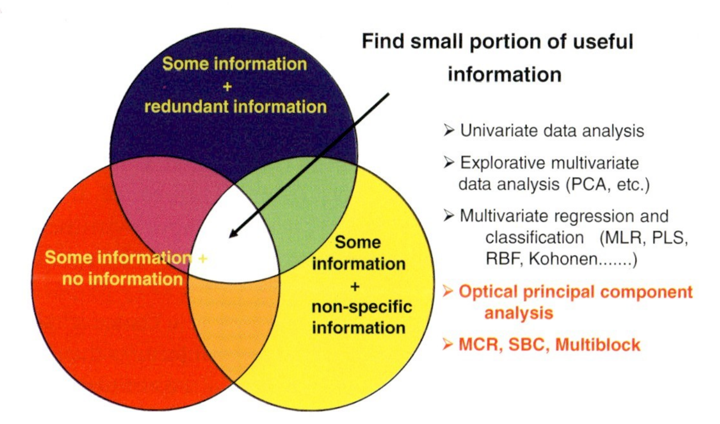

Figure 1.18 Extraction of information from a large data set.

In common data there is high redundancy of the information, overlapping with no-information (white noise) and information which is of no use for the specific problem. Besides univariate data analysis, the chemometric toolbox includes explorative data analysis like principal component analysis (PCA); multivariate regression analysis like partial least square analysis (PLS); and nonlinear approaches like neural networks. It is important to emphasize that the first step for a proper calibration and modeling is always the correct optical set-up of the spectrometer device and the appropriate definition of the measurement procedure. As will be shown later, complex systems are preferably analyzed and described by using multiple spectroscopic methods (multimodal spectroscopy) which may be addressed as “optical principle component” analysis. Hybrid models like multivariate curve resolution (see below) or science-based calibration (SBC) allow the introduction of knowledge into the modeling.

A standard procedure to extract information and transform this information into knowledge may be:

- Standardization and calibration of the instrument
- Spectral data pretreatment
- Data cleaning
- Principal component analysis
- Regression analysis
- Evaluation and figures of merit

The procedure for standardization and calibration of the instrument was described in the previous section.

1.3.5.2 Spectral Data Pretreatments and Data Cleaning

As common in spectroscopy, the measured dark spectra, reference spectra and sample spectra are used to calculate the corrected sample spectra. Data transformation may then involve the conversion of the raw data into, for example, absorbance. The measured diffuse reflectance spectra can be transformed to absorbance or Kubelka-Munk units in order to linearize the correlation to chemical constituent concentrations.

Changes in sample surface, sample orientation, particle size distributions, compaction of loose samples like powders and external changes in the illumination or the detector response (for instance by temperature drift) may result in unwanted spectral signals, which are added or subtracted throughout the whole spectral range. To reduce these additive effects a first and second derivative can be carried out. If the spectra are very noisy they have to be smoothed before calculating the derivatives. A detailed overview of standard pretreatment procedures and their effects on the optical spectra is given in [34].

Spectra normalization, either to length one or area one, can be a choice if the interesting information is more related to the shape of the spectral features than to changes in absorbance intensity due to concentration variations of a constituent. In such cases when classification (qualitative information) is aimed at, normalization is a very helpful pretreatment method since the spectra will become independent of their global intensity.

To correct for particle size or other scattering effects a multiplicative signal correction (MSC) can be applied [35, 36]. Several methods have been described in the scientific literature, ranging from simple MSC to more sophisticated methodologies such as extended MSC [37] and stepwise multiplicative scatter correction (PMSC) [38]. Alternatively, the standard normal variate (SNV) correction for scatter effects can be used. SNV is a simpler but purely mathematical-based procedure to correct for scatter.

In order to correct for baseline curvature or other nonlinear effects across the NIR spectral range a de-trending algorithm can be applied subsequently after an SNV transformation. Barnes *et al.* [39] have shown that MSC and SNV give more or less the same results.

1.3.5.3 **Chemometrics**

Chemometrics offers the possibility to extract relevant information from multiple wavelengths and methods instead of using single-wavelength channels only. Additionally, chemometrics reduces this relevant information into one or a few quality defining parameters (underlying entities) by applying either multivariate classification or regression models to the data.

There has been constant development in chemometrics and a number of good reviews and useful tutorials have been published [34, 40–42]. The advent of modern computer systems in the past decades has boosted widespread use of chemometric software packages, and has also had a very positive effect on the broader distribution of mathematics-intensive spectroscopic on-line methods.

Principal component analysis is a chemometric method that decomposes a two-dimensional data Table X into a bilinear model of latent variables, the so-called principal components, according to the following expression:

$$X = TP^T + E \tag{1.1}$$

where T is the scores matrix and P^T the transposed loadings matrix. The matrix E is the residual matrix and accounts for the experimental error (noise), which is not part of the model. The principal components are calculated so that they explain as much variance of the data as possible. The first principal component captures most of the variance in the data set. This information is then removed from the data and the next principal component is calculated, which again captures the major part of the remaining variance; this procedure is continued until a pre-defined stopping criterion is met, which is based on falling below a lower limit of variance explained by an addition of another principal component. All principal components are linearly independent (orthogonal); that means there is no correlation among them and they can, therefore, serve as a new coordinate system with reduced dimensions.

A so-called loading plot shows the relative importance of the individual variables (here: absorbance at different wavelengths). It can be used to assign the spectral classification to molecular structures of the chemical components. The objects can also be represented by their scores in the new principal components space. This allows clustering and structuring the samples quantitatively.

The fact that the principal components have no correlation among each other, as they are calculated to be orthogonal, results in negative scores and loadings. This makes it often difficult to interpret the underlying chemistry. To overcome this deficiency, MCR can be applied instead, where non-negativity is one of the basic prerequisites for calculation. Such MCR methods have been introduced to image analysis only recently, with growing attention and success. More details can be found in [9, 40].

1.3.5.4 **Regression Analysis**

Regression The target of PCA is more explorative but it is well possible to build regression models with the PCA scores regressed on target values. This is called principal component regression (PCR). However, as in traditional spectroscopy, the most commonly used algorithm for multivariate regression is partial least squares (sometimes also called projection to latent structures). The PLS algorithm builds an inverse calibration model for the spectra X and the target value Y according to the following regression equation:

$$Y = XB \tag{1.2}$$

The matrix X contains the spectra and Y holds the corresponding target values, which are the properties to be predicted. Y can be a matrix, but especially in process control it often has only one *y*-variable. PLS uses latent variables, similar to the principal components in PCA, to build the calibration model. The latent variables are calculated so that they explain as much as possible of the covariance between X and Y.

The model size (number of latent variables) is determined by the internal validation data set and is checked for correctness with an external data set. The figures of merit are given as bias and root mean square error of prediction as a measure of accuracy and precision. They are calculated separately for the different data sets according to the following formulae:

$$bias = \sum_i (y_i - \hat{y}_i) \tag{1.3}$$

$$RMSEP = \sqrt{\frac{\sum_i (y_i - \hat{y}_i)^2}{n}} \tag{1.4}$$

where y_i is the reference concentration for the ith sample (given by the reference method), $\hat{y}_i$ is the predicted concentration by the calibration model and n the number of samples. When the model has been validated, it can be used to predict y values (e.g., concentration) for measured spectra.

Evaluation of the Calibration The reliability of a method includes accuracy and precision. "*Accuracy*" in testing means "closeness to the true value". Especially in biotechnology, this is hard to define because usually the relevant constituents cannot be prepared in a pure state and their spectral characteristics depend strongly on the interfering matrix material. Within a laboratory, accuracy can be established by repeated analysis. Between laboratories, accuracy can be assessed by using the mean results of collaborative studies (ring tests) among all of the laboratories belonging to the same organization. "*Precision*" in any testing means obtaining the same result every time the measurement is repeated. It includes repeatability and reproducibility. "*Reproducibility*" includes all features of the test, including sub-sampling, sample preparation and presentation to the instrument, and testing by all of the operators that are likely to be involved in the testing. It is determined by repeated analysis of the same sample, including all of the steps involved in the analytical procedure and all of the operators likely to be involved in future testing. "*Repeatability*" includes all features of the test except sub-sampling and sample preparation. It is determined by performing duplicate or replicate tests on the same sample, after sub-sampling and sample preparation, and is a test of the actual method on a single sample after sample preparation. It is important to emphasize that PLS can be misleading if it is not used with care [43, 44]. To select the correct validation technique is the key for causality (see Section 3.1).

1.3.6
Process Control (How to Control a Process)

As described in the "Guidance for Industry: Process Validation: General Principles and Practices" and other papers, process analysis provides quantitative and qualitative information about a chemical process in real time, using on-line and in-line analyzers [29, 45]:

"The information given by these systems is used to control the process. The control strategy is defined as the input material controls, process controls and monitors, and finished product tests, as appropriate, that are proposed and justified in order to ensure product quality. The control strategy will ensure the product is manufactured within the Design Space to meet all Critical Quality

Attributes and other business-driven quality attributes (e.g., that affect cost or manufacturability)".

Process control is important for economic, safety and environmental reasons. Improved process control allows more efficient use of feedstock and energy, giving better product quality and ensuring consistency of quality. It also enables improved treatment of waste products and effluent to meet continually more stringent environmental legislation.

The key to good process control lies in the ability to measure fluctuations in the system behavior (e.g., changes in feedstock composition or temperature build-up). This information is then used to compensate for these changes and to optimize process parameters. How representative measurements are of a system, how long the interpretation of the data takes and how quickly this information can be acted on, are important factors of process control.

In this section, the strategy of a modern manufacturing using feed-back and feed-forward control is described. Figure 1.19 visualizes the strategy for process control and how to manage variability.

Process control summarizes all measures to keep quality within certain limits. *On-line statistical process control* involves actions to monitor deviations from a desired state while the process is actually running and manufacturing takes place. Hence, in order to be able to react promptly in response to observed deviations, real-time monitoring techniques are advantageously employed. Even when a process is well designed, statistical process control measures are always useful, since they allow one to detect and act upon unforeseen effects of immediate or abrupt changes in the process

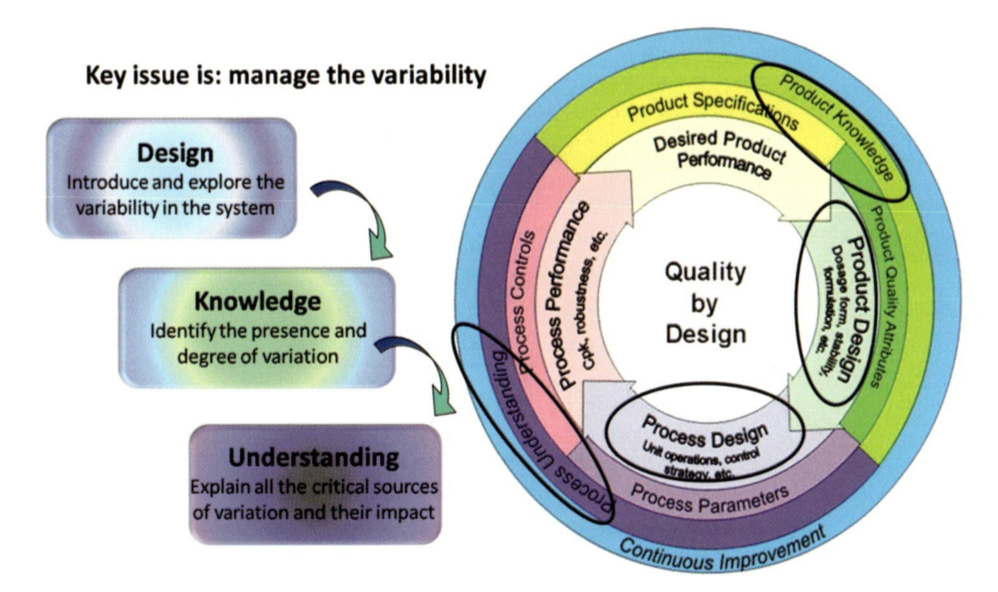

Figure 1.19 Managing variability (modified after FDA).

conditions, and to correct the process for statistical variations that occur, for instance, in the raw material, which is always likely to happen. Moreover, processes when first implemented are usually not yet optimized with regard to the highest possible level of quality they might produce; on-line statistical process control may well assist in such optimization and continuous improvement. Regarding the information that is used for statistical process control, laboratory data or process data may be used. *Laboratory data* typically comprise physical or chemical tests performed on the incoming raw material to control the input quality, or technological tests for the application behavior of the final products to control the output quality. Especially in the early phases of industrial manufacturing, this was the main mode of controlling the overall performance of a process, the major indicator of process quality being the amount of waste production or sorted-out parts (Figure 1.2). Regarding process-related data two different types of *process data* are available: the first includes machine data, which are accessible by recording parameter settings of the machine equipment, or general, unspecific process data, like temperature, pH or conductivity, which may be recorded by various sensors throughout the whole course of the production. The second type includes quality influencing, material specific data which are directly related to and measured on the produced goods by means of process analytical methods which are applied at numerous intermediate product stages. Traditional laboratory data alone, especially when only applied as end-of-pipeline tests of product performance, only allow *feed-back control* loops, that is, after having detected that a fraction of the final

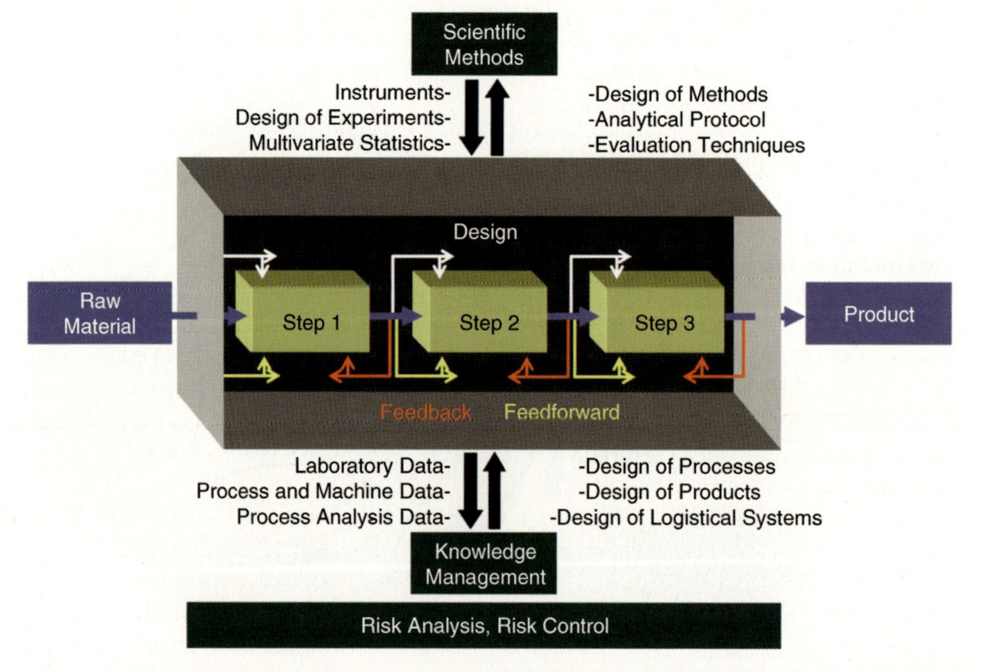

Figure 1.20 Process control by feed-back and feed-forward control requires a holistic view of the process and the trans-disciplinary interplay of numerous techniques and methods [46].

product is unusable and must either be downgraded or even discarded, the process is modified until subsequent batches may again be within specifications. In contrast, process data (in combination with a causal understanding of the overall process) may more intelligently be used in so-called *feed-forward control* loops which allow anticipation of product properties before they have actually manifested. In this way, correcting measures may be undertaken in advance in the case of expected negative deviations and inferior or waste production may be totally avoided. Figure 1.20 (adapted from [46]) summarizes these concepts schematically.

1.4
Specific Problems Encountered in Industrial Process Analytics

In this section, selected process analytical problems that are of crucial importance to many applications throughout various branches of industry are addressed. Since they are often encountered in an industrial environment, this section is dedicated to principle problem solutions that deal with moisture determination, the process analytics of solids and surfaces, strongly scattering systems, and the spatially resolved spectroscopy of samples using spectral imaging.

1.4.1
Moisture Measurements (NIR, MW)

Moisture is an important quantity which influences processability and shelf-life. Moreover, delivering a product of defined moisture content not only saves energy but also increases profit. Off-line moisture measurements are time-consuming and cannot be used for a feed-back or a feed-forward control in manufacturing. Therefore, there is a strong demand in industry for on-line and in-line control of water in a substrate [47].

1.4.1.1 NIR Spectroscopy
As described in the previous chapter, MIR and NIR spectroscopy show strong absorption for water and, therefore, can be used for in-line analysis of moisture content. NIR spectroscopy is based on measurements of light absorbed by the sample when it is exposed to electromagnetic radiation in the range from 780 nm (12 820 cm^{-1}) (short-NIR, s-NIR) to 2500 nm (4000 cm^{-1}) (NIR). As described in the previous chapter, water absorption in the NIR occurs mainly at wavelengths around 1445 and 1900 nm (Figure 1.21).

There are numerous investigations which deal with the determination of surface water and intrinsic water (bulk water). To separate these two kinds of water spectroscopically, it is generally postulated that surface water predominantly absorbs around 1900–1906 nm and intrinsic water, respectively bound water, at 1936 nm. However, bands in the NIR are broad and, in most cases, spectrometers were used which were limited in their optical resolution. Thus care must be taken to interpret spectra obtained from measurements on real life components.

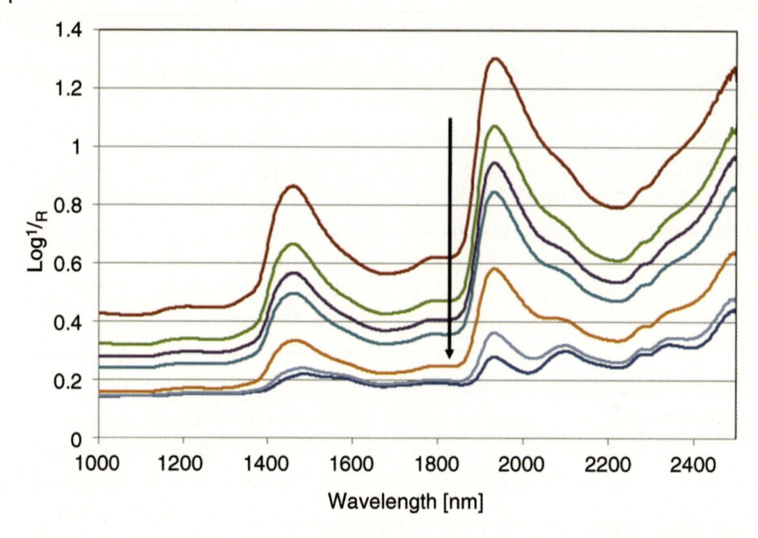

Figure 1.21 NIR spectra showing the typical absorbance pattern of water with maxima at 1450 and 1950 nm. The spectra show the evaporation of water from cellulose-based filter paper during drying. Measurements were performed with a Lambda 1050 NIR spectrometer (Perkin Elmer) in diffuse reflectance using an integrating sphere (150 mm).

Furthermore, as described in the next chapters, the cross sections for absorption, and also for scattering, are usually in a range of low penetration depths of photons. At strong absorptions in the NIR (e.g., combination vibration), penetration and, hence, information depth may, therefore, be only a few hundred microns or even less; in the third overtone of water (s-NIR), penetration into the substrate may go up to several millimeters. Therefore, the measurement of moisture using NIR combination bands is usually restricted to the determination of surface water. The typically used reference measurement techniques for calibration of the spectroscopic signal, such as Karl Fischer titration or gravimetric methods ("loss on drying"), however, determine the overall bulk moisture content, which does not necessarily correlate to the surface moisture that is measured by NIR spectroscopy. Hence, calibration of NIR spectra may be erroneous. Moreover, NIR spectra often also contain spectral contributions from other components which may overlap with water peaks. To make things even more complicated, signals may also be perturbed by scattering due to particles of different (and unknown) size distribution, resulting in nonlinear mean free path length variations of the photons. In multivariate calibrations, SNV, MSC or EMSC have been successfully used to eliminate baseline offsets present in the raw spectra and can compensate for differences in thickness and light scattering of the analyzed samples.

1.4.1.2 Microwave Resonance Technique (MWR)

In the case of microwave resonance (MWR) measurements the specific absorption of water in the microwave wavelength range of 2–3 GHz is used. Sensors are designed from cavity resonators or stray field resonators. The penetration depth of the microwaves is in the several-cm range and, therefore, bulk moisture measurements

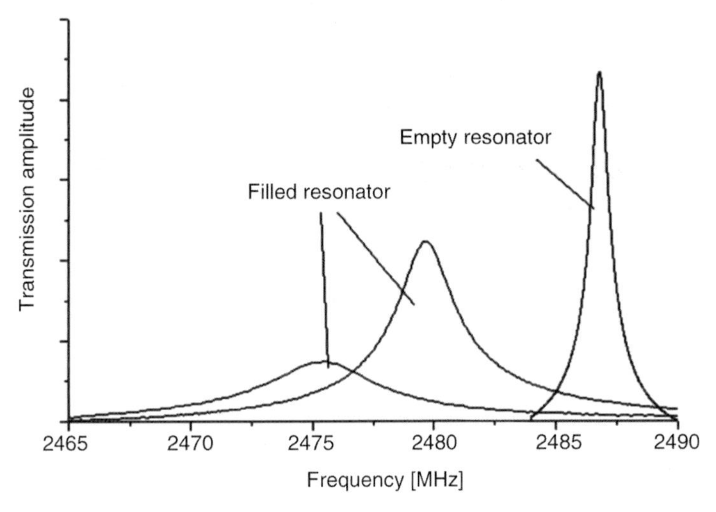

Figure 1.22 Microwave spectra of an empty and filled resonator.

can easily be correlated. Figure 1.22 shows an example of the microwave spectrum of an empty resonator and a resonator filled with a substrate.

Usually, the spectrum of the empty resonator is stable and can be used as an "absolute" reference which is specific for a given system and system set-up under certain temperature-controlled conditions. When a moisture-containing substrate is measured (surface and bulk), the spectrum of the empty resonator changes in two ways: First, the resonance spectrum shifts towards lower frequencies due to the dielectric losses and, secondly the spectral bandwidth increases in response to the density, respectively, the mass of the substrate. This information can be used to compensate for fluctuations of the measured path lengths due to particle size variations or other density changes (like, for instance, compaction) during on-line measurements. Hence, besides being a very precise and reproducible measurement method for moisture content, MWR also allows mass or density determination of the bulk material which is corrected for the moisture content (dry mass determination). Being a very rapid measurement method, the MWR technique allows up to 500 single measurements per second of powders, granules, fibers and even solids. This allows easy averaging of the spectral information. Unlike other dielectric techniques, such as capacitive techniques, conductivity measurements or microwave transmission measurements, unperturbed information is obtained, even in cases where substrates are used that contain high amounts of ionic material. Numerous applications can be found in the literature for food and feed control [48, 49]. Figure 1.23 shows examples for the set-up of on-line measurements in the food industry.

1.4.2
Process Analytics of Solids and Surfaces: Specular and Diffuse Reflectance

In real-life samples of solids or surfaces, reflectance spectra are composed of contributions from specular and diffuse reflected light. Pharmaceutical tablets show

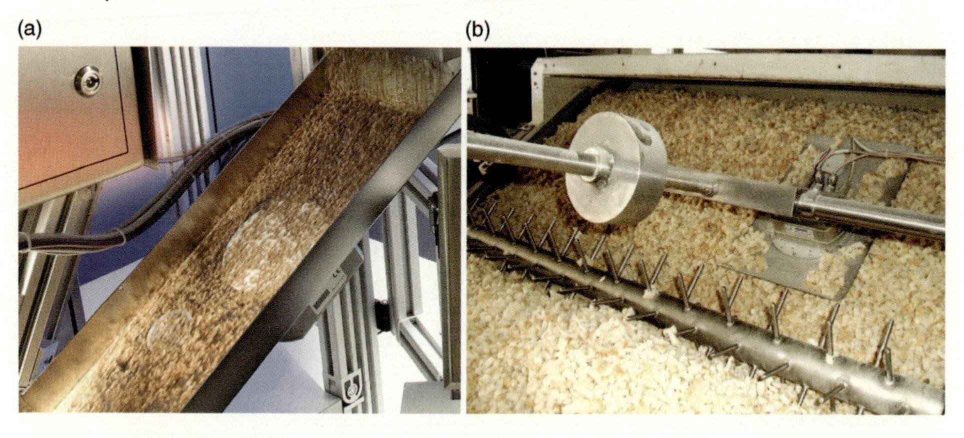

Figure 1.23 On-line control of wheat grains by a microwave stray field sensor (a) and a slide sensor measuring moisture of bread crumbs (b), (courtesy of Sartorius).

primarily diffuse reflectance but, due to different degrees of compaction, specular surface reflectance is also observed. Decisive changes in the scattering coefficient occur during compaction and relaxation and have a great impact on the measured signal. Metal surfaces, on the other hand, exhibit mainly specular reflectance and interference; however, they may also show some diffuse reflectance due to defects within the layer, surface roughness, or contaminants on the surface. In a more complex system like wood chips, on-line control allows correlation of the spectral signature not only to a single target value like lignin, but to a complex quality definition like functionality.

The Fresnel equations are the basis for the calculation and interpretation of the portion of light which is specular reflected from optically smooth surfaces and consists of wavelengths that are comparable to those of the incident light. Diffuse scattering originates from surface irregularities that are of the same order of magnitude (or slightly smaller) as the wavelength of the irradiating light source, as shown schematically in Figure 1.24. Diffuse reflected light is described by the

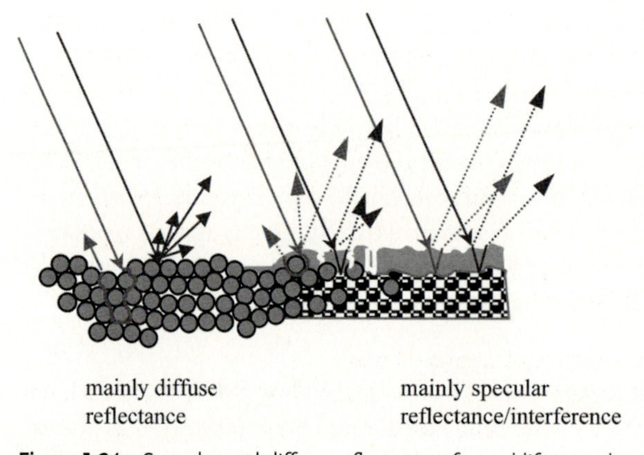

Figure 1.24 Specular and diffuse reflectance of a real-life sample surface [50].

Kubelka Munk function [50]. In this case, diffuse reflectance is influenced by both the absorption and scattering coefficients, which are usually denoted as k and s. In an ideal case, the absorption coefficient k and the scattering coefficient s can be separated by the measurement of a sample with defined thickness in diffuse reflectance on a highly scattering white or black background, or by means of diffuse transmission and reflectance spectroscopy [50].

In real-life samples, specular and diffuse reflectance contribute simultaneously to the overall spectrum, which is then a mixture of both effects. A detailed mathematical description of the resulting solutions of the Fresnel and Kubelka Munk function is given in [51].

The ideal experimental set-up would be to measure the reflectance by means of an integrating sphere. But integrating spheres are not suitable for in-line or on-line control. Moreover, the analytical deconvolution of the spectral information into the contributions of the pure specular and diffuse components is not a simple task. Thus, efforts must be made to focus the measurement on either diffuse reflectance or specular reflectance in order to optimize the response, depending on the required physical or chemical information.

An even stronger discrimination between diffuse and specular reflectance is feasible using polarized light. Polarized light is depolarized by scatter centers and therewith the information on defect sites is emphasized.

A specific target of the PAT initiative of the Food and Drug Administration of the United States of America is to identify and quantify an active pharmaceutical ingredient (API) in a complex formulation. Thus a direct relation between the measured spectrum and the concentration of the API must be established at high precision. The objectives of on-line control are, therefore, a fast and robust noninvasive measurement protocol which is not perturbed by artefacts. Pharmaceutical tablets are made from small particles which can act as ideal scatter centers for diffuse reflectance. Spectra, therefore, simultaneously show (i) wavelength-dependent scattering, and (ii) specific absorption due to their chemical composition. Since the process of tablet formation leads to a smooth surface after compaction, the tablets additionally exhibit specular reflectance which perturbs the diffuse reflectance measurement. For on-line control, the amount of specular reflected light being transmitted to the detector should be minimized. A possible set-up to bring this about is the diffuse illumination of the sample and a detection perpendicular to the illuminated surface. Alternatively, the illumination can be performed at an angle of 45° while the detection of the diffuse reflected light is again done at an angle of 90° to the surface (assignment: 45R0). A detailed description is given in [52].

1.4.3 Working in Multiple Scattering Systems: Separating Scatter from Absorbance

1.4.3.1 Basics in the Measurement of Opaque Systems

In scattering systems, the interaction of light (photons) is complex and includes refraction, specular and diffuse reflectance and/or transmission, as well as absorption and scattering simultaneously. Due to the diffusion of photons, even the spatial identification and attribution to a defined spatial coordinate in x and y may diminish. Hence, the spectroscopic investigation of samples that contain phase boundaries and, therefore, simultaneously display absorption and scattering effects show several restrictions regarding the experimental procedure (methodological approach) and the substrate:

Methodological influences

- Angle of illumination and detection (specular and diffuse light)
- Wavelength range: for example, UV–VIS–NIR–IR
- Polarization of the light (illumination and detection)
- Illumination and detection area
- Focal planes of illumination and detection

Substrate influences

- Differences in particle and matrix refractive indices
- Particle size
- Particle size distribution
- Volume concentration – compaction
- Scattering and absorption coefficients

Figure 1.25 shows schematically the different effects occurring when photons interact with an opaque substrate.

Particles produce scattered light. The intensity of the scattered light is dependent on the size of the particle and the wavelength of the interacting photons. Smaller particles show higher scatter and light of shorter wavelengths is more strongly scattered than light of longer wavelengths. This results in a difference in mean free path lengths and penetration depths of the photons. In combination with absorption phenomena, the overall spectral response will be significantly different when the same sample is measured in diffuse reflectance or in diffuse transmission. Also, the thickness of the measured sample will have a strong influence on the obtained spectra due to photon loss at the rear side of the sample. These photons cannot be

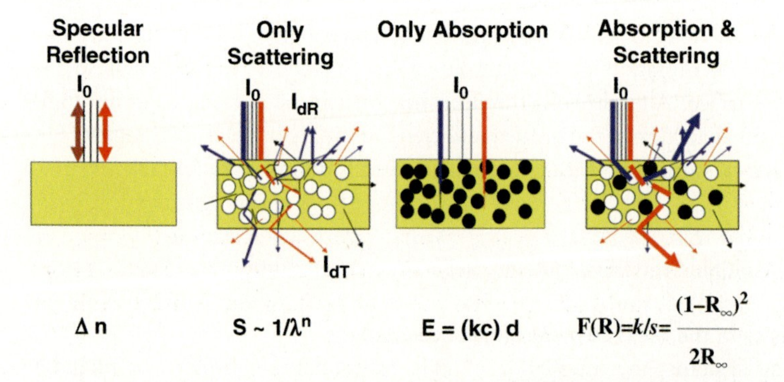

Figure 1.25 Schematic representation of the different possibilities for the interaction of photons with opaque systems.

scattered back again as in the case of infinite sample thickness. Thus, the geometrical set-up of the measurement is an important factor and changes in particle sizes and size distribution must be compensated during calibration.

It is also important to emphasize that scattering also changes the direction of the photons. Theoretical calculations have shown that the travel distance of the photon may even be expressed in millimeters and more. Therefore, in highly scattering systems, information from one spatial region may be influenced by information from another region, resulting in a mixing of the spatially separated spectral information. As a result, there is a trade-off between spatial resolution and chemical (quantitative) information [6].

1.4.3.2 **Separation of Scatter from Absorption**

In many chemical, pharmaceutical and biotechnological processes not only the chemical changes are of importance but also the morphological variation of the particulates. When spectroscopic in-line control is applied to such complex systems very often the scatter in the spectra is regarded as unwanted and, therefore, eliminated by chemometric methods instead of using it as a supplementary source of information on the morphology of the substrate. One of the most appropriate theories to describe multiple scattering and absorption is the radiative transfer equation (RTE). A summary of RTE is given in [53]. A survey of the different techniques is described in [54].

Using this equation, three separate and independent measurements are necessary to separate scatter from absorption. Kortüm [50] and coworkers have demonstrated extensively the power of diffuse reflectance spectroscopy in quantitative measurements of turbid systems. The simplest approach is to use the Kubelka Munk (KM) function, where $F_{(R\infty)}$ describes the reflectance of an optically infinitely thick sample.

Linearity between the spectral response and concentration is only obtained (i) with a specific optical set-up for diffuse illumination and (ii) when a defined preparation procedure of the samples is used with special consideration for reproducible grinding and dilution of the sample with powders that do not absorb. In practice, the alternative log $1/R_\infty$, that is called "absorbance", prevails over $F_{(R\infty)}$ of the KM function in most publications. The KM function is shown in Eq. (1.5). Here K stands for absorption and S for scattering

$$E(R_\infty) = \frac{K}{S} = \frac{(1-R_\infty)^2}{2R_\infty} \tag{1.5}$$

Several models exist to determine the wavelength-dependent scattering and absorption coefficients in diffuse reflectance and diffuse transmission spectroscopy [19]. K and S can be calculated for example, from the simplified solution of the differential equations of the Kubelka Munk function of a light flux into and from the sample. Eqs. (1.6) and (1.7) describe the relation between S and K with the measurements of a sample in diffuse reflectance.

$$S = \frac{2.303}{d} \cdot \frac{R_\infty}{1-R_\infty^2} \cdot \log\left(\frac{R_\infty(1-R_\infty \cdot R_0)}{R_\infty - R_0}\right) \tag{1.6}$$

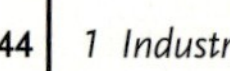

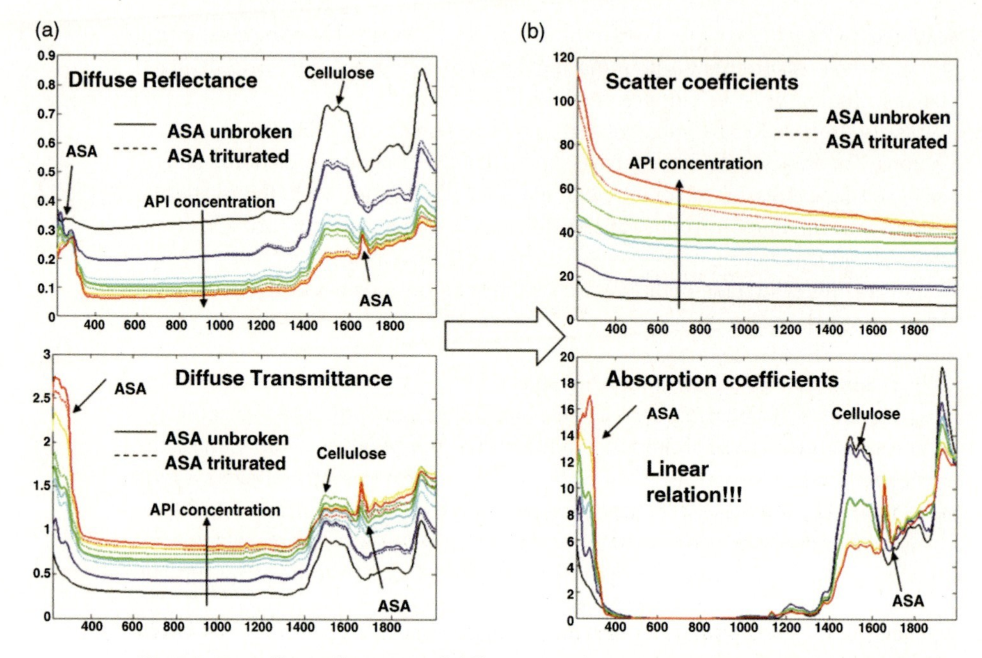

Figure 1.26 Diffuse reflectance and diffuse transmittance spectra of Aspirin in cellulose (a) and calculated wavelength-dependent scatter and "pure" absorption spectra (b) [52].

$$K = \frac{2.303}{2 \cdot d} \cdot \frac{1 - R_\infty}{1 + R_\infty} \cdot \log\left(\frac{R_\infty(1 - R_\infty \cdot R_0)}{R_\infty - R_0}\right) \qquad (1.7)$$

Here, R_0 denotes the spectrum measured at a definite sample thickness with a black, strongly absorbing background and R_∞ denotes the measurement of the same sample with an ideal white scatter (= nonabsorbing material for example, barium sulfate) as a background. Other approaches are described by Oelkrug, Dahm and Dahm [55, 56].

Figure 1.26 shows the result of a spectrum where the contributions of scatter and absorption have been separated. For demonstration, an aspirin tablet with cellulose as excipient is measured at different compactions in diffuse reflectance with infinite and definite thickness. For details see [6, 9, 52].

As can be seen clearly, the absorption portion of the spectrum in the longer wavelength region containing chemical information is slightly perturbed by scatter, whereas in the shorter wavelength range, scatter is the dominating information source on the morphology of the sample.

This deconvolution can also be used with advantage in chemical imaging. Figure 1.27 shows the image of an aspirin (ASS) tablet as measured at the specific absorption wavelength of ASS in diffuse transmission and diffuse reflectance. The S and K values were calculated at the specified wavelengths.

The figure shows nicely that the same particle of ASS shows different regions of absorption depending on the geometrical set-up of the measurement and the used

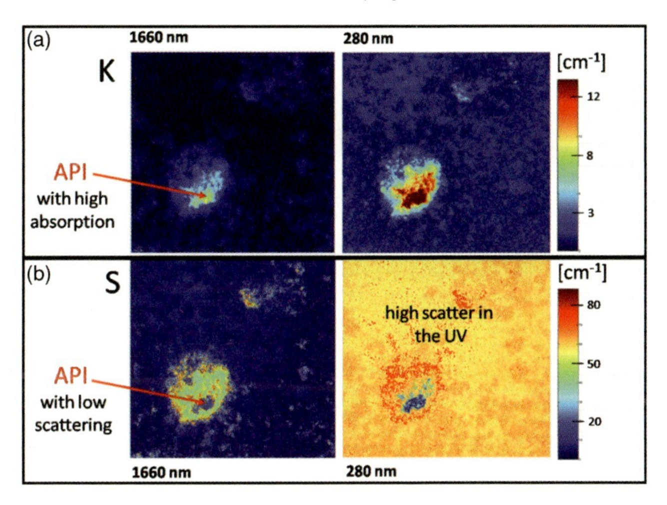

Figure 1.27 Spectral imaging of an aspirin particle in cellulose: (a) "pure" absorption of the aspirin at different wavelengths, (b) "pure" scatter of the particle at different wavelengths (for more details, see [52].

wavelength. UV absorption is significantly stronger than the absorption in the NIR range. This is also true for the scattering, which at lower wavelengths is much stronger than in NIR. It can also be seen that scattering is especially strong at the phase boundaries of the particle.

Figures 1.26 and 1.27 clearly show that the overall spectrum is the superposition of the contributions of absorption and scattering effects. Therefore, the mean free path lengths of a photon into and through a tablet will depend on the number of scattering centers and the scattering characteristics of the particles. The probability of interaction of the photon, and thus the signal intensity, also depends on the portion of photons that is absorbed during their motion through the particulate system. Certainly, the measured intensity also depends on the layer thickness since photons are lost by transmission when the layers are of finite thickness. In theory, if a photon is not absorbed it will travel for an infinite distance within the particulate system.

1.4.3.3 **Optical Penetration Depth**

The optical penetration can be defined as the depth δ, at which the intensity of the radiation inside the material I falls to 1/e (about 37%) of the value of the incident beam, I_0. For a semi-infinite medium, in the range of validity of the K-M theory, the intensity at a distance z from the surface can be approximated[2] as:

$$\delta = \frac{1}{\sqrt{K(K+2S)}} \tag{1.8}$$

2) D. Oelkrug, B. Boldrini, and R. W. Kessler, unpublished results.

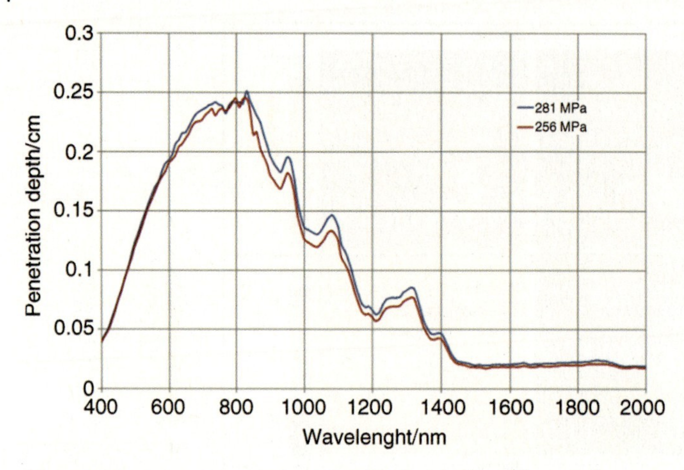

Figure 1.28 Penetration spectrum for the theophyllin tablet in cellactose at different degrees of compaction.

This is a generalized form of the penetration depth integrating absorption and scattering. Details will be described elsewhere.

Figure 1.28 shows the result of a calculated penetration spectrum of the theophyllin tablet example shown in Figure 1.4 at two different compactions. Because of the high absorbance at longer wavelengths the penetration depth is low. On the other hand, due to the strong absorption, the spatial resolution at higher wavelengths may be higher because the number of scattered photons is lower and specular reflectance has a lower probability. In the short NIR range (below 1100 nm), high penetration takes place, mainly due to the low absorption and mean scattering properties of the sample. As demonstrated, the penetration depth and, therefore, the scale of scrutiny can be determined, when the scatter and absorption coefficients are known. Similar approaches with similar results are described in [53, 57].

A much simpler approach is described in [58, 59]. Here the spectral response of an absorber is measured when this absorber is covered with different layers of scattering particles. The description of the determined path length is, however, strongly dependent on the geometrical set-up of the system, the sensitivity of the instrument and the properties of the particulates and, therefore, cannot be generalized.

1.4.4 Spectral Imaging and Multipoint Spectroscopy

Spectral or chemical imaging combines the physical and chemical characterization of samples by spatially resolved spectroscopy. The techniques may either be classified according to the used wavelength ranges or, more generally, into mapping or imaging techniques. *Mapping* usually means to get an image by exploiting the full spectrum (λ) of a local point through point measurements (at spatial x and y coordinates). The term *imaging* is used, when two-dimensional pictures (x, y) are obtained by a camera

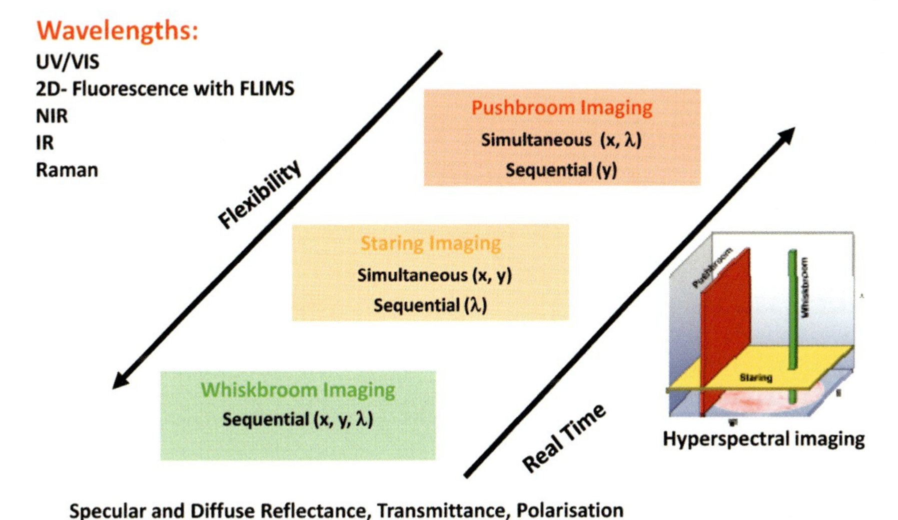

Figure 1.29 Visualization of the different imaging technologies: whiskbroom imaging, staring imaging, pushbroom imaging.

at different wavelengths (λ). In addition, *line scans* are defined as measured spectra along a line [60]. A more straightforward taxonomy is used in remote sensing. The different applied technologies are labeled as:

- Whiskbroom imaging
- Staring ("stardown") imaging
- Pushbroom imaging

Figure 1.29 visualizes the different techniques. All techniques can use the full optical wavelength range from the UV to the IR as well as Raman and fluorescence.

In process control, the spectroscopic method is often applied to moving samples and, therefore, should be as fast as possible. Hence, in this case the pushbroom acquisition mode is usually the best suited technique. Along a line of the sample, which represents the first spatial direction (x-axis), several spectra are measured at the same time. The pixel spatial resolution of this x-axis is determined by the number of pixels used and the distance of the sample from the spectrometer objective lens. The pixel resolution of the spectral λ-direction (z-axis,) is determined by the spectral range of the spectrometer (camera) and the number of pixels that are combined together to yield one spectrum. The second spatial dimension (y-axis) is correlated with time. If the sample is moved before the next line-scan starts, the time and the second spatial dimension on the sample (y-axis) are correlated to the moving speed of the sample. The data cube of the measurement thus represents the two spatial axes and the wavelength axis.

Figure 1.30 shows an example of the data cube of a hyperspectral image of a wood chip on which the letters RRI (Reutlingen Research Institute) are written with a UF (urea formaldehyde) resin binder that is commonly used in the production of

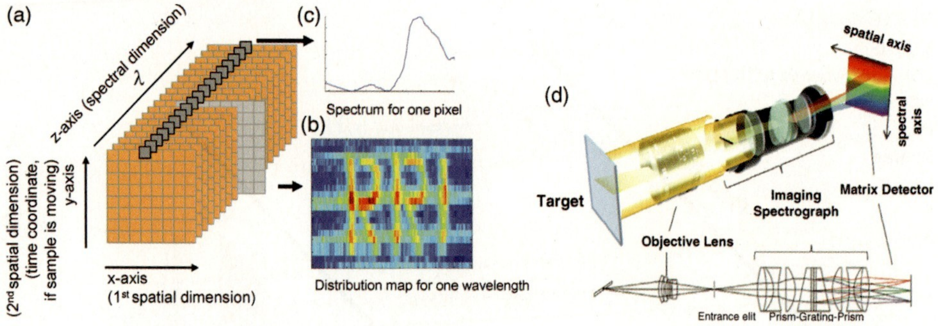

Figure 1.30 left: Sketch of three-dimensional hypercube (a) and distribution map for one selected wavelength $\lambda = 1506$ nm (b) and one spectrum along the spectral direction for one pixel $x = 90$, $y = 10$ (c). (d) Schematic drawing of a pushbroom imager (courtesy of Specim, Finland).

oriented strand boards (OSB). The objective of the chemometric treatment is to separate the information "wood chip" from the information "binder" and is intended finally to determine the thickness of the binder layer. A hyperspectral image was recorded for this sample with a pushbroom imager from Specim Finland. The first spatial resolution (x-axis) contains 180 pixels and each pixel represents 0.1 mm on the sample. For the second spatial direction (y-axis) 20 line-scans were performed after positioning the sample on a conveyer belt another 0.5 mm further in the y-direction from the starting position. This means 20 pixels for this y-direction with a spatial pixel resolution of 0.5 mm. The chemical information for each pixel is contained in 214 NIR absorption values in the wavelength range from 900 to 1700 nm (spectral pixel resolution is 3.7-nm). The resulting three-dimensional data cube ($180 \times 20 \times 214$) is sketched in Figure 1.30 together with a distribution map for one selected wavelength (1506 nm) and one spectrum along the spectral direction for pixel $x = 90$, $y = 10$. The full hyperspectral image has ($180 \times 20 = 3600$) spectra with 214 spectral readings in each spectrum.

To get the full chemical information out of the hyperspectral image it is necessary to look at the full spectral information contained in the image. This can be done by applying multivariate data analysis methodologies to extract the relevant information and at the same time reduce the dimensionality of the data. For images it is called multivariate image analysis (MIA) and for a qualitative analysis it is very often linked to principal component analysis (PCA). If a quantitative value for a quality or process parameter is required it is called multivariate image regression (MIR) and is mostly related to the partial least squares algorithms (PLS) [61].

When an optical fiber bundle is coupled to a pushbroom imager, multi-point spectroscopy at different locations in a reactor will be possible. These fibers can, however, also be coupled to various probes, thereby enabling simultaneous measurements like reflection, transmission or ATR measurements (see Section 6.2).

Pushbroom imaging technology is, for example, used to identify plastics in waste management systems. Figure 1.31 shows an example of a commercial application in separation of the different polymers automatically on a conveyor belt.

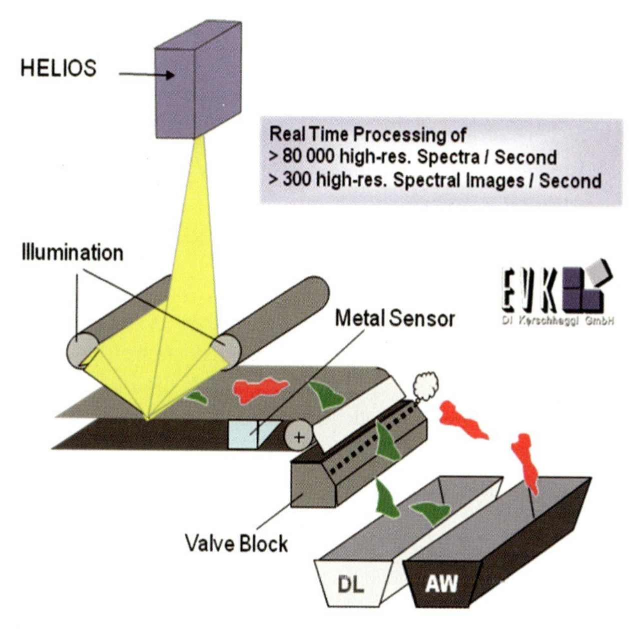

Figure 1.31 Real time characterization of polymers in plastic waste separation (courtesy of EVK company).

1.5
Survey Through Industrial Applications

1.5.1
Selection of Applications

This short survey through selected industrial applications of process analytical technology in the industry should illustrate the potential and widespread use that such methods already experience in the current industrial landscape. It should stimulate interest in potential problem solutions and with several thousand publications on the subject only in the course of the last five years is by far not intended to be comprehensive.

The objective of this section is to give also a short overview of the different applications of hyperspectral imaging in industry and science. It should provide some ideas on the heritage of information which is gained from hyperspectral images. Historically, the pharmaceutical industry uses mainly staring imaging technology, whereas food and agriculture is more focused on pushbroom imaging technology due to its background in remote sensing. Whiskbroom imaging technology will not be included in the selected applications as this technique is more restricted to scientific interests [62].

It is by far beyond the scope of this chapter to mention all relevant techniques and methods in detail, in the following sections only a brief summary of important applications of the various optical spectroscopic techniques is given. For an in-depth summary of new developments in PAT the reader is referred to the review articles on process analytical chemistry by Workman and coworkers, for instance Refs [2, 63] which are published every couple of years and reflect the most important new developments in the field. Recent accounts of the state of the art of process analytical technology are also to be found, for example, in [6, 33, 64, 65].

1.5.2 Pharmaceutical Industry

The most important applications of process analytical technology in the pharmaceutical industry cover on-line measurements in raw material control, reaction monitoring and control, crystallization monitoring and control, drying, milling, cleaning validation, granulation and blending, drug polymorphism detection, particle size, moisture content determination, and tablet analysis, as well as packaging.

1.5.2.1 UV/Vis Spectroscopy

Spectroscopic analysis in the pharmaceutical industry is very often concerned with the analysis of drug formulations such as tablets. For instance, dissolution testing of tablets is mandatory and there exist guidelines of how to perform these tests. Diode-array spectrometer-based UV/Vis techniques can be employed in this context to monitor the dissolution behavior of tablets [66]. Another important application of UV/Vis spectroscopy is the controlling of cleaning operations. The cleanliness of vessels intended for pharmaceuticals, beverage or food production has to be guaranteed when production shifts from one batch to another with an intermittent cleaning step. Fiber-optic UV/Vis spectrometers have been used for this purpose. UV/Vis spectroscopy has also been used to determine the active ingredients content in tablets, for example paracetamol, ibuprofen, and caffeine [67].

1.5.2.2 NIR Spectroscopy

NIR spectroscopy is an established measurement technology that has been employed for several decades under *in situ* conditions. In comparison to the chemical industry, the diversity of industrial operations is much smaller. Hence, a number of standard applications of NIR spectroscopy have been developed that are established and widely used by numerous pharmaceutical companies. Typical standard applications are, for example, the NIR spectroscopic control of raw material identity by using fingerprint analysis of the incoming substances, the NIR monitoring of mixing processes, the NIR monitoring of granulation, or the NIR analysis of tablet formation and compaction. While raw material control in many cases is accomplished by laboratory equipment, in-process applications are of importance, especially in the control of the granulation process. An important target signal here is the time-dependent moisture content of the aggregate in the course of particle compaction. Besides monitoring the kinetics of the granulate water content, the kinetics of changes in the particle size distribution can be determined [68]. NIR spectroscopy is also used in the characterization of the mixture homogeneity of powder preparations. Here, two major strategies have been followed: (i) on-line NIR monitoring of the mixing process [69–72] by implementing either stationary or moving NIR-probes directly in the mixing machinery, and (ii) imaging approaches, which will be described below. In the NIR spectroscopic analysis of tablets, the recent development of very rapid measurement devices based on diode array spectrometers allows 100% control of all manufactured tablets [73, 74], which represents an important step forward from random sampling. A typical problem in tablet manufacturing is not only the absolute amount of active ingredient but also its distribution within the particular formulation. While chromatographic analysis upon homogenization of the tablet by dissolution reveals the actual overall content of the drug in the tablet, the information generated by spectroscopic measurements is more complicated to unravel. Since NIR spectroscopy of tablets is typically carried out in diffuse reflectance the simultaneous contributions of scattering and absorption phenomena have to be considered and the arguments presented in Section 4.3 need to taken into account. Besides photon diffusion spectroscopy, here, too, imaging methods will play an increasingly important role in process analysis. Besides the active ingredient content and its distribution, in the automated analysis and quality control of tablets, physical parameters such as mass, diameter, thickness and hardness of the tablets are targeted [75, 76]. More examples on the application of NIR spectroscopy in combination with chemometric techniques can be found in the recently published review by Roggo [77] and in [78].

1.5.2.3 **Raman Spectroscopy**

Since interpretation of Raman bands is much more straightforward than qualitative interpretation of NIR bands and, hence, with Raman less chemometric effort is required, information deduction from Raman spectroscopy provides great potential also for the pharmaceutical industry. Monitoring of pharmaceutical blend homogeneity has been successfully accomplished using Raman based on a calibration-free approach by mean square difference between two sequential spectra; see, for example, [79]. In this study it was demonstrated that particle size and mixing speed significantly influence the time required to obtain homogenous mixing. Raman spectroscopy has also been used to follow the kinetics of polymorph conversion from one form to another [80]. Automated analysis of micro-titer plates can also be performed using Raman spectroscopy. Since it is possible to distinguish between crystal forms Raman spectroscopy has been suggested as a tool for high-throughput screening for different crystal structures [81]. Raman spectroscopy can also be used in combination with NIR data to determine correlations between physical pharmaceutical properties of tablets, such as hardness, porosity, and crushing strength [82]. Among others, the review published by Barnes [83] summarizes more recent examples of the application of Raman spectroscopy in the monitoring of pharmaceutical processes.

1.5.2.4 **Imaging Techniques**

There are still only a limited number of real life on-line applications for chemical imaging in process control in the pharmaceutical industry. Exhaustive reviews of different applications can be found in [33, 84]. Focus is mainly on three different uses: blend uniformity of powders and tablets, composition and morphological features of coated tablets and granules, spatial changes during hydration, degradation and active release.

Pharmaceutical solid dosage forms are ideally suited for NIR chemical imaging. The systems are chemically complex and the distribution of the components affects dramatically the product quality and performance [85]. Recently, Lewis and coworkers compared wet granulation, direct compression and direct compression together with a micronized API. Besides the evaluation of the homogeneity of the mixing, they could also qualify the appearance and the morphology of API hot spots and agglomeration [86]. It is also possible to relate this information to the performance of the final product [87].

In another work, the utility of NIR chemical imaging in measuring density variations within compacts is demonstrated. The data are also used to relate these variations to tableting forces which are controlled by frictional properties and the quantity of the Mg stearate concentration during production [88]. This is possible as the spectral information also includes the percentage of specular reflected light at the surface of the tablet and the change in penetration depth of the photons which change after compaction.

Counterfeit pharmaceutical products are a real threat to the health of patients. NIR chemical imaging provides a rapid method for detecting and comparing suspected counterfeit products without sample preparation. The advantage of imaging is that the discrimination of the tablets is not only caused by changes in the chemical composition, but also from its spatial distribution. This reflects the use of different raw materials as well as the distribution of the components in the tablet which depends strongly on processing conditions [89]. Thus the combined chemical and morphological information provides an individual measure for the characterization of tablets.

Applications of chemical imaging in the pharmaceutical industry are mainly related to NIR imaging. Some papers describe also the use of terahertz imaging. Terahertz is ideally suited to the identification of chemical components with strong phonon bands, thus components with high crystallinity can easily be distinguished from amorphous systems [90].

1.5.3
Food and Agriculture

Spectroscopic quality monitoring, and especially NIR spectroscopy, has a very long tradition in the food and agricultural industries. The main areas of application here regard the quality control of raw materials, intermediate products and final products. Many different sample types have to be dealt with, as reflected by the vast diversity of possible food stuff, ranging from liquids, powders, slurries to solid materials. One major problem often encountered in the application of NIR in food technology is the preparation of appropriate reference materials – soft matter such as food usually is connected with complex and varying matrix systems. Typical applications include the transmission, transflection and diffuse reflection measurement of crop grains and seeds, fruits and vegetables, livestock products, beverages, marine foods and processed foods.

1.5.3.1 UV/Vis Spectroscopy

UV/Vis spectroscopy is still an under-represented technique in process analysis. The advantages of UV/Vis spectroscopy are the low cost and high reproducibility of the equipment and its precision, especially in quantitative analysis. In recent years, UV/Vis analysis has been successfully implemented in the beverage industry and for wine color analysis [91–93]. The main application is the use of color cameras in food processing, which is also a sort of spectroscopic application.

1.5.3.2 NIR Spectroscopy

Water is an important issue in the food industry and can be measured using NIR spectrometers integrated in the food processing facility [94]. Typically, a rotating filter wheel is used and the characteristic absorbance peak of water at 1940 nm is used for calibration. By this method, water has been determined in a wide variety of dry products, such as instant coffee powder, potato chips, bonbons, tobacco, wheat, noodles, cookies, or dry milk powder [95]. Another important food constituent is fat, which may also be measured using infrared spectroscopy; in aqueous systems, however, it is preferable to use measurements in the mid-infrared range to avoid interfering water bands [96].

Although determination of major components in foodstuff, such as acid or sugar content, is still an important task, increasing attempts are made to determine complex quality parameters governing the expected sensory perception or storage stability of a certain product instead of only measuring one defined component. For comprehensive and recent reviews on the subject, see, for instance [97–101].

Assuring food safety is still a key issue in the food industry. Bacteria usually show broad bands in the NIR region even on taxonomically unrelated bacteria. However, using MVA it is possible to differentiate most of the bacteria [102].

1.5.3.3 Raman Spectroscopy

Raman spectroscopy is also a versatile method in applications related to food and agricultural products [103]. Food components, such as carbohydrates, edible oils, cyanogenic components, or proteins can readily be identified due to their characteristic absorption pattern by means of univariate characterization and peak assignment. For instance, proteins can be studied by comparing changes in the amide I (carbonyl stretch) band, which shifts from 1655 cm^{-1} for the α-helix to 1670 cm^{-1} for the anti-parallel β-sheet structures and, hence, allows detailed studies of food stuff. Cyanides show a very distinctive sharp band for the $C \equiv N$ triple bond at 2242 cm^{-1}. Inorganic food constituents, such as calcium carbonate, can also be identified due to their very sharp spectra. It was, for instance, possible to characterize the pre-freezing treatments of shrimp by studying the relative differences in Raman signals obtained from different crystal structures of calcium carbonate (ratio between calcite and vaterite), respectively, by studying the dehydration of the hexahydrate (ikaite, absorbance at 1070 cm^{-1}) to the anhydrous form (vaterite, 1089 and 1075 cm^{-1}).

Certainly, the potential of Raman spectroscopy can be enhanced by subjecting the spectral information to chemometric post-treatment, opening the field for rather quantitative applications of Raman methods. For example, alginate powders have been studied quantitatively with respect to their characteristic β-D-mannuronic acid to α-L-guluronic acid (M : G) ratio. It is recognized that Raman spectroscopy must no longer be regarded as an exotic tool for routine analysis of soft matter like food but has its rightful place in the PAT toolbox available for the food and agricultural industries [103].

1.5.3.4 **Imaging Techniques**

On-line chemical imaging in agriculture is mainly remote sensing. Satellite or aerial remote sensing (RS) technology uses nowadays pushbroom imaging technology in the Vis, s-NIR and NIR-range. Vegetation images show crop growth from planting through to harvest, changes as the season progresses and abnormalities such as weed patches, soil compaction, watering problems, and so on. This information can help the farmer make informed decisions about the most feasible solution.

The differences in leaf colors, textures, shapes, or even how the leaves are attached to plants, determine how much energy will be reflected, absorbed or transmitted. The relationship between reflected, absorbed and transmitted energy is used to determine spectral signatures of individual plants. Spectral signatures are unique to plant species [104].

Remote sensing is used to identify stressed areas in fields by first establishing the spectral signatures of healthy plants. For example, stressed sugar beets have a higher reflectance value in the visible region of the spectrum from 400–700 nm. This pattern is reversed for stressed sugar beets in the non-visible range from about 750–1200 nm. The visible pattern is repeated in the higher reflectance range from about 1300–2400 nm. Interpreting the reflectance values at various wavelengths of energy can be used to assess crop health. The comparison of the reflectance values at different wavelengths, called a vegetative index, is commonly used to determine plant vigor (for details, see [104]).

In the food industry, numerous on-line controls are still made by human vision, especially for sorting bad looking products. Cameras can perform this task more efficiently and, using RGB cameras, a limited certain spectral differentiation is possible in machine vision [105]. However, due to the great variety of states (solid, fragmented, etc.) of the food, shapes, color and chemical composition, as well as seasonal variations it is difficult to monitor and control food in an unbiased manner. The driving force to introduce NIR imaging was to qualify food not only on its appearance but also on its chemical composition, such as water or starch content. An extensive demonstration of different applications is given in [105], which includes on-line characterization of chemical composition, detection of external contamination, surface and subsurface nonconformities and defects in food.

The airborne or on-line information for process control in food qualification can be complemented with a higher resolution by hyperspectral imaging techniques on a laboratory scale. The ripeness of tomatoes was qualified using Vis-spectroscopy [106], moreover, apples were qualified on the quantification of starch and other chemical compounds [107, 108]. Differences in phenol-typing in plant breeding development can also be visualized by NIR imaging as shown in [105].

Chemical imaging in food and agriculture can also be used to identify diseases, rot and contamination by insects, for example, larvae. A method to detect animal proteins in feed is described in [108]. The objective was to determine the limits of detection, specificity and reproducibility.

1.5.4 Polymers

Process analytics is even more widespread in the chemical industry and such diverse processes as polymerization, halogenation, calcination, hydrolysis, oxidation, corrosion, purification, or waste disposal, to name just a few, can be advantageously monitored using PAT tools. As an example for the chemical industry, a few application examples from polymer manufacturing are given.

1.5.4.1 UV/Vis Spectroscopy

Applications of UV/Vis spectroscopy in the polymer industry are numerous; for example it can be successfully used to determine the color of extruded plastics [109]. Color is a key quality criterion for the customer and automated on-line analysis of color is of great importance and relatively easily accomplished, and various experimental set-ups have been described. Another example of the industrial application of UV/Vis spectroscopy is determination of the thickness of polymer films.

1.5.4.2 NIR Spectroscopy

NIR spectroscopy is widely used in laboratory and industrial applications for material classification [65]. One of the most important processes in the industrial synthesis of polymeric materials is polymer extrusion. NIR has been employed to control the extrusion process, for example in terms of characterization of the starting formulation and of the final polymeric product [110–112]. Moreover, the additive content of polymer blends can be determined by using NIR [113, 114]. For that kind of measurement, on-line probes can be installed before or after the actual extrusion unit. More sophisticated application of NIR process analytical technology uses the direct implementation of NIR sensors in the extrusion machinery. Here, a flow cell with an integrated NIR sensor may be located after the mixing unit, directly before the hot polymer melt enters the mold, and spectra may be recorded either in transmission or diffuse reflectance, depending on the transparency of the polymer. Since the absorbance of the glass fiber optics has been shown to be temperature dependent [115], the temperature needs to be carefully controlled in the measurement chamber and the measured spectra have to be baseline corrected. Important requirements for reliable extrusion monitoring are reproducibility, accuracy and long-term stability of the NIR system [116]. For each measurement position (for example, at the end of the extruder, in the transport zone of the extruder, or at the cooling, extruded polymer material) separate calibration has to be performed that, besides polymer composition, also includes parameters like color or particle content and particle size distribution [117].

The moisture content in polymers has also been determined by NIR spectroscopy [116]. Besides composition, physical parameters of polymers such as density [118] and melting flow index [119] have been determined by NIR, and even more complex rheological parameters have been correlated with NIR spectra [120]. When NIR spectroscopy was used in combination with an ultrasonic treatment, even the bubble formation in the transformation of polystyrene was observed, which demonstrated the versatility of the method [121]. A review of process analytical applications in the polymer industry has recently been published by Becker [116]. Siesler also gives a very good overview of the potential of NIR in polymer analysis [122].

1.5.4.3 Raman Spectroscopy

Raman spectroscopy has been shown to have great potential in polymer production, for example for monitoring the emulsion polymerization of various polymers [123–126] or polymer curing reactions [127, 128], and also for the determination of the residence time of TiO_2-containing polypropylene in an extruder [129]. Polymer films can also be characterized in real-time during the blown film extrusion process with respect to composition, crystallinity and their microstructure development [130]. Besides polymer bulk and film characterization, application of Raman spectroscopy as a PAT tool may also be very promising in the characterization of synthetic and natural polymer fibers and their composites [131], foams [132], and even liquid crystals [133].

However, so far not so many applications of Raman spectroscopy in industrial polymer processing have been described in comparison to NIR applications. NIR spectroscopy is by far the most commonly employed method here. One major reason for this is the much higher cost of Raman equipment. Recent developments towards lower investment costs will certainly broaden the scope of potential applications.

1.5.4.4 Imaging Techniques

While standard spectrometers only allow measurement at one sampling point at a time, NIR spectral imaging techniques can identify, in real-time, both the size and shape of an object, as well as the material it is made from. The robust classification of materials, such as polymers, is based on their characteristic reflectance spectra. Sorting, for example, requires the correct material, size and shape of the entire object to be known for reliable separation [134–136].

1.6 Perspectives

1.6.1 Technology Roadmap 2015 +

Regarding the various PAT, the following trends in future development of process analytical equipment have been identified and summarized in the roadmap 2015 + for process sensors published by NAMUR/GMA [137].

1) It is expected that sensor systems will display much greater robustness and long-time stability, that is, they will be required to depend on much less efforts in maintenance and, hence, will have a significantly lower cost-of-ownership
2) New in-process sensors will not only be incorporated in the design of newly erected industrial plants but they will also be employed in the optimization of existing processes
3) The new requirements for process analysis systems will not be exhausted by the pure collection and storage of process data. Information on the current state of intermediates while production is still operating and predictions on expected end-product properties will be increasingly used for controlling and process management purposes.
4) For specific purposes, an increased accuracy of the gathered process information will be required
5) As a most important trend, spatially resolved information of process and material parameters will become of increasing importance.
6) Since interfaces play an important role for practically all industrial and engineering areas, the localization, identification and characterization of interfaces will play an increasingly important role in process analysis applications
7) There is a growing trend toward manufacturing products in an environmentally compatible way. This means that a growing number of synthetic processes currently performed in a classic chemical pathway will be translated into biotechnological approaches. As a consequence, process monitoring systems suitable for biotechnological applications, and specifically fermentation processes, will gain in importance and will have to be further developed and refined
8) In bioprocess technology, an industrially feasible and robust in-process analysis of specific targeted proteins would be a revolutionary achievement.
9) In order to comply with an ever increasing cost-pressure imposed on industrial processes, the substitution of sophisticated and expensive process analysis methods carrying a significant cost of ownership during maintenance by low-cost, ideally single-use devices that are disposed of after having delivered the required information is an important task for future research and development activities in the field of process analyzers.
10) The growing use of renewable resources or recycled materials as environmentally compatible starting materials for new products imposes novel analytical problems on process analytical devices which will have to be addressed in future development and applications of process analysis
11) The sensitivity of process analysis devices in many cases will have to be significantly increased. For instance, in the analysis of potentially harmful or catalytically effective trace components in gases, continuously lower detection limits will be required
12) In the context of a holistic process management, there is also an increasing demand in developing non-invasive analysis techniques in the field of warehouse logistics and product organization

13) An overall trend towards in-line measurements is evident, which in turn means that intensification of research on the translation from laboratory analytical methods into stable, processable and robust in-line methods that can routinely be applied will be required

Some of the new challenges and developments in PAT are discussed in the following sections.

1.6.2
Medical Applications and Tomographic Imaging

The next step in imaging is to integrate the *z*-spatial axis into the data cube. Especially for medical applications, it is important to feature the spatial distribution in all three directions.

Techniques which are available for 3D-imaging often lack the spectral information or are difficult to introduce into in-line or high throughput applications like optical coherent tomography (OCT) and laser scanning confocal microscopy. As stated in the introduction, medical ultrasonography, magnetic resonance imaging (MRI) and confocal microscopy are not suited to morphological or spectral imaging: the first two have poor resolution; the last lacks millimeter penetration depth and suitable applications in process control. On the other hand, in process analysis, electrical resistance tomography (ERT) is very popular nowadays and involves measurements around the periphery of an object for example, a process vessel or a patient. The instrument operates by taking data from an array of electrodes in contact with the process media. The electrical field lines "communicate" and are, therefore, affected by the presence of conducting and nonconducting regions [138]. However this technique does not provide any information on a molecular basis.

New developments in laser scanning confocal microscopy allow the use of different lasers at selected wavelengths; thus spectral imaging may then be realized. Also, when fluorescence is excited during scanning, for example, acousto-optical tuneable filters can be used to analyze the fluorescence spectrum at each focal point. However, there is still a long way to go for on-line or at-line applications in process control. This is also true for fluorescence imaging techniques, including lifetime imaging systems without using fluorescent markers [139]. On the other hand, multiphoton tomography is now at the edge of a broad "at-line" application in clinical high throughput *in vivo* studies [140].

Diffuse optical imaging (DOI) is a new emerging technique for functional imaging of biological tissues. It involves generating images using measurements in the visible or s-NIR-light scattered across large and thick tissues (about several centimeters) [141, 142]. As shown in Figure 1.7, penetration depth in the s-NIR range is high, therefore the objective of DOI is to provide low cost sensors for *in vivo* applications including imaging for example, of breast and brain cancer. The key issue, however, is the image reconstruction which is difficult due to the scattering of the material and, therefore, low spatial information of the data. Progress has been made on developing more realistic and efficient models of light transport in tissue and solving the *ill-posed inverse* problem in an increasingly rigorous way [141].

Another way to solve the diffusion of photons problem and to localize spatially resolved features is time-resolved NIR-spectroscopy. This technique enables the separation of the absorption properties of the sample from the scattering properties. This improves the localization of changes in the physical parameters, for example, in a tablet [143].

Until recently, researchers did not extensively explore the material interactions occurring in the terahertz spectral region, the wavelengths that lie between 30 μm and 1 mm. Terahertz spectroscopy can be used to image through materials yielding high spatial resolution in the x, y and z directions. The technique also has the ability to resolve both time and amplitude information. Terahertz has the potential to be applied to many applications, including: on-line non-destructive testing for quality control of products and packaging for industrial markets; weapons and explosives detection for homeland security and defense markets; topical imaging for the medical market; quality control of external skins on space vehicles for the aerospace and military markets and many more. The technique is still quite expensive but will find its way into process control in the future [144].

1.6.3 Multi- Point-Information Systems in Manufacturing

The main objective of any process control is to guarantee a constant and specified product quality. Especially, maximum yield at a minimum of manufacturing costs are more and more important parameters, even in the pharmaceutical industry. Complex manufacturing with many different successive production steps must introduce a feed-back and a feed-forward control in the future. This requires specific efforts to integrate at each step of the production line an appropriate quality control for example, by spectroscopic methods.

Instead of using a single spectrometer at each individual production step, pushbroom imaging systems with numerous attached fiber bundles allow individual control of the quality at each intermediate and the final step. In this context, the pushbroom imager is used as a multipoint information source and can substitute a moving multiplexer. Many fibers per spectrometer can be used for simultaneous measurements. In addition, different spectrometer technologies for UV–Vis–NIR, fluorescence or Raman can also be combined. Thereby the probe becomes a multi-information system, which describes the sample in an ideal and complete way.

Figure 1.32 shows a schematic diagram of some production steps in tablet production with the integration of a multipoint spectroscopic information system using push-broom imaging technology.

In this example the pushbroom technology is able to substitute several single spectrometer systems. For example, at the fluid bed dryer, the moisture or the homogeneity of the blending process is controlled on a molecular level. Combined with multivariate data analysis science-based production can be realized.

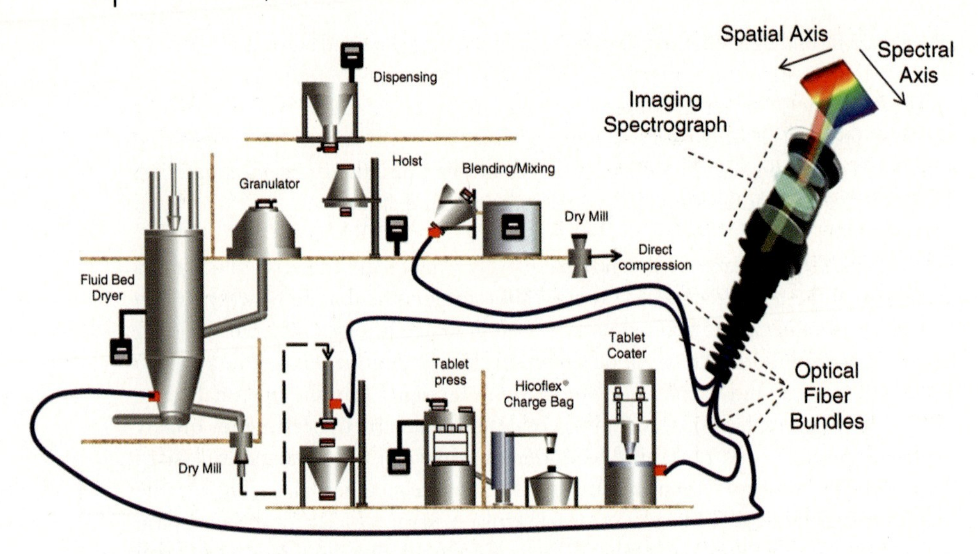

Figure 1.32 Schematic diagram of a tablet production process with different possibilities for pushbroom imaging.

1.6.4 Multimodal Spatially Resolved Spectroscopy

In recent years multimodal methods have matured to an accepted new technology, especially in the field of medicine [145]. The term multimodal relates to the possibility of collecting information on chemical and physical parameters at the same time. Not only chemical changes at the molecular level but also morphological changes in the sample can be detected. Complex interrelations of the system can be reduced and described in an easy and complete way. Several different imaging techniques are frequently used to realize the multimodal concept.

Process control is subject to a continuous change and the approach of multimodal methods will be extended in the future. The increasing complexity of high quality products and the cost pressure demand new strategies. In the future, the focus will not only be on process optimization but also on tailoring property profiles of products according to specifications and individual desires of customers. This will allow the creation of operating instructions, including the choice of materials and process parameters, to design a product with a given preference profile. Multimodal methods will then also enable the realization of multipurpose optimization [146].

Optical molecular spectroscopy, and in particular chemical imaging, will play an important role in implementing these objectives. By combining different wavelength ranges and techniques multimodal spectroscopy synergistically provides much more useful information than each technique on its own. It produces complementary chemical and morphological information about reaction products. Measurements can be carried out quickly, sensitively, selectively and economically at a reasonable price. In principle, the implementation of the multimodal approach can be achieved in three different ways.

The first implementation strategy uses different wavelength ranges. The UV range is more suitable for scattering effects, while the short near-infrared range has a much higher penetration depth because of the low absorption probability in organic samples. Fluorescence techniques for excitation and emission spectroscopy can be used to separate molecules due to their different spectral signatures. Raman spectroscopy as a light scattering technique can be used as a fingerprint method to identify molecular structures or bonding effects.

The second implementation strategy involves using different technologies within the same wavelength ranges. For example, diffuse reflectance and transmission in the UV or NIR can be used to separate morphological scattering from chemical effects [9]. In addition attenuated total reflection (ATR) spectroscopy can be used to analyze highly concentrated samples without prior dilution and avoiding interference by scatter.

The third implementation strategy deals with laterally resolved measurements to achieve the desired differentiation. Angular resolved spectral measurements or line scans with a pushbroom imaging system lead to different penetration depths, which are highly specific for particulate systems. Besides the chemical information, parameters like homogeneity, particle size, particle distribution and density can also be detected.

1.6.5 Microreactor Control and Reaction Tomography

Microreactor technology is a powerful tool and has become indispensable over a wide application range from organic synthesis to enzymatic controlled reactions [147, 148]. A miniaturization with the possibility of reaction tomography, and thus a significant reduction in costs, will be the next step in the foreseeable future. Many chemical reactions, especially organic and fine chemical synthesis, could already be transferred to continuous microreaction processes [149]. The small geometric dimensions result in an intensified mass and heat transfer which often leads to increased yield and selectivity compared to the classical batch approach. However, microreaction technology today is still at the threshold between academia and industry.

A crucial factor for the successful implementation of microstructured production processes in industry is a suitable process analytical technology. Time and spatially-resolved on-line analysis must be implemented directly in the microfluidic channels. Thus parallel and multiplexed measurement technologies are needed to reduce costs and increase the robustness of information. To date, no commercially available solutions exist.

Typical state of the art procedures to study chemical processes use flow cells positioned after the microstructured environment or off-line methods. However, these approaches have several disadvantages. A critical point is the creation of distorted results due to changed geometrical proportions with measurements

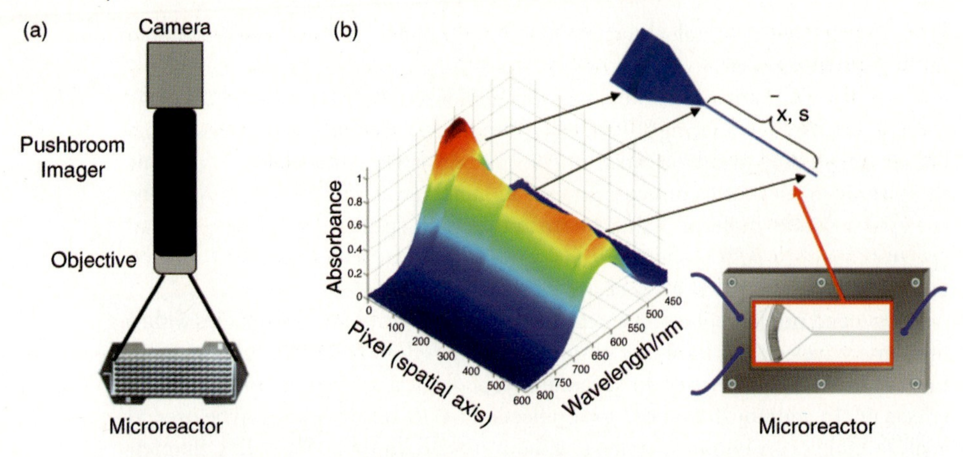

Figure 1.33 (a) Schematic diagram of the experimental set-up for an on-line spectrometer system. (b) Example of reaction monitoring of different concentration levels of an aqueous blue solution flowing through a microreactor system.

downstream of the microstructured environment. Moreover, no information about the actual reaction process inside the microreactor is generated. The major disadvantage, however, especially for industry, is the increasing costs for analytical devices required to assure constant product quality. The production of high added value chemicals by means of a microstructured process requires several hundred microreactors in order to produce sufficiently large amounts of final product. Therefore, large numbers of flow cells, each attached to a separate spectrometer (= pushbroom imager), are needed to meet these unique requirements.

An alternative technology is to analyze several microchannels simultaneously or to analyze a single microchannel spatially resolved along the reaction path using pushbroom imaging technology. This type of optical on-line spectroscopy is an ideal tool to characterize chemical reactions in a fast and reliable way, even on a molecular

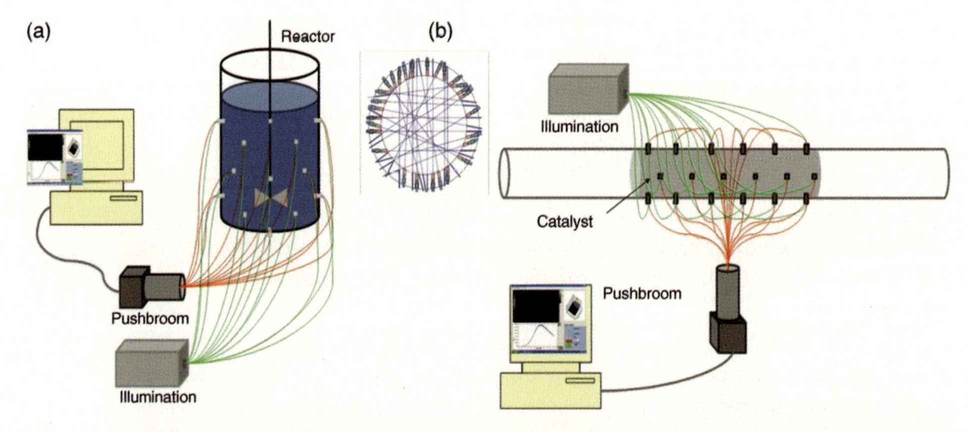

Figure 1.34 Pushbroom imager as a multipoint spectrograph for tomography of a batch reactor (a) and a continuous reactor (b).

level. Transmission or reflectance spectra are registered through a fixed prism–grating–prism optics with a two-dimensional CCD camera attached to it. Thereby, the x-axis of the CCD array corresponds to the spatial resolution and the y-axis of the camera provides the full spectrum of the sample. Figure 1.33 shows the experimental set-up for an on-line spectrometer system as well as an example for reaction monitoring.

Figure 1.34 shows a sketch of a pushbroom imager connected to a fiber bundle where a batch reactor can be analyzed, and another example is shown where spectroscopy is applied to a continuous reactor.

This multipoint spectroscopy can also be applied to microreaction systems to analyze along a reactor or as a multiplexed microreactor analyzing tool.

References

1 U. S. Department of Health and Human Services, Food and Drug Administration, *Guidance for Industry "PAT – a Framework for Innovative Pharmaceutical Development, Manufacturing and Quality Assurance"*, FDA, Silver Spring (2004) http://www.fda.gov/cvm/guidance/published.html.

2 Workman, J., Koch, M., Lavine, B., and Chrisman, R. (2009) Process analytical technology. *Anal. Chem.*, **81** (12), 4623–4643.

3 Jochem, R. (2010) *Was kostet Qualität? Wirtschaftlichkeit von Qualität ermitteln*, Carl Hanser Verlag, München, Germany ISBN 978-3-446-42182-0.

4 Garwin, D. (1984) What does product quality really mean? *Sloan Manag. Rev.*, **25** (3), 25–43.

5 Zollondz, H.D. (2002) *Grundlagen Qualitätsmanagement. Einführung in Geschichte, Begriffe, Systeme und Konzepte*, 2nd edn, Oldenbourg Verlag, München Wien, Austria, ISBN 978-3-486-57964-2.

6 Kessler, R.W. (ed.) (2006) *Prozessanalytik. Strategien und Fallbeispiele aus der industriellen Praxis*, Wiley–VCH, Weinheim, ISBN 3-527-31196-3.

7 Kano, N., Seraku, N., Takahashi, F., and Tsuji, S. (1984) Attractive quality and must-be quality. *Quality*, **14** (2), 39–48.

8 Cohen, L. (1995) *Quality Function Deployment: How to Make QFD Work for You*, Addison Wesley., Reading, MA.

9 Kessler, W., Oelkrug, D., and Kessler, R.W. (2009) Using scattering and absorption spectra as MCR-hard model constraints for diffuse reflectance measurements of tablets. *Anal. Chim. Acta*, **642**, 127–134.

10 Mitra, A. (2008) *Fundamentals of Quality Control and Improvement*, 3rd edn, John Wiley & Sons, Inc., Hoboken, New Jersey, USA, ISBN 978-0-470-22653-7.

11 ISPE " ISPE Product Quality Lifecycle Implementation (PQLI) Guide: Overview of product design, development and realization: A Science- and risk based approach to implementation" http://www.ispe.org/ispepqliguides/overviewofproductdesign.

12 Woelbeling, C. (2008) Creating Quality by Design/Process Analytical Technology management awareness. *Pharm. Eng.*, **28** (3), 1–9.

13 Rehorek, A. (2006) Prozess-Flüssigchromatographie, in *Prozessanalytik. Strategien und Fallbeispiele aus der Industriellen Praxis* (ed. R.W. Kessler), Wiley-VCH, Weinheim, pp. 429–474.

14 Becker, T. and Krause, D. (2010) Softsensorsysteme-Mathematik als Bindeglied zum Prozessgeschehen. *Chem. Ing. Tech.*, **82** (4), 429–440.

15 BASF Publication.

16 Schmidt-Bader, T. (2010) PAT und QbD im regulatorischen Umfeld der Pharmazeutischen Industrie. *Chem. Ing. Techn.*, **82** (4), 415–428.

17 Maiwald, M. (2010) Prozessanalytik als Instrument des

Informationsmanagements in der Chemischen und Pharmazeutischen Industrie. *Chem. Ing. Tech.*, **82** (4), 383–390.

18 ICH website http://www.ich.org.

19 ICH Guideline Q8 (R2) (August 2009) Pharmaceutical Development, step 4 http://www.ich.org/fileadmin/Public_Web_Site/ICH_Products/Guidelines/Quality/Q8_R1/Step4/Q8_R2_Guideline.pdf.

20 ICH Guideline Q6A (October 1999) Specifications: Test procedures and acceptance criteria for new drug substances and new drug products: Chemical substances (including decision trees), step 5 http://www.ema.europa.eu/docs/en_GB/document_library/Scientific_guideline/2009/09/WC500002823.pdf.

21 Box, G.E.P. and Draper, N.R. (1987) *Empirical Model-Building and Response Surfaces*, John Wiley & Sons, New York, ISBN: 0-471-81033-9.

22 Ghosh, S. (ed.) (1990) *Statistical Design and Analysis of Industrial Experiments*, Marcel Dekker Inc., New York, ISBN: 0-8247-8251-8.

23 Pyzdek, T. (2004) *The Six Sigma Handbook. A Complete Guide for Green Belts, Black Belts and Managers at all Levels*, McGraw-Hill, New York.

24 Herwig, C. (2010) Prozess Analytische Technologie in der Biotechnologie. *Chem. Ing. Tech.*, **82** (4), 405–414.

25 Behrendt, S. (2006) Integrierte Technologie-Roadmap Automation 2015 + http://www.zvei.org/fileadmin/user_upload/Fachverbaende/Automation/Nachrichten/1_Roadmap_Behrendt.PDF.

26 http://en.wikipedia.org/wiki/Causality.

27 Esbensen, K.H. and Paasch-Mortensen, P. (2010) Process sampling: Theory of sampling –the missing link in process analytycal technlogies (PAT), in *Process Analytical Technology*, 2nd edn (ed. K.A. Bakeev), Wiley, UK, pp. 37–79.

28 http://www.cpac.washington.edu/NeSSI/NeSSI.htm.

29 U. S. Department of Health and Human Services, Food and Drug Administration, Guidance for Industry, "Process Validation: General Principles and Practices" (Revision January 2011) http://www.fda.gov/Drugs/GuidanceComplianceRegulatoryInformation/Guidances/default.htm.

30 Myers, R.H. and Montgomery, D.C. (2002) *Response Surface Methodology*, John Wiley and Sons, New York.

31 VDA Band 4: Sicherung der Qualität vor Serieneinsatz, VDA, Frankfurt, Germany (1986)

32 VDA Band 4: Sicherung der Qualität vor Serieneinsatz, Teil 2: System FMEA, VDA, Frankfurt, Germany (1996)

33 Bakeev, K.A. (ed.) (2010) *Process Analytical Technology*, 2nd edn, Wiley, UK.

34 Kessler, W. (2007) *Multivariate Datenanalyse für die Pharma-, Bio- und Prozessanalytik*, Wiley-VCH, Weinheim, ISBN 13: 978-3-527-31262-7.

35 Geladi, P., McDougall, D., and Martens, H. (1985) Linearisation and scatter correction for near infrared reflectance spectra of meat. *Appl. Spectrosc.*, **39**, 491.

36 Burger, J. and Geladi, P. (2005) Hyperspectral NIR Image Regression Part 1: Calibration and Correction. *J. Chemom.*, **19**, 355–363.

37 Martens, H., Nielsen, J.P., and Engelsen, S.B. (2003) Light Scattering and Light Absorbance Separated by Extended Multiplicative Signal Correction. *Anal. Chem.*, **75**, 394–404.

38 Isaksson, T. and Kowalski, B. (1993) Piece-Wise Multiplicative Scatter Correction Applied to Near-Infrared Diffuse Transmittance data from Meat Products. *Appl. Spectrosc.*, **47**, 702–709.

39 Barnes, R., Dhanoa, M., and Sister, S. (1998) Standard Normal Variate Transformation and Detrending fo Near Infrared Diffuse Reflectance. *Appl. Spectrosc.*, **43**, 772–777.

40 Brown, S.D., Tauler, R. and Walczak, B. (eds) (2009) *Comprehensive Chemometrics*, vol. 4, Elsevier, Amsterdam.

41 Jaumot, J., Gargallo, R., de Juan, A., and Tauler, R. (2005) A graphical user-friendly interface for MCR-ALS: a new tool for multivariate curve resolution in MATLAB. *Chemom. Intell. Lab. Sys.*, **76**, 101–110.

42 Ulmschneider, M. and Roggo, Y. (2008) *Pharmaceutical Manufacturing Handbook* (ed. S.C. Gad), John Wiley & Sons, Hoboken NJ, pp. 353–410.

43 Small, G.W. (2006) Chemometrics and near infrared spectroscopy: Avoiding the pitfalls. *Tr. Anal. Chem.*, **25**, 1057–1066.

44 Kjeldahl, K. and Bro, R. (2010) Some common misunderstandings in chemometrics. *J. Chemom.*, **24**, 558–564.

45 Smith, C.L. (2009) *Basic Process Measurements*, John Wiley & Sons, New York.

46 Kandelbauer, A., Gronalt, M., Lammer, H., Penker, A., and Wuzella, G. (2007) Kosteneinsparungen in der Holzindustrie durch Logistikoptimierung, Prozessanalytik und Wissensmanagement, Teil 1: Allgemeine Aspekte und Logistik-Management. *Holztechnol.*, **48** (6), 5–9.

47 Kappes, R. and Grimm, C. (2010) NIR Feuchtemesstechnik mit vereinfachter Kalibrierfunktion. *Techn. Mess.*, **77**, 293–304.

48 Hauschild, T. (2005) Density and moisture measurements using microwave resonators, in *Electromagnetic Aquametry, Electromagnetic Wave Interaction with Water and Moist Substances* (ed. K. Kupfer), Springer Verlag, Heidelberg.

49 Kappes, R., Grimm, C., and Scholz, J. (2010) Feuchtemesstechnik vom Labor bis in den Prozess. *Pharm. Ind.*, **72** (7), 1231–1238.

50 Kortüm, G. (1969) *Reflectance Spectroscopy*, Springer Verlag, New York, USA.

51 Kessler, R.W., Böttcher, E., Füllemann, R., and Oelkrug, D. (1984) *Fresenius Z. Anal. Chem.*, **319**, 695–700.

52 Rebner, K., Kessler, W., and Kessler, R.W. (2010) in *Science Based Spectral Imaging: Combining First Principles with New Technologies, Near Infrared Spectroscopy: Proceedings of the 14th International Conference, 7.-16.11 Bangkok, Thailand* (eds S. Saranwong, S. Kasemsumran, W. Thanapase and P. Williams,), IMPublications, Chichester, UK, pp 919–927, ISBN 978-1-906715-03-8.

53 Bentsson, O., Burger, T., Folestad, S., Danielsson, L.-G., Kuhn, J., and Fricke, J. (1999) Effective sample size in diffuse reflectance Near –IR spectrometry. *Anal. Chem.*, **71**, 617–623.

54 Shi, Z. and Anderson, C.A. (2010) Pharmaceutical Applications of Separation of Absorption and Scattering in Near-Infrared Spectroscopy (NIRS). *J. Pharm. Sc.*, **99**, 4766–4783.

55 Oelkrug, D., Brun, M., Hubner, P., and Egelhaaf, H.-J. (1996) Optical parameters of turbid materials and tissues as determined by laterally resolved reflectance measurements. *SPIE 2445*, **248**.

56 Dahm, D.J. and Dahm, K.D. (2007) *Interpreting Diffuse Reflectance and Transmittance*, NIR Publications, Chichester, UK.

57 Thenadil, S. (2008) Relationship between the Kubelka-Munk scattering and radiative transfer coefficients. *J. Opt. Soc. Am. A*, **25**, 1480–1485.

58 Clarke, F.C., Hammond, S.V., Jee, R.D., and Moffat, A.C. (2002) Determination of the information depth and sample size for the analysis of pharmaceutical materials using reflectance NIR microscopy. *Appl. Spectrosc.*, **56**, 1475–1483.

59 Hudak, S.J., Haber, K., Sando, G., Kidder, L.H., and Lewis, E.N. (2007) Practical limits of spatial resolution in diffuse reflectance NIR chemical imaging. *Nir News*, **18**, 6–8.

60 Griffiths, P.R. (2009) Infrared and Raman Instrumentation for Mapping and Imaging, in *Infrared and Raman Spectroscopic Imaging* (eds R. Salzer and H.W. Siesler), Wiley-VCH, Weinheim, Germany, pp. 3–64, ISBN 978-3-527-31993-0.

61 Lewis, E.N., Schoppelrei, J.W., Makein, L., Kidder, L.H., and Lee, E. (2010) Near infrared chemical imaging for product and process understanding, in *Process Analytical Technology* (ed. K.A. Bakeev), Wiley, pp. 245–278.

62 Kessler, R.W. (2011) *Handbook of Spectroscopy*, 2nd edn (eds G. Gauglitz and T. Vo-Dinh), Wiley-VCH.

63 Workman, J. Jr., Koch, M., and Veltkamp, D. (2007) Process Analytical Chemistry. *Anal. Chem*, **79**, 4345–4364.

64 Chalmers, J.A. (ed.) (2000) *Spectroscopy in Process Analysis*, Sheffield Academic Press, Sheffield, UK.

65 Burns, D.A. and Ciurczak, E.W. (eds.) (2001) *Handbook of Near-Infrared Analysis*, 2nd edn, Marcel Dekker, New York Basel.

66 Lu, X., Lozano, R., and Shah, P.(November (2003)) *In-situ Dissolution Testing Using Different UV Fiber Optic Probes and Instruments*, Dissolution Technologies, pp. 6–15.

67 Khoshayand, M.R., Abdollahi, H., Shariatpnahi, M., Saadatfard, A., and Mohammedi, A. (2008) *Spectrochim. Acta, Part A*, **70A** (3), 491–499.

68 Saal, C. (2006) Prozessanalytik in der pharmazeutischen Industrie, in *Prozessanalytik. Strategien und Fallbeispiele aus der Industriellen Praxis* (ed. R.W. Kessler), Wiley-VCH, Weinheim, pp. 499–512.

69 Ufret, C. and Morris, K. (2001) Modelling of powder blending using on-line near infrared measurements. *Drug Dev. Ind. Pharm.*, **27**, 719.

70 El-Hagrasy, A.S. and Drennen, J.K. (2004) PAT Methods provide process understanding: characterization of intra-shell versus inter-shell mixing kinetics in V-blenders. *NIR News*, **15**, 9–11.

71 Duong, N.-H., Arratia, P., Muzzio, F., Lange, A., Timmermans, J., and Reynolds, S. (2003) A homogeneity study using NIR spectroscopy: tracking Magnesium stearate in Bohle Bin Blender. *Drug Devel. Ind. Pharm.*, **29**, 679.

72 Winskill, N. and Hammond, S.,An industry perspective on the potential for emerhng process analytical technologies, Pfizer Global manufacturing services http://www.fda.gov/ohrms/dockets/ac/ac/02/briefing/3841B1_05_PFIZER.pdf.

73 http://www.uhlmann-visiotec.de/html/dinline-wirkstoffanalyse.html.

74 Herkert, T., Prinz, H., and Kovar, K.A. (2001) One hundred percent on-line identity check of pharmaceutical products by NIR spectroscopy of the packaging line. *Eur. J. Pharm. Biopharm.*, **51**, 9.

75 http://www.brukeroptics.de/ft-nir/tandem.html.

76 http://www.brukeroptics.com/downloads/AF512E_Tablet_Uniformity.pdf.

77 Roggo, Y., Chalus, P., Maurer, L., Lema-Martinez, C., Edmont, A., and Jent, N.J. (2007) *Pharma. Biomed. Anal.*, **44** (3), 683–700.

78 Doherty, S.J. and Kettler, C.N. (2005) On-line applications in the pharmaceutical industry, in *Process Analytical Technology* (ed. K.A. Bakeev), Blackwell Publishing, Oxford, UK, pp. 329–361.

79 Vergote, G.J., De Beer, T.R.M., Vervaet, C., Remon, J.P., Baeyens, W.R.G., Diericx, N., and Verpoort, F. (2004) Inline monitoring of a pharmaceutical blending process using FT Raman spectroscopy. *Eur. J. Pharm. Sci.*, **21**, 479–485.

80 Starbuck, C., Spartalis, A., Wai, L., Wang, J., Fernandez, P., Lindemann, C.M., Zhou, G.X., and Ge, Z. (2002) Process optimization of a complex pharmaceutical polymorphic systemvia in situ Raman spectroscopy. *Cryst. Growth Des.*, **2**, 515–522.

81 Hilfiker, R., Berghausen, J., Blatter, F., DePaul, S.M., Szelagiewicz, M., and Von Raumer, M. (2003) Hiigh throughput screening for polymorphism. *Chem. Today*, **21**, 75.

82 Shah, R.B., Tawakkul, M.A., and Khan, M.A. (2007) *J. Pharm. Sci.*, **96** (5), 1356–1365.

83 Barnes, S., Gillian, J., Diederich, A., Barton, D., and Ertl, D. (2008) *Am. Pharm. Rev.*, **11** (3), 80–85.

84 Reich, G. (2005) Near Infrared spectroscopy and imaging: Basic principles and pharmaceutical applications. *Adv. Drug Delivery Rev.*, **57**, 1109–1143.

85 Fernandez Pierna, J.A., Baeten, V., Dardenne, P., Dubois, J., Lewis, E.N., and Burger, J. (2009) Spectroscopic Imaging, in *Comprehensive Chemometrics*, vol. **4** (eds S.D. Brown, R. Tauler, and B. Walczak), Elsevier, Amsterdam, The Netherlands, pp. 173–197.

86 Makeln, L.J., Kidder, L., Lewis, E.N., and Valleri, M. (2008) Non-destructive evaluation of manufacturing process changes using near infrared chemical imaging. *NIR News*, **19**, 11–15.

87 Clarke, F. (2004) Extracting process related information from pharmaceutical dosage forms using near infrared microscopy. *Vibr. Spectrosc.*, **34**, 25–35.

88 Ellison, C.D., Ennis, B.J., Hamad, M.L., and Lyon, R.C. (2008) Measuring the distribution of density and tabletting force in pharmaceutical tablets by chemical imaging. *J. Pharm. Biomed. Anal.*, **48**, 1–7.

89 Dubois, J., Wolff, J.-C., Warrack, J.K., Schoppelrei, J., and Lewis, E.N. (2007) NIR chemical imaging for counterfeit pharmaceutical product analysis. *Spectrosc.*, **22**, 40–50.

90 Cogdill, R.P., Short, S.M., Forcht, R., Shi, Z., Shen, Y., Taday, P.F., Anderson, C.A., and Drennen, J.K. III (2006) An efficient method-development strategy for quantitative chemical imaging using teraherz pulse spectroscopy. *J. Pharm. Innov.*, **1**, 63–75.

91 Ghosh, P.K. and Jayas, D.S. (2009) Use of spectroscopic data for automation in food processing industry. *Sens. & Instrumen. Food Qual.*, **3**, 3–11.

92 Acevedo, F.J., Jiménez, J., Maldonado, S., Domínguez, E., and Narváez, A. (2007) Classification of Wines Produced in Specific Regions by UV-Visible Spectroscopy Combined with Support Vector Machines. *J. Agric. Food Chem.*, **55**, 6842–6849.

93 Yu, H.Y., Ying, Y.B., Sun, T., Niu, X.Y., and Pan, X.X. (2007) Vintage year determination of bottled rice wine by Vis–NIR spectroscopy. *J. Food Sci.*, **72**, 125–129.

94 Reh, C. (2006) Prozessanalytik in der Lebensmittelindustrie, in *Prozessanalytik. Strategien und Fallbeispiele aus der Industriellen Praxis* (ed. R.W. Kessler), Wiley-VCH, Weinheim, pp. 539–549.

95 Reh, C., Bhat, S.N., and Berrut, S. (2004) Determination of water content in powdered milk. *Food Chem.*, **86**, 457–464.

96 Lanher, B.S. (1996) Evaluation of Aegys MI 600 FTIR milk analyzer for analysis of fat, protein, lactose and solid non-fat: a compilation of eight independent studies. *J. AOAC Int.*, **79** (6), 1388–1399.

97 Kawang, S. (2006) Application of NIR spectroscopy to agricultural products and foodstuffs, in *Near–Infrared Spectroscopy: Principles, Instruments, Applications* (eds H.W. Siesler, Y. Ozaki, S. Kawata, and H.M. Heise), Wiley-VCH, Weinheim, pp. 269–288.

98 Kington, L.R. and Jones, T.M. (2001) Application for NIR analysis of beverages, in *Handbook of Near-Infrared Analysis*, 2nd edn (eds D.A. Burns and E.W. Ciurczak), Marcel Dekker, New York Basel, pp. 535–542.

99 Shenk, J.S., Workman, J.J. Jr., and Westhouse, M.O. (2001) Aplication of NIR spectroscopy to agricultural products, in *Handbook of Near-Infrared Analysis*, 2nd edn (eds D.A. Burns and E.W. Ciurczak), Marcel Dekker, New York Basel, pp. 419–474.

100 Osborne, B.G. (2001) NIR analysis of bakery products, in *Handbook of Near-Infrared Analysis*, 2nd edn (eds D.A. Burns and E.W. Ciurczak), Marcel Dekker, New York Basel, pp. 475–498.

101 Frankhuizen, R. (2001) NIR analysis in dairy products, in *Handbook of Near-Infrared Analysis*, 2nd edn (eds D.A. Burns and E.W. Ciurczak), Marcel Dekker, New York Basel, pp. 499–534.

102 Dubois, J., Lewis, E.N., Fry, F.S., and Calvey, E.M. (2007) Near infrared chemical imaging for high throughput screening of food bacteria. *NIR News*, **18**, 4–6.

103 Viereck, N., Salomonsen, T., van den Berg, F., and Engelsen, S.B. (2009) Raman applications in food analysis, in *Raman Spectroscopy for Soft Matter Applications* (ed. M.S. Amer), John Wiley & Sons, Hoboken NJ, pp. 199–223.

104 Campbell, J.B. (ed.) (2007) *Introduction to Remote Sensing*, 4th edn, Guildford Press, N.Y.

105 Bellon-Maurel, V. and Dubois, J. (2009) Near infrared hyperspectral imaging in food and agriculture science, in *Infrared and Raman Spectroscopic Imaging* (eds R. Salzer and H.W. Siesler), Wiley-VCH, Weinheim, Germany, ISBN: 978-3-527-31993-0, pp. 259–294.

106 Polder, G., Heyden, G.W.A.M.v.d., and Young, I.T. (2003) Tomato sorting using

independent component analysis on spectral images. *Real Time Imaging*, **9**, 253–259.

107 Peirs, A., Scheerlinck, N., de Baerdemaeker, J., and Nicolai, B.M. (2003) Starch index determination of of apple fruit by means of a hyperspectral near infrared hyperspectral imaging system. *J. Near Infrared Spectrosc.*, **11**, 379–389.

108 Fernandez Pierna, J.A., Baeten, V., and Dardenne, P. (2006) Screenig of compound feeds using NIR hyperspectral data. *Chemom. Intell. Lab. Syst.*, **84**, 114–118.

109 Baylor, L.C. and O'Rourke, P.E. (2005) UV-Vis for On-line Analysis, in *Process Analytical Technology* (ed. K.A. Bakeev), Blackwell Publishing, Oxford, UK, pp. 170–186.

110 McPeters, H.L. and Williams, S.O. (1992) 1 In-line monitoring of polymer processes by near-infrared spectroscopy. *Process Contr. Qual.*, **3**, 75–83.

111 Hansen, M.G., and Khettry, A. (1994) In-line monitoring of titanium dioxide content in poly(ethylene terephthalate) extrusion. *Polym. Engin. Sci.*, **34** (23), 1758.

112 Rohe, T., Becker, W., Krey, A., Nägele, H., Kölle, S., and Eisenreich, N. (1998) In-line monitoring of polymer extrusion processes by NIR spectroscopy. *J. Near Infrared Spectrosc.*, **6**, 325–332.

113 Fischer, D., Bayer, T., Eichhorn, K.J., and Otto, M. (1997) In-line process monitoring on polymer melts by NIR-spectroscopy. *Fresen. J. Anal. Chem.*, **359**, 74–77.

114 Rohe, T. and Kölle, S. (2000) *GIT Lab. Fachzeitschr.*, **44**, 1444.

115 Rohe, T. (2001) Inline Nahinfrarot (NIR Spektroskopie bei der Kunststoffextrusion PhD Thesis, University Stuttgart.

116 Becker, W. (2006) Prozessanalytik in der Kunststoffindustrie, in *Prozessanalytik. Strategien und Fallbeispiele aus der Industriellen Praxis* (ed. R.W. Kessler), Wiley-VCH, Weinheim, pp. 551–570.

117 Reshadat, R., Desa, S., Joseph, S., Mehra, M., Stoev, N., and Balke, S.T. (1999) In-line near-infrared monitoring of polymer processing, Part I: Process/monitor interface development. *Appl. Spectrosc.*, **53** (11), 1412.

118 Nagata, T., Oshima, M., and Tanigaki, M. (2000) On-line NIR sensing of CO2 concentration for polymer extrusion foaming processes. *Polym. Eng. Sci.*, **40** (8), 1843–1844

119 Hansen, M.G. and Vedula, S. (1998) In-line fiber-optic near-infrared spectroscopy: Monitoring of rheological properties in an extrusion process. Part I. *J. Appl. Polym. Sci.*, **68** (3), 859–872.

120 Vedula, S. and Hansen, M.G. (1998) In-line fiber-optic near-infrared spectroscopy: Monitoring of rheological properties in an extrusion process. Part II. *J. Appl. Polym. Sci.*, **68** (3), 873–889.

121 Thomas, Y., Cole, K.C., and Daigneault, L.E. (1997) n-line NIR monitoring of composition and bubble formation in polystyrene/blowing agent mixtures. *J. Cell. Plast.*, **33**, 516–527.

122 Siesler, H.W., Ozaki, Y., Kawata, S., and Heise, H.M. (2006) *Near-Infrared Spectroscopy. Principles, Instruments, Applications*, Wiley VCH, Weinheim.

123 van den Brink, M., Pepers, M., and Van Herk, A.M. (2002) Raman spectroscopy of polymer latexes. *J. Raman Spectrosc.*, **33**, 264–272.

124 Bauer, C., Amram, B., Agnely, M., Charmot, D., Sawatzki, J., Dupuy, N., and Huvenne, J.-P. (2000) On-line monitoring of a latex emulsion polymerization by fiber-optic FT Raman spectroscopy, Part 1: Calibration. *Appl. Spectrosc.*, **54**, 528–535.

125 Wenz, E., Buchholz, V., Eichenauer, H., Wolf, U., Born, J.-R., Jansen, U., and Dietz, W. (2001) Process for the production of graft polymers, US Patent Application Publication 2003/0130433 A1, Assigned to Bayer Polymers LLC, Filed in 2002. Priority number DE 10153534.1.

126 Elizalde, O. and Leiza, J.R. (2009) Raman application in emulsion polymerization systems, in *Raman Spectroscopy for Soft Matter Applications* (ed. M.S. Amer), John Wiley & Sons, Hoboken NJ, pp. 95–144.

127 Schrof, W., Horn, D., Schwalm, R., Meisenburg, U., and Pfau, A. (1998) Method for optimizing lacquers, US 6447831 B1, Assigned to BASF

Aktiengesellschaft, filed in 1999, priority number DE 19834184.

128 Van Overbeke, E., Devaux, J., Legras, R., Carter, J.T., McGrail, P.T., and Carlier, V. (2001) Raman spectroscopy and DSC determination of conversion in DDS-cured epoxy resin: application to epoxy copolyethersulfon blend. *Appl. Spectrosc.*, **55**, 540–551.

129 Ward, N.J., Edwards, H.G.M., Johnson, A.F., Fleming, D.J., and Coates, P.D. (1996) *Appl. Spectrosc.*, **50**, 812.

130 Gururajan, G. and Ogale, A.A. (2009) Raman applications in polymer films for real-time characterization, in *Raman Spectroscopy for Soft Matter Applications* (ed. M.S. Amer), John Wiley & Sons, Hoboken NJ, pp. 33–62.

131 Young, R.J. and Eichhorn, S.J. (2009) Raman applications in synthetic and natural polymer fibers and their composites, in *Raman Spectroscopy for Soft Matter Applications* (ed. M.S. Amer), John Wiley & Sons, Hoboken NJ, pp. 63–94.

132 Amer, M.S. (2009) Raman applications in foams, in *Raman Spectroscopy for Soft Matter Applications* (ed. M.S. Amer), John Wiley & Sons, Hoboken NJ, pp. 181–198.

133 Hayashi, N. (2009) Raman applications in liquid crystals, in *Raman Spectroscopy for Soft Matter Applications* (ed. M.S. Amer), John Wiley & Sons, Hoboken NJ, pp. 145–180.

134 Leitner, R., Mairer, H., and Kercek, A. (2003) Real-time classification of polymers with NIR spectral imaging and blob analysis. *Real Time Imaging*, **9**, 245–251, (Special Issue on Spectral Imaging).

135 Kulcke, A., Gurschler, C., Spöck, G., Leitner, R., and Kraft, M. (2003) On-line classification of synthetic polymers using near infrared spectral imaging. *J. Near Infrared Spectrosc.*, **11**, 71–81.

136 Pourdeyhimi, B. (ed.) (1999) *Imaging and Image Analysis Applications for Plastics*, William Andrew Publishing/Plastics Design Library.

137 Maiwald, M., Gerlach, G., and Kuschnerus, N. (2009) Prozess-Sensoren 2015 +. Technologie-Roadmap für Prozess Sensoren in der chemisch-pharmazeutischen Industrie, VDI-VDE – NAMUR, November.

138 Stanley, S.J. and Bolton, G.T. (2008) Review of Recent Electrical Resistance Tomography (ERT) Applications for Wet Particulate Processing. *Particle Particulate Syst. Charact.*, **25**, 207–215.

139 Bearman, G. and Levenson, R. (2003) Chapter 8, *Biological Imaging Spectroscopy, Biomedical Photonics Handbook*, CRC Press, Boca Raton.

140 König, K. (2008) Clinical multiphoton tomography. *J. Biophoton.*, **1**, 13–23.

141 Gibson, A.P., Hebden, J.C., and Arridge, S.R. (2005) Recent advances in diffuse optical imaging. *Phys. Med. Biol.*, **50**, R1–R43.

142 Boas, D.A., Dale, A.M., and Franceschini, M.A. (2004) Diffuse optical imaging of brain activation: approaches to optimizing image sensitivity, resolution and accuracy. *Neuro Image*, **23**, 275–288.

143 Abrahmasson, C., Johansson, J., Anderson-Engels, S., Svanberg, S., and Folestad, S. (2005) Time-resolved NIR spectroscopy for quantitative analysis of intact pharmaceutical tablets. *Anal. Chem.*, **77**, 1055–1059.

144 Dexheimer, S.L. (ed.) (2007) *Terahertz Spectroscopy – Principles and Applications*, CRC Press.

145 Salomatina, E., Muzikansky, A., Neel, V., and Yaroslavsky, A.N. (2009) Multimodal optical imaging and spectroscopy for the intraoperative mapping of nonmelanoma skin cancer. *J. Appl. Phys.*, **105**, 102–110.

146 Merz, T. and Kessler, R.W. (2007) On-line Prozesskontrolle mittels 2D-Fluoreszenzspektroskopie. *Process*, **9**, 44–45.

147 Ahmed-Omer, B., Brandta, J.C., and Wirth, T. (2007) Advanced organic synthesis using microreactor technology. *Org. Biomol. Chem.*, **5**, 733–740.

148 Tisma, M., Zelic, B., Vasic-Racki, D., Znidarsic-Plazl, P., and Plazl, I. (2009) Modelling of laccase-catalyzed l-DOPA oxidation in a microreactor. *Chem. Eng. J.*, **149**, 383–388.

149 Hessel, V. and Löwe, H. (2004) Organische Synthese mit mikrostrukturierten Reaktoren. *Chem. Ing. Tech.*, **76**, 535–554.

2
Applications of Spectroscopy and Chemical Imaging in Pharmaceutics

Aoife A. Gowen and José M. Amigo

2.1
The New Paradigm of Process Control

The US Food and Drug Administration (FDA)-led process analytical technology (PAT) initiative is transforming approaches to quality assurance in the pharmaceutical industry. Core to the PAT initiative is increased process understanding by monitoring of critical performance attributes, leading to better process control and ultimately improved drug quality [1]. Potential advantages of PAT implementation include reduced production cycle times, prevention of rejects, reduction of human error and facilitation of continuous processing to improve efficiency. With adequate knowledge of the ingredient – process relationship, the FDA will allow a pharmaceutical company to adjust their manufacturing process to maintain a consistent output without notification to the FDA [2]. The real-time release paradigm assures that when the last manufacturing step is passed, all final release criteria are met [3]. Adoption of such a quality by design approach would allow significant benefits to be gained by pharmaceutical companies: labor- and time-intensive finished product testing would be avoided if the product could be released based on in-process data.

This possibility is driving pharmaceutical companies to gain a more complete understanding of their processes and input variable relationships to realize the benefits of PAT registrations. Summarizing, pharmaceutical companies are encouraged to:

1) Implement real-time process analyzers and control tools
2) Develop new multivariate tools for design, data acquisition and analysis
3) Utilize (1) and (2) for continuous improvement and knowledge of the process.

Process monitoring and control are necessary at all stages of pharmaceutical processing, from raw material to packaged product characterization (Figure 2.1).

Traditional quality control methods (e.g., high performance liquid chromatography (HPLC) or mass spectroscopy (MS)) are time consuming, destructive, expensive, require lengthy sample preparation and give no information about the distribution of components within a sample. Due to the destructive and time-consuming nature of these methods, only small samples of drugs may be tested from given production

Handbook of Biophotonics. Vol.3: Photonics in Pharmaceutics, Bioanalysis and Environmental Research, First Edition.
Edited by Jürgen Popp, Valery V. Tuchin, Arthur Chiou, and Stefan Heinemann

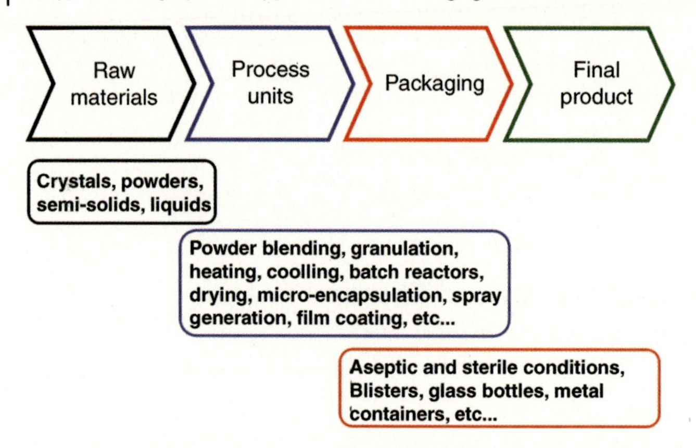

Figure 2.1 Flow-chart of the main steps in the production of pharmaceutical products.

batches [4]. Motivated by the PAT initiative, numerous techniques have been developed to provide real-time process control in order to improve understanding of the chemical and physical processes that occur during pharmaceutical unit operations. This chapter will focus on the major spectroscopic and chemical imaging techniques that have been applied in this context.

2.2
Overview of Spectroscopic Techniques Commonly Used

Spectroscopy concerns the scattering, absorption and emission of light originating from various regions of the electromagnetic spectrum. The ultraviolet (UV), visible (Vis), near-infrared (NIR) and mid-infrared (MIR) wavelength regions have been demonstrated as the most important areas of the electromagnetic spectrum for the collection of information about pharmaceutical processes. UV–Vis light is energetic enough to cause electronic transitions in molecules, while light in the NIR and MIR wavelength ranges results in molecular vibrations. A summary of the key features of various spectroscopic techniques employed in pharmaceutical analysis is presented in Table 2.1.

UV–Vis spectroscopy has a long history of use in pharmaceutical quality control, and is regarded as one of the most important techniques in analytical chemistry due to its low cost, ruggedness of spectrometers, rapidity and lower sensitivity to environmental factors (e.g., temperature) than the vibrational spectroscopies. This method is usually applied to samples in the liquid state; chromophore-containing materials require dilution by factors of 10^3–10^5 in a nonabsorbing solvent prior to spectral acquisition. In order to overcome this limitation, attenuated total reflectance (ATR), a technique more commonly associated with MIR spectroscopy, has been coupled with UV–Vis spectroscopy. ATR-UV–Vis spectroscopy is suitable for monitoring crystallization and for quantitative analysis of solid dosage forms containing chromophores [5, 6].

Table 2.1 Summary of main features of UV–Vis, NIR, MIR and Raman (RS) spectroscopies.

	UV–Vis	NIR	MIR	RS
Electromagnetic region	190–700 nm.	2500–7500 nm.	10000–2500 nm.	LASER radiation in UV–Vis or NIR
Low cost[a)]	X	x		
High speed	X	XX	x	X
Rugged instrumentation	X	X		X
Selective spectra			x	X
Portability	x	XX (probes)		
Possibility of plugging into a reactor[b)]	x	XX		
Nondestructive		x		X

a) The cross denotes the degree of the property in relationship to their implementation in PAT methodologies. For instance, in "low cost" property, the X denotes that there are devices affordable at a reasonable price. The x denotes that there are devices, but with higher cost.
b) The availability of portable probes makes it possible to plug them into reactors, blenders, or any device for real-time acquisition of spectra.

NIR spectroscopy (NIRS) is well recognized as a nondestructive tool in pharmaceutical analysis for raw material testing, quality control and process monitoring, mainly due the advantages it allows over traditional methods, for example, speed, little/no sample preparation, capacity for remote measurements (using fiber optic probes), prediction of chemical and physical properties from a single spectrum. Räsänen and Sandler [7] provide a comprehensive review of the applications of NIRS in the solid dosage form development, documenting the usefulness of this technique for each of the manufacturing steps, from raw material screening to assessment of the final dosage form. Absorption bands in the NIR are overtones and combinations of bands originating in the MIR region. As a consequence, they are relatively weak and overlapped, making molecular structural elucidation difficult. Nevertheless, multivariate data analysis has facilitated the development of robust calibration models from NIRS data.

MIR light excites fundamental molecular vibrations; therefore, MIR spectra are more selective than NIR, facilitating the extraction of information on the structure of molecules. Fourier transform infrared (FT-IR) spectroscopy is the standard spectroscopic method currently used for acquisition of MIR spectra. However, the high absorption of water in the MIR range means that, in order to measure aqueous samples, very thin samples are required. About 10–100 μm path length is necessary to avoid signal saturation, and at the same time to provide a sufficiently strong signal. Such thin samples are difficult to prepare, however, attenuated total reflection (ATR) - FTIR, also known as evanescent wave spectroscopy (EWS) provides a solution to this problem.

Raman spectra are more distinct and less overlapped than NIR spectra. Low signal to noise ratio is, however, a problem for RS, which arises due to the very small proportion of scattered photons that make up the Raman signal, and the need to use filters to remove the excitation line from the collected radiation. Rantanen [8] recently published a review documenting the process analytical applications of RS in the area of pharmaceutical dosage forms. Its potential application in dosage formulation, from chemical synthesis to tablet coating was described, along with a discussion of the main challenges facing its implementation in process analysis, which include: interfacing with the process environment, sample heating during measurement and small sampling area. Šašić provides a good introduction to Raman spectroscopy and applications in pharmaceutical process monitoring in his recent book [9].

Some spectrometers integrate spatial information to give an average spectrum for each sample studied. The inability of NIR and Raman spectrometers to capture internal constituent gradients within samples may lead to discrepancies between predicted and measured composition. Furthermore, point-source spectroscopic assessments do not provide information on spatial distribution of different constituents, which is important in many pharmaceutical applications such as blending. Chemical imaging (CI) integrates conventional imaging and spectroscopy to attain both spatial and spectral information from an object (Figure 2.2). The terms *spectral*

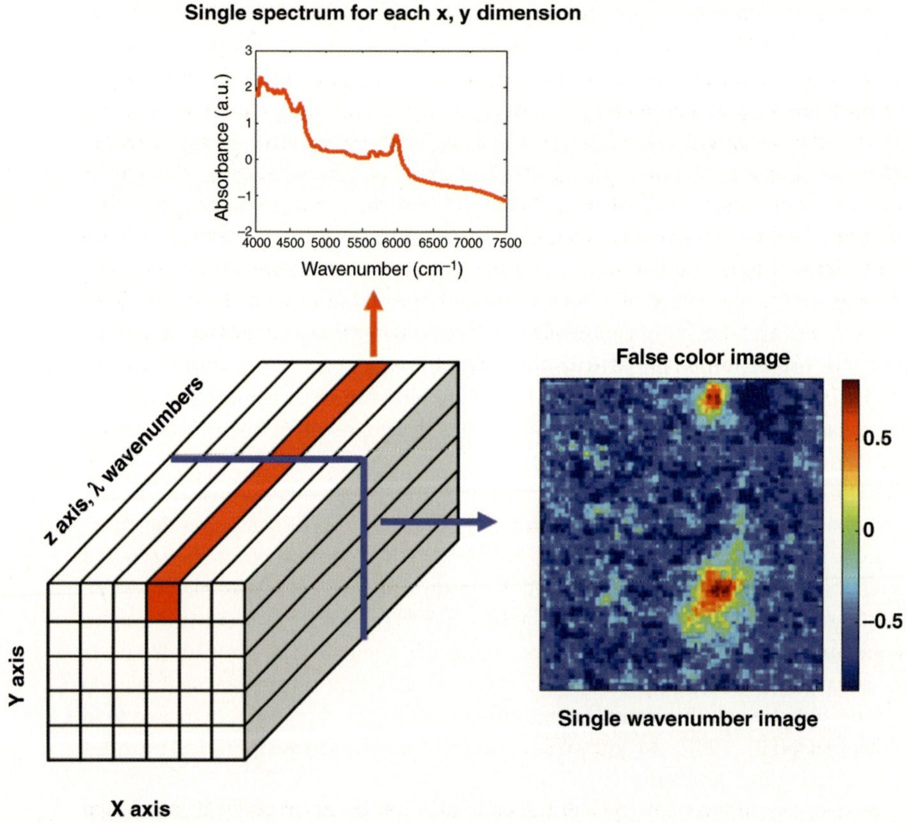

Figure 2.2 Hyperspectral data cube obtained with a chemical imaging device. In this case, the device is a NIR system.

and *chemical* are often interchanged to describe this type of imaging: Spectral imaging is an umbrella term that can be applied to almost any optical spectroscopic technique (e.g., IR, Raman, fluorescence, UV), while chemical imaging generally refers to the vibrational spectroscopies (IR, NIR, Raman). By combining the chemical selectivity of vibrational spectroscopy with the power of image visualization, CI enables a more complete description of ingredient concentration and distribution in inhomogeneous solids, semi-solids, powders, suspensions and liquids.

What basically is obtained by using chemical imaging devices is a hyperspectral data cube of information. That is, for each pixel of the surface scanned (*X* and *Y*-axes in Figure 2.2) a spectrum is recorded.

The attractiveness of surface information provided by, mainly MIR, NIR and Raman imaging devices, together with proper data analysis methods, has opened new research pathways in pharmaceutics, providing excellent possibilities toward obtaining reliable and accurate information in different stages of the process (e.g., correct distribution of components in tablets and powdered mixtures, monitoring mixtures being blended, studies of segregation in granulated preparations, study of the interface between the coating and the core of the tablet, detection of minor compounds, counterfeits, contaminants, etc.). This is confirmed by the high quality publications and book chapters, generated in a few years, encompassing the fundamentals of the new technology and its applications in pharmaceutical research and development [10–18]. Moreover, advances in the production of low cost array detectors have resulted in faster data collection and cheaper chemical imaging systems [19]. Recently, a novel chemical imaging instrument has been developed [48], in which the broadband light source has been replaced by a tunable laser source. The laser source (based on optical parametric oscillator, OPO technology) generates a single wavelength light source at a time. The wavelength can be varied on demand over the wavelength range of interest. This eliminates the need for filters, since only one wavelength is generated at a time. Moreover, this technology reduces the time required for chemical image acquisition to a few seconds.

2.3 The Need for Multivariate Data Analysis

As in classical spectroscopy, the use of multivariate data analysis techniques is becoming a crucial part of the data analysis in CI. The basic steps of the data analysis are summarized in Figure 2.3.

2.3.1 Pre-Processing

The spectral patterns collected are usually affected by noise or by instrumental variations that may alter the analysis and conclusions that may be drawn with further data processing [10, 20]. The most common de-noising techniques are based on the

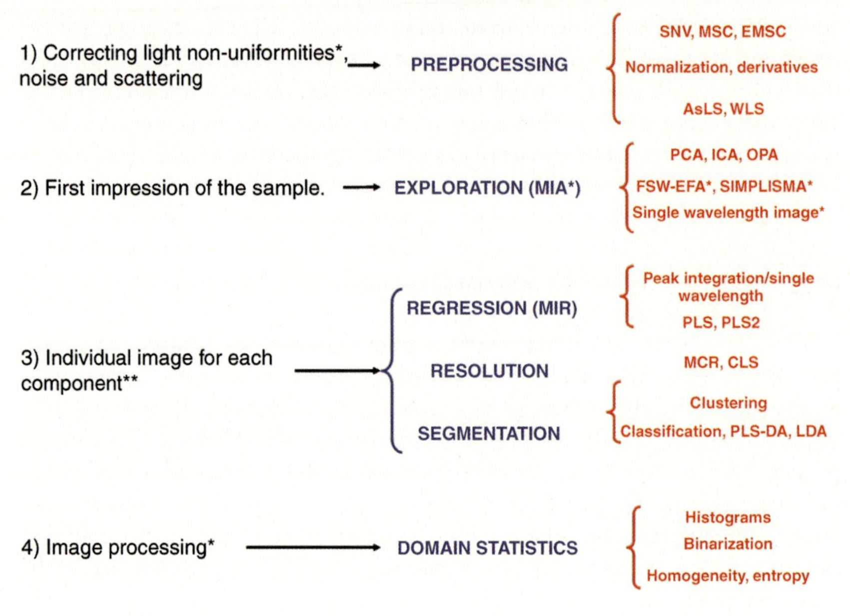

Figure 2.3 Data processing steps. The steps marked with (*) are exclusively applied on hyperspectral image data cubes. (**) The third step is also applied in classical spectroscopy (single measurement/sample). SNV standard normal variate, MSC multiplicative scatter correction, EMSC extended multiplicative scatter correction, AsLS asymmetric least squares, WLS weighted least squares, PCA principal component analysis, ICA independent component analysis, OPA orthogonal projection approach, FSW fixed-size windows, EFA evolving factor analysis, SIMPLISMA simple-to-use interactive self-modeling mixture analysis, PLS partial least squares, PLS2 partial least squares 2, MCR multivariate curve resolution, CLS classical least squares, DA discriminant analysis, LDA linear discriminant analysis.

Savitzky–Golay approach, which fits the spectral pattern to a polynomial function (second-order polynomial) in a step-wise manner [10]. Less used is wavelet-based filtering or principal component analysis. Pre-processing is a convenient and often necessary step. Nevertheless, it can introduce data artifacts or generate the loss of significant spectral information if the method is not properly selected. Consequently, minimal and careful data pre-processing is usually preferred.

2.3.2
Exploration Techniques

Principal components analysis (PCA) (also called multivariate image analysis MIA) is arguably the most useful exploratory technique in spectroscopy. This technique was designed with the aim of explaining the main sources of variability in bi-dimensional experimental datasets (e.g., classical spectroscopy). That is, PCA gives information about the main compounds (chemical compounds and physical artifacts) and their distribution on the measured surface. This has generated the adaptation of the general PCA method to methods more focused on considering the spatial information gathered, by incorporating the spatial information in the analysis. For instance, methods like fixed-size windows-evolving factor analysis (FSW-EFA) or enhancing contrast methods try to explore the analyzed surface, merging surface and spectral information [21–24].

2.3.3 Regression, Resolution and Classification Techniques

Regression, resolution and classification techniques are demanded when more precise information is required from the samples (e.g., active pharmaceutical ingredient (API) concentration or micro-domain segmentation), Regression methods aim to extract quantitative information from spectroscopic data. The setting-up of calibration models is usually performed by laboratory-made tablets that cover the whole concentration range of the components. In chemical imaging datasets, owing to pixel-to-pixel heterogeneity the quantitative model is usually built using the average spectrum of the regions of interest (ROI) in the surface [25, 26] and the bulk concentration.

On the contrary, resolution techniques (e.g., multivariate curve resolution (MCR) [22, 27–29]) do not usually require a complete calibration set. The feasibility of MCR models relies on the assumption that the spectral information in each sample is the weighted sum of the spectral influence of each component in the mixture. Therefore, the concentration values obtained are fractions or percentages of the pure component sample.

Classification techniques are very useful when the target is to find groups with similar chemical composition, as a function of their similarity in the spectral profiles. They are especially useful in CI to locate and identify impurities or physical defects, coating thickness, or counterfeit detection.

2.3.4 Image Processing Techniques

Up to now, the abovementioned methods have been adapted from classical multivariate analysis to the study of three-way arrays found in CI experiments. Image processing techniques, though, answer the need to extract additional information from the images derived from resolution/regression/classification techniques. That is, study of the distribution (surface distribution) of elements in an image, texture, histogram statistics, and particle domain statistics. These tools are especially useful in the monitoring of blending and granulation processes (Figure 2.4).

Binarization plays a major role in the calculation of domain size statistics. The binary images are created by setting a threshold value, and converting the resulting image into a mixture of two colors denoting the presence or absence of the component of interest. Selection of the proper threshold is critical, since this selection

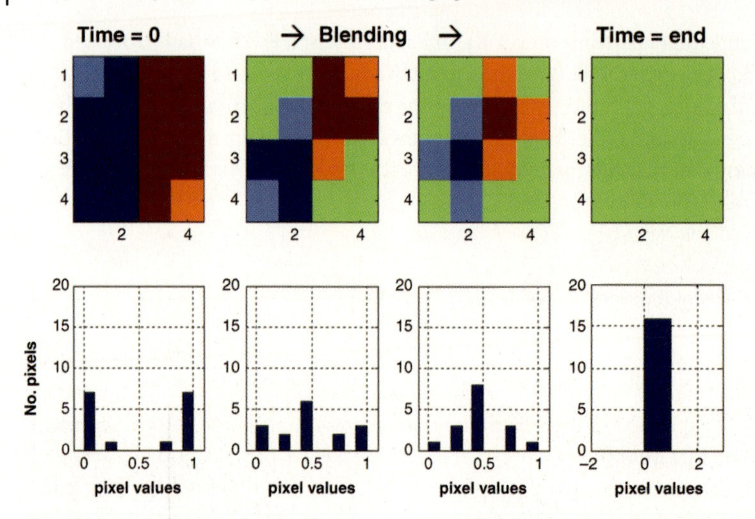

Figure 2.4 Simulation of a blending process monitored by looking at the histograms.

is subjective and will always depend on the final spatial resolution, the correct performance of the data analysis and the criteria of the operator (Figure 2.5).

2.4
Applications in Pharmaceutical Process Monitoring and Quality Control

Rising awareness of spectroscopic techniques and their potential role in the realization of PAT has promoted an increase in the number and diversity of research carried out on their application in manufacturing, development and real implementation of pharmaceutical products. New perspectives and challenges emerge to deeply understand the process or to control the final quality of the preparation. A selection of important applications is discussed below.

2.4.1
Spray Formulations

Doub *et al.* [30] reported the application of Raman-CI for particle size characterization of nasal spray formulations. In order to analyze the formulations, it was necessary to spray aqueous suspensions onto inverted aluminum-coated glass microscope slides which were dried before imaging. A Raman imaging particle size distribution (PSD) protocol was developed and validated using polystyrene (PS) microsphere size standards. Good statistical agreement was obtained between reported PS particle sizes and those estimated using Raman-CI. Moreover, Raman-CI was demonstrated as being suitable for distinguishing between active samples and placebos. It was noted, however, that vigorous validation of Raman-CI for PSD is required, due to the high variability associated with replicate measurements.

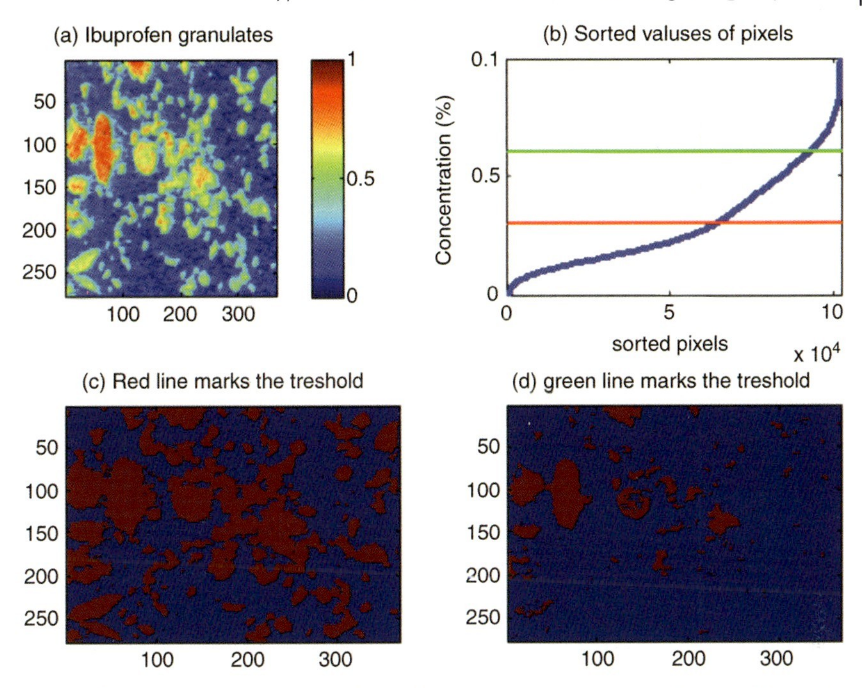

Figure 2.5 (a) Concentration surface obtained for an ibuprofen granulate. (b) Sorted values of the pixels to select a threshold. (c) and (d) Binary images obtained with different thresholds (red and blue denote presence and absence, respectively).

2.4.2 Powders

Powder blend homogeneity in final dosage product poses a significant problem in pharmaceutical quality assurance; poor blending tends to adversely affect tablet hardness, appearance and dissolution kinetics. Lyon *et al.* in 2002 [31] compared the ability of NIR-CI and NIRS in distinguishing blending grade for a selection of final dosage forms. While NIRS was limited in its ability to evaluate drug product homogeneity, NIR-CI provided the opportunity to investigate localized microdomains of ingredients within a drug. It was noted that physical and chemical abnormalities that made minimal contribution to the bulk tablet would be undetected by traditional NIRS but detected by NIR-CI. Compared to conventional "thief" sampling, traditionally used in UV analysis of powder samples, imaging multiple sections of a powder blend offers improved blend characterization [32].

2.4.3 Polymorphism

Characterization of identical chemicals of different crystalline form (polymorphs) is important for quality control in drug development, since polymorphs have different physical properties such as solubility. Current methods for polymorph screening, such as X-ray diffraction, are time consuming and expensive. Raman spectroscopy coupled with hot-stage microscopy has been demonstrated for rapid screening of polymorphism [33]. More recently, transmission RS has been applied to quantify the polymorphic content of pharmaceutical formulations [34]. This method has the advantage of bulk measurement as compared with backscattering RS methods.

2.4.4 Solid Dosage Forms

Many new drug release systems require complex tablet architecture. In 2001, Lewis *et al.* [35] showed how NIR-CI could be employed for the qualitative description of composition and architecture of a solid dosage form. Bell *et al.* [36–39] published findings on the development of quantitative analytical methods for Raman analysis of solid dosage forms using point mapping. Another more recent article discusses the impact of spatial sampling density in Raman imaging [40]. It was demonstrated that the coarse measurement could be useful for selection of regions which required further fine measurement. Although the fine grid method is time consuming, it would allow visualization of the chemical distribution of the sample, showing greater details of crystal morphology. Li *et al.* [42] applied NIR-CI for characterization of swellable matrix tablets. Tablets were immersed in water for various periods of time and dissected along the radial direction prior to image acquisition. Single wavelength images were used for estimation of the thickness of the gel layer, while PCA images were interpreted in terms of the glass–rubber phase transition.

2.4.5 Root Cause Analysis

Pharmaceutical products undergo various chemical and physical changes during processing. Characterizing the nature and sources of these changes is a major challenge for PAT. Clarke [43] reported the utilization of NIR-CI microscopy for identification of tablets which experienced problems during processing, and for investigation of pre-tabletting blends with dissolution issues. The cluster size of ingredients was measured from chemical images to investigate the effect of compaction force on dissolution properties. In another study, the potential of NIR-CI for investigating the effects of processing on tablet dissolution was examined [44]. It was shown that NIR-CI could be used for the prediction of coating time and to understand batch differences due to different processing conditions.

2.4.6 API Distribution

Henson and Zhang [45] published an article on drug characterization of low dosage pharmaceutical tablets using Raman microscopic mapping. Raman images were obtained using a Raman line-mapping instrument. Domain size and spatial distribution of the API and major excipients were estimated from Raman maps of the tablets. It was demonstrated that the domain size of the API was dependent upon the particle size distribution of the ingoing API material; therefore, the Raman maps could be used to indicate the source of API used in tablet manufacturing. The potential utility of Raman microscopic mapping in manufacturing process optimization or predictive stability assessments was also outlined.

2.4.7 Coatings

Lewis *et al.* [46] applied NIR-CI for quantification of coating thickness in a single time release microsphere. NIR chemical images demonstrated the differences in chemical structure of the tablet core and coating, enabling quantitative information on coating thickness and homogeneity to be determined. Uneven coating could be clearly identified by visualization of the total intensity image at 2080 nm. A group of 135 microspheres containing two different microsphere types was also imaged in this study. Discrimination of microsphere type was visualized using a spectral difference image.

2.4.8 Counterfeit Identification

Rodionova *et al.* [48] reported advances in the feasibility of detecting counterfeit drugs using both NIRS and NIR-CI. The authors remarked that NIR-CI is useful for identifying small differences in ingredient distributions, which is important for identification of so-called "high quality" counterfeit samples. Westenberger *et al.* [49] compared quality assessment of pharmaceutical products purchased via the internet using a variety of traditional (HPLC-based) and non-traditional (NIR, NIR-CI) techniques. The NIR-CI technique had an added advantage over traditional techniques in its ability to detect suspect manufacturing issues, such as blending and density pattern distribution of components within a drug product, allowing direct qualitative comparison with control products. Lopes *et al.* have developed novel multivariate data analysis techniques for detection and classification of counterfeit drugs using NIR-CI [50–52].

2.4.9 High Throughput Analysis

The huge volume of production encountered in the pharmaceutical industry necessitates high throughput quality analysis techniques. NIR-CI may be adapted to perform analysis on multiple samples in a single field of view by increasing the distance between the camera and target. Fast and nondestructive identification of active ingredients and excipients in whole tablets can be achieved, even through packaging. Hamilton and Lodder [53] reported that a typical NIR chemical imaging system could simultaneously analyze around 1300 tablets in blister packaging. In another, more recent, paper [54] high throughput analysis of pharmaceutical tablet content uniformity was reported using NIR-CI. A large field of view NIR-CI device was used to take simultaneous chemical images of 20 tablet samples at 61 wavelengths. The total time for image acquisition was less than 2 min, while the conventional UV method took approximately half a working day for sample analysis. It was observed that the mean intensity of images at 1600 nm was proportional to API concentration; the authors suggested that a multispectral instrument based on a small number of wavelengths around 1600 nm could be built to rapidly identify API concentration of these tablets in packaged form.

2.5
Issues Facing the Implementation of Spectroscopic Techniques in the Pharmaceutical Industry

A wealth of evidence suggests that advances in the control of pharmaceutical processes are possible through the on-line implementation of PATs, such as the spectroscopic techniques described in the previous section. These advances would include decreased batch variability, fewer rejects, faster response, lower costs, improved productivity and better process understanding. However, many emerging PAT methods have not yet been implemented in routine process control. The hesitancy to implement PATs in industry may be due in part to resistance to change in some processes and also, as has been suggested by Winskill and Hammond [55], due to concerns over delays in regulatory approval. Reich [37] reviewed the regulatory status of NIR analysis in the pharmaceutical industry, noting that although many pharmaceutical companies have successfully implemented NIR systems in QC labs for raw material identification and qualification, only a few quantitative NIR methods have yet gained regulatory approval.

Apart from regulatory issues, despite the benefits associated with chemical imaging, measurements and subsequent data analysis are not exempt from issues and drawbacks that should be considered to extract accurate and meaningful information and to relate the results obtained to the objective pursued. Key issues and limitations associated with the quality and interpretation of chemical imaging data are discussed below.

2.5.1
Sampling

Nondestructive techniques, such as NIR, RS and their imaging counterparts, offer the possibility of sampling each tablet in a manufacturing process. However, consideration of within-object sampling is an important, if often overlooked, issue. The most common way of measuring NIR and mid-IR radiation in pharmaceutical research is diffuse reflectance. When the incident radiation, composed of millions of photons, reaches the sample surface, it may be immediately reflected (specular reflectance) or enter the sample, depending on the radiation wavelength and the nature of the sample [16, 35–37]. Some of the photons will be absorbed by the material, whereas others will pass through the entire sample or be scattered/reflected to such a degree that they will not be captured and detected by the instrument. Thus, the photons that ultimately arrive at the detector (the diffusely reflected ones) will contain mixed information about the sample components at various depths and locations in the sample (Figure 2.6a). The final result is a spectral pattern influenced not only by the components of the measured pixel, but also by neighboring pixels and by the depth of penetration. Raman scattering measurements, in contrast, are less affected by this last parameter since the laser beam source, with which the sample is irradiated, penetrates only a few micrometers into the sample.

2.5.2
Spatial Resolution

An important consideration in chemical imaging experiments of pharmaceutical experiments is the domain size of components (e.g., APIs) in comparison to the spatial resolution of the system. Spatial resolution is typically calculated by dividing

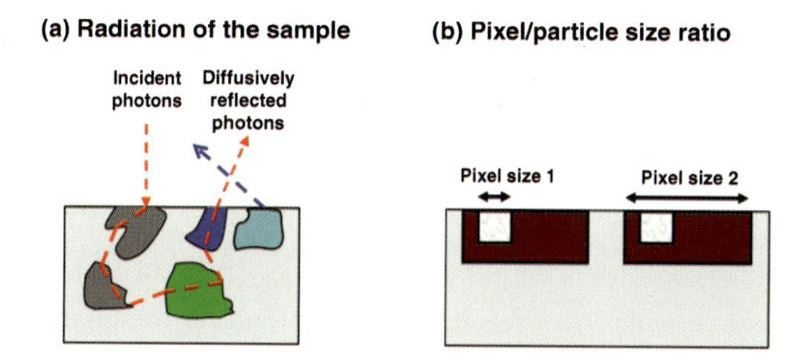

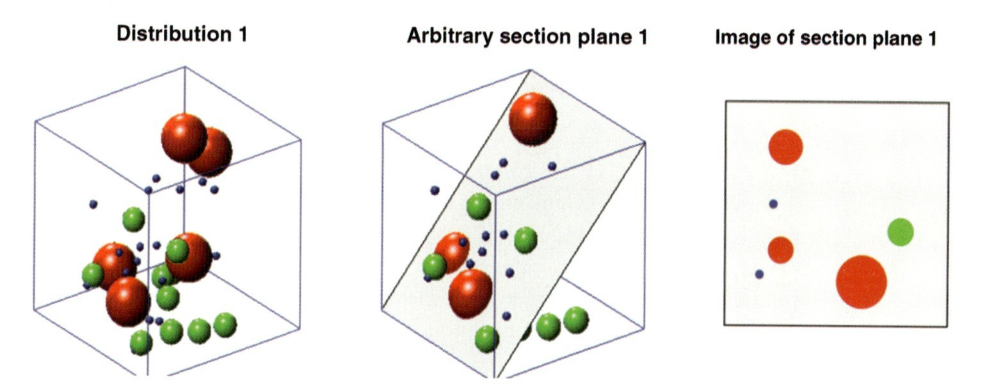

Figure 2.6 Some spatial issues in the analysis of pharmaceutical surfaces [18].

the width and length of the field of view by the number of pixels in the x- and y- dimensions of the detector. Should the domain size be smaller than the nominal spatial resolution, it is necessary to apply spectral unmixing algorithms (e.g., MCR) to estimate the relative abundance of a species in a given pixel. This kind of situation, where pixels contain contributions from numerous sources, is endemic in remote-sensing hyperspectral imaging where spatial resolution can range from meters to kilometers. The particle sizes of pharmaceutical products vary in nature, depending on the formulation and the product itself, but it is common to find particles with an effective diameter less than 10 μm. This plays a fundamental role in hyperspectral measurements, since the pixel size can usually vary from 10 to 100 μm, depending on the configuration. Figure 2.6b illustrates a schematic model of a pharmaceutical sample where a small particle (gray) is embedded in a larger one (red). The sample is analyzed using two different pixel sizes. In the first case, the pixel size is small enough to obtain selective information about the gray particle. The second case represents a pixel size much bigger than the smallest particle. Thus, depending on the absorptivity of the red particle, the gray one will be difficult to detect because it may be overwhelmed by the signal of the former.

Moreover, spatial resolution obtainable by a given CI system is not just a function of the size of the detector and image magnification but fundamentally limited by photon diffusion through the sample. Sub-surface scattering often distorts the image obtained and thus limits the spatial resolution that can be obtained. In order to address this issue, special spatial resolution targets have been developed that mimic pharmaceutical materials [57].

2.5.3 Representativeness of the Measured Surface

Another concept to bear in mind is that chemical imaging techniques are surface-based techniques. This is especially relevant in tablets, which are usually presented as homogeneous volumetric samples. However, tablets are usually formed from components with different physical and chemical properties, as well as diverse particle sizes, promoting nonhomogeneous distributions of the components in the tablet. Figure 2.6c presents one "volumetric" tablet with the same quantitative composition (equal amount, size, and shape of red, green, and blue spheres) [38–41]. If a section is cut through the tablet (arbitrary section plane) intersecting the components, then an image of that section plane will show two important aspects:

1) The concentration of the components obtained in the surface plane is not representative of their bulk concentration in the tablet.
2) The sizes of the particles in section plane 1 in Figure 2.6c do not correspond to the real sizes of the components in the tablet (clearly visible in the red component). This may lead to the conclusion that the sample is composed of differently sized particles of the same component.

Therefore, special care must be taken in the interpretation of a chemical image obtained from a cross-section of one tablet. A possible solution is to measure several samples of a product (different tablets of the same batch) or to measure a number of cross-sections of each sample, to guarantee the representativeness of the cross-section with respect to the bulk content of the components.

2.5.4 Irregularities in the Measured Surface

One of the advantages of chemical imaging devices is that almost no sample preparation is needed prior to data acquisition. However, flat-surface samples are preferable owing to the focusing nature of the devices. Moreover, the surface must be parallel to the image plane. Since tablets often have a convex shape, the use of a bevel-blade is necessary to mill flat samples after fixing the tablet to a crystal slide with acrylonitrile adhesive [58]. The microtoming milling technique may present some problems as some component materials may smear or be displaced during sample preparation [59]. It is also convenient to attain a surface as plane as possible when powders or mixtures being blended are measured, to avoid the scattering of the light source owing to the roughness of the surfaces.

The intensity of the light source is one of the key points to consider when obtaining spectral patterns of high quality. Powerful light sources are usually needed in mid-IR and NIR devices to irradiate polychromatic light in the spectral range selected. Laser beam radiation devices working at high resolution are also needed in Raman measurements. These illumination sources generate high levels of energy that induce strong heating of the samples. This may reveal several problems in samples, such as structural changes in granulates and effervescent tablets, and melting or even burning of some organic dyes used in coatings. To overcome this difficulty, cooling systems (e.g., using liquid nitrogen) are usually added to the device. Still, new advances in detectors and dispersive systems are continuously occurring, mainly intended to avoid cooling needs [60].

2.6 Conclusions

Attempts to apply spectroscopy and chemical imaging to real processes have demonstrated high potential to provide useful information in routine pharmaceutical tasks. However, further development in instrumentation technology, speed, and data processing is needed to fully employ chemical imaging in manufacturing process environments. The cost of equipment is commonly stated as a barrier to the implementation of chemical imaging in routine process monitoring. However, the non-destructive, rugged and flexible nature of chemical imaging makes it an attractive PAT for identification of critical control parameters that impact on finished product quality, and time savings associated with its implementation would result in lower production costs over the long term. Judging by the continuing emphasis on process analytical technologies to provide accurate, rapid, nondestructive analysis of pharmaceutical processes, it is likely that CI will be increasingly adopted for process monitoring and quality control in the pharmaceutical industry, as has been the case with NIR spectroscopy. Future innovations in CI equipment manufacture are likely to depress purchase costs and encourage more widespread utilization of this emerging technology in the pharmaceutical sector.

Acknowledgements

A.A. Gowen acknowledges funding from the EU FP7 under the Marie Curie Outgoing International Fellowship.

References

1. www.fda.gov/cder/OPS/PAT.htm.
2. http://www.emea.europa.eu/Inspections/PAThome.html.
3. Schoneker, D. (2006) Impact of Excipient Variability on PAT Applications: The Need for Good Qualification Processes, AAPS Annual Meeting and Exposition, October 28 - November 3, San Antonio, TX.
4. Skibsted, E.T., Westerhuis, J.A., Smilde, A.K., and Witte, D.T. (2007) Examples of NIR based real time release in tablet manufacturing. *J. Pharm. Biomed. Anal.*, **43**, 1297–1305.
5. Florence, A.J. and Johnston, A. (2011) Applications of ATR UV/vis spectroscopy in physical form characterisation of pharmaceuticals, *Spectrosc. Eur.*, 24–27.
6. Doyle, W.M. and Tran, L. (1999) Analysis of strongly absorbing chromophores by UV-visible ATR spectroscopy. *Spectroscopy*, **14** (4), 46–54.
7. Räsänen, E. and Sandler, N. (2007) Near infrared spectroscopy in the development of solid dosage forms. *J. Pharm. Pharmacol.*, **59**, 147–155.
8. Rantanen, J. (2007) Process analytical applications of Raman spectroscopy. *J. Pharm. Pharmacol.*, **59**, 171–177.
9. Šašić, S. (2007) *Pharmaceutical Applications of Raman Spectroscopy*, Wiley.
10. Amigo, J.M., Cruz, J., Bautista, M., Maspoch, S., Coello, J., and Blanco, M. (2008) *Trends Anal. Chem.*, **27**, 696–713.
11. Gendrin, C., Roggo, Y., and Collet, C. (2008) *J. Pharm. Biomed. Anal.*, **48**, 533–553.
12. Clark, D., Henson, M.J., LaPlant, F., Sasic, S., and Zhang, L. (2007) *The Handbook of Vibrational Spectroscopy, Applications in life, Pharmaceutical and Natural Sciences, Pharmaceutical Applications* (eds D.E. Pivonka, J.M. Chalmers, and P.R. Griffiths), Wiley, Hoboken, pp. 1–27.
13. Gowen, A.A., O'Donnell, C.P., Cullen, P.J., and Bell, S.E.J. (2008) *Eur. J. Pharm. Biopharm.*, **69**, 10–22.
14. Hammond, S.V. and Clarke, F.C. (2002) *The Handbook of Vibrational Spectroscopy, Sampling Techniques, Microscopy* (eds J.M. Chalmers and P.R. Griffiths), Wiley, Hoboken, pp. 1405–1431.
15. Lewis, E.N., Dubois, J., Kidder, L.H., and Haber, K.S. (2007) *Techniques and Applications of Hyperspectral Image Analysis* (eds H. Grahn and P. Geladi), Wiley, Hoboken, pp. 335–361.
16. Lewis, E.N., Schoppelrei, J., Lee, E., and Kidder, L.H. (2005) *Process Analytical Technology*, Blackwell, Cambridge.
17. Reich, G. (2005) *Adv. Drug. Delivery Rev.*, **57**, 1109–1143.
18. Amigo, J.M. (2010) Practical issues of hyperspectral imaging analysis of solid dosage forms. *Anal. Bioanal. Chem.*, **398** (1), 93–109.
19. Kim, M.S., Chao, K., Chan, D., and Lefcourt, A. (2009) Line-scan imaging for high-speed food safety inspection. *SPIE Newsroom*. doi: 10.1117/2.1200903.1564
20. Levin, I.W. and Bhargava, R. (2005) *Annu. Rev. Phys. Chem.*, **56**, 429–474.

21 Cairós, C., Amigo, J.M., Watt, R., Coello, J., and Maspoch, S. (2009) *Talanta*, **79**, 657–664.

22 de Juan, A., Tauler, R., Dyson, R., Marcolli, C., Rault, M., and Maeder, M. (2004) *Trends Anal. Chem.*, **23**, 70–79.

23 de Juan, A., Maeder, M., Hancewicz, T., and Tauler, R. (2005) *Chemom. Intell. Lab. Syst.*, **77**, 64–74.

24 de Juan, A., Maeder, M., Hancewicz, T., and Tauler, R. (2008) *J. Chemom.*, **22**, 291–298.

25 Burger, J. and Geladi, P. (2006) *Analyst*, **131**, 1152–1160.

26 Kolomiets, O., Hoffmann, U., Geladi, P., and Siesler, H.W. (2008) *Appl. Spectrosc.*, **62**, 1200–1208.

27 Amigo, J.M. and Ravn, C. (2009) *Eur. J. Pharm. Sci.*, **37**, 76–82.

28 Awa, K., Okumura, T., Shinzawa, H., Otsuka, M., and Ozaki, Y. (2008) *Anal. Chim. Acta*, **619**, 81–86.

29 Cruz, J., Bautista, M., Amigo, J.M., and Blanco, M. (2009) *Talanta*, **80**, 473–478.

30 Doub, W.H., Adams, W.P., Spencer, J.A., Buhse, L.F., Nelson, M.P., and Treado, P.J. (2007) Raman chemical imaging for ingredient-specific particle size characterization of aqueous suspension nasal spray formulations: a progress report. *Pharm. Res.*, **24**, 934–945.

31 Lyon, R.C., Lester, D.S., Lewis, E.N., Lee, E., Yu, L.X., Jefferson, E.H., and Hussain, A.S. (2002) Near-infrared spectral imaging for quality assurance of pharmaceutical products: analysis of tablets to assess powder blend homogeneity. *AAPS PharmSciTech.*, **3**, 17.

32 Ma, H. and Anderson, C.A. (2008) Characterization of pharmaceutical powder blends by NIR chemical imaging. *J. Pharm. Sci.*, **97**, 3305–3320.

33 Al-Dulaimi, S., Aina, A., and Burley, J. (2009) Rapid polymorph screening on milligram quantities of pharmaceutical material using phonon-mode Raman spectroscopy. *CrystEngComm*, **12**, 1038–1040.

34 Aina, A., Hargreaves, M.D., Matousek, P.A.U., and Burley, J.C. (2010) Transmission Raman spectroscopy as a tool for quantifying polymorphic content of pharmaceutical formulations. *Analyst*, **135**, 2328–2333

35 Lewis, E.N., Carroll, J.E., and Clarke, F.M. (2001) A near-infrared view of pharmaceutical formulation analysis. *NIR News*, **12**, 16–18.

36 Bell, S.E.J., Beattie, J.R., McGarvey, J.J., Peters, K.L., Sirimuthu, N.M.S., and Speers, S.J. (2004) Development of sampling methods for Raman analysis of solid dosage forms of therapeutic and illicit drugs. *J. Raman Spectrosc.*, **35**, 409–417.

37 Bell, S.E.J., Barrett, L.J., Burns, D.T., Dennis, A.C., and Speers, S.J. (2003) Tracking the distribution of "ecstasy" tablets by Raman composition profiling: a large scale feasibility study. *Analyst*, **128**, 1331–1335.

38 Bell, S.E.J., Burns, D.T., Dennis, A.C., and Speers, J.S. (2000) Rapid analysis of ecstasy and related phenethylamines in seized tablets by Raman spectroscopy. *Analyst*, **125**, 541–544.

39 Bell, S.E.J., Burns, D.T., Dennis, A.C., Matchett, L.J., and Speers, J.S. (2000) Composition profiling of seized ecstasy tablets by Raman spectroscopy. *Analyst*, **125**, 1811–1815.

40 Lee, E., Adar, F., and Whitley, A. (2006) The impact of spatial sampling density in Raman imaging. *Spectrosc.*, **21S**, 21.

41 Lee, E., Adar, F., and Whitley, A. (2006) The impact of spatial sampling density in Raman imaging. *Spectrosc*, **21**, 21.

42 Li, W., Woldu, A., Araba, L., and Winstead, D. (2010) Determination of Water Penetration and Drug Concentration. Profiles in HPMC-Based Matrix Tablets by Near Infrared. Chemical Imaging. *J. Pharm. Sci*, **99** (7), 3081–3088.

43 Clarke, F. (2004) Extracting process-related information from pharmaceutical dosage forms using near infrared microscopy. *Vib. Spectrosc.*, **34**, 25–35.

44 Roggo, Y., Jent, N., Edmond, A., Chalus, P., and Ulmschneider, M. (2005) Characterizing process effects on pharmaceutical solid forms using near-infrared spectroscopy and infrared imaging. *Eur. J. Pharm. Biopharm.*, **61**, 100–110.

45 Henson, M. and Zhang, L. (2006) Drug Characterization in Low Dosage Pharmaceutical Tablets Using Raman Microscopic Mapping. *Appl. Spectrosc.*, **60**, 1247–1255.

46 Lewis, E.N., Kidder, L., and Lee, E. (2005) NIR Chemical Imaging as a Process Analytical Tool. *Innov. Pharm. Tech.*, **17**, 107–111.

47 Spectral imaging device with tunable light source (2007) US patent 7,233,392.

48 Rodionova, O., Houmøller, L., Pomerantsev, A., Geladi, P., Burger, J., Dorofeyev, V., and Arzamastsev, A. (2005) NIR spectrometry for counterfeit drug detection: A feasibility study. *Anal. Chim. Acta*, **549**, 151–158.

49 Westenberger, B.J., Ellison, C.D., Fussner, A.S., Jenney, S., Kolinski, R.E., Lipe, T.G., Lyon, R.C., Moore, T.W., Revelle, L.K., Smith, A.P., Spencer, J.A., Story, K.D., Toler, D.Y., Wokovich, A.M., and Buhse, L.F. (2005) Quality assessment of internet pharmaceutical products using traditional and non-traditional analytical techniques. *Int. J. Pharm.*, **306**, 56–70.

50 Lopes, M.B., Wolff, J.-C., Bioucas-Dias, J.M., and Figueiredo, M.A.T. (2010) Near-infrared hyperspectral unmixing based on a minimum volume criterion for fast and accurate chemometric characterization of counterfeit tablets. *Anal. Chem.*, **82**, 1462–1469.

51 Lopes, M.B., Wolff, J.-C., Bioucas-Dias, J.M., and Figueiredo, M.A.T. (2009) Determination of the composition of counterfeit HeptodinTM tablets by near-infrared chemical imaging and least squares estimation. *Anal. Chim. Acta*, **641**, 46–51.

52 Lopes, M.B. and Wolff, J.-C. (2009) Investigation into classification/sourcing of suspect counterfeit HeptodinTM tablets by near infrared chemical imaging. *Anal. Chim. Acta*, **633**, 149–155.

53 Hamilton, S.J. and Lodder, R.A. (2002) Hyperspectral imaging technology for pharmaceutical analysis. *Proc. SPIE*, **4626**, 136–147.

54 Malik, I., Poonacha, M., Moses, J., and Lodder, R.A. (2001) Multispectral imaging of tablets in blister packaging. *AAPS PharmSciTech*, **2**, 9.

55 Winskill, N. and Hammond, S., An industry perspective on the potential for emerging process analytical technologies http://www.fda.gov/ohrms/dockets/ac/02/briefing/3841B1_05_PFIZER/sld001.htm.

56 Reich, G. (2005) Near-infrared spectroscopy and imaging: Basic principles and pharmaceutical applications. *Adv. Drug Deliv. Rev.*, **57**, 1109–1143.

57 Kauffman, J.F., Gilliam, S.J., and Martin, R.S. (2008) Chemical imaging of pharmaceutical materials: fabrication of micropatterned resolution targets. *Anal. Chem.*, **80** (15), 5706–5712.

58 Sasic, S. (2008) *Anal. Chim. Acta*, **611**, 73–79.

59 LaPlant, F. (2004) *Am. Pharm. Rev.*, **7**, 16–24.

60 Onat, B.M., Carver, G., and Itzler, M. (2008) *Proc. of SPIE*, **7310**, 731004-1/11.

3
Quality Control in Food Processing

Colette C. Fagan

3.1
Introduction

Food processors have come under increased pressure to ensure the large-scale production of consistently high quality food products with specific functional properties. This has been driven by the requirement for consumer acceptability of products as either food ingredients or consumer products, as well as increased concerns over the safety and, to some degree, authenticity of food products. If food processors are to meet these demands, they must implement and maintain procedures which will ensure this outcome. However, there are a number of challenges which the food industry must overcome to achieve this, for example most unit operations are batch or semi-continuous operations, there is often a large inherent variation in the raw material used, and food products and processes are complex, heterogeneous and involve the interaction of a number of factors. Therefore, the food industry has looked towards strategies, for example process analytical technology (PAT), to achieve consistently high quality products. PAT is a systems for designing, analyzing and controlling manufacturing through timely measurements of critical quality and performance attributes of raw and in-process materials and processes [1]. However, this requires the control of the manufacturing process through real-time analysis of critical quality parameters. Therefore, techniques and technologies, that is, PAT tools, which can rapidly, accurately and preferably nondestructively, assess the quality, and functional properties of food products, are essential for the modern food processing industry. These tools should also facilitate increased process understanding, development of risk-mitigation strategies, and continuous improvement. Optical and imaging technologies have been developed which have the potential to be ideal PAT tools. These have primarily been applied to quality and safety assessment, authenticity determination, and process control applications. These include near-infrared (NIR) spectroscopy, mid-infrared (MIR) spectroscopy, fluorescence spectroscopy, hyperspectral imaging (HSI), and Raman spectroscopy (Table 3.1).

Handbook of Biophotonics. Vol.3: Photonics in Pharmaceutics, Bioanalysis and Environmental Research, First Edition.
Edited by Jürgen Popp, Valery V. Tuchin, Arthur Chiou, and Stefan Heinemann

Table 3.1 Examples of the application of optical and imaging technologies to food quality and processing.

Principle	Application	Product	Reference
NIR	Sensory quality	Cheese	[111]
FTIR, FT-NIR and FT-Raman	Authenticity	Oils and fats	[59]
FTIR	Geographic origin	Honey	[112]
NIR	Process monitoring	Hamburger meat	[113]
Fluorescence and MIR	Geographic origin	Cheese	[114]
Spectral imaging	Defect detection	Fruit	[115]
Hyperspectral imaging	Contaminant detection	Poultry	[116]

Integration of such strategies and technologies results in the generation of large amounts of diverse data. Therefore, appropriate data management and processing approaches are critical if such data rich technologies are to be employed in the food processing sector.

Advances in chemometrics, which applies optimal mathematical and statistical methods to process data, has been key to the adoption of PAT tools. Chemometric approaches are vital in understanding, diagnosing real-time processes, and keeping them under multivariate statistical control. Multivariate calibration and multivariate classification are two of the most common chemometric methodologies. Chemometric approaches can be used to overcome problems in spectroscopy such as collinearity that is, where variables in the calibration have high correlations between them, or where it is difficult to select a specific wavelength for calibration as infrared spectra frequently contain data points carrying overlapping information [2]. Calibration approaches such as principal components analysis (PCA) and partial least squares regression (PLSR), and classification approaches such as cluster analysis or discriminant analysis (DA) are commonly used and have been explained in detail elsewhere [3, 4].

Recent advances in instruments have widened the potential applications of photonics in the food industry. Developments in fiber optics have assisted the expansion of NIR reflectance and transmission sensors for online analysis, while the move towards miniaturization of spectrometers has opened up new opportunities for food processors to monitor critical control point. The development and improvement in MIR instrumentation, in particular Fourier transform infrared (FTIR) spectrometers, has increased the feasibility of applying MIR spectroscopy to the precise control and monitoring of a manufacturing process. Emerging platform technologies such as hyperspectral imaging (HSI), which integrates conventional imaging and spectroscopy to attain both spatial and spectral information from an object, have also shown potential for widespread adoption in the areas of food quality, safety, and product development. Table 3.2 outlines the advantages and disadvantages of some of technologies most commonly applied to food products.

Table 3.2 Advantages and disadvantages of the near infrared (NIR), Fourier transform infrared (FTIR) and fluorescence spectroscopy from [117].

Techniques	Sensitivity	Information content	Absence of interferences	Repeatability	Absence of light scatter
NIR	**	**	*	**	**
FTIR	***	***	*	***	**
Fluorescence	***	*	***	**	*

Sensitivity and information content: *, low; ***, high; Absence of interferences: *, many interferences; ***, few interferences; Repeatability: *, poor; ***, good; Absence of light scatter: *, severe light scatter; ***, no light scatter.

3.2 Quality Applications

Food quality cannot be considered as a single, well-defined attribute. In fact it encompasses a number of properties or characteristics, which are often referred to as the quality indices of the product under test [5]. Examples of such indices may be chemical composition, sensorial and textural attributes, aroma, and appearance. While the use of photonics offers a solution to the challenge of determining such complex attributes, one must ensure that the basis for the prediction of quality is fully understood, as well as its inherent limitations.

The advantages of employing FTIR spectroscopy in conjunction with attenuated total reflectance (ATR) as a quantitative quality control tool for the food industry have been previously outlined [6]. Villé *et al.* [7] developed a method for the determination of total fat and phospholipid content in intramuscular pig meat using FTIR spectroscopy. Meat samples were extracted with chloroform and methanol and FTIR spectra were recorded in transmission mode. They developed a linear regression equation to predict total fat using the spectral region associated with the C=O band (1785–1697 cm^{-1}). The resulting model had a determination coefficient of 99%. FTIR spectroscopy has also been employed to predict the fatty acid content in fat extracts of pork meat [8]. Samples were again extracted using chloroform and methanol and the sample accessory was an ATR cell. Models were developed which predicted four fatty acids with R^2 values between 0.91 and 0.98.

Lefier *et al.* [9] employed FTIR spectroscopy to predict milk fat, crude protein, true protein and lactose content. Due to the success of the FTIR measuring principle for the analysis of milk this technology has been successfully commercialized with products such as the MilkoScan™ FT 120 (Foss, Denmark), which employs this principle in compliance with IDF and AOAC standards.

The quality of any given type of cheese is related to a large extent to its texture, which in turn is influenced by moisture and other composition components, and processing conditions. Therefore MIR spectroscopy has been investigated as a tool for predicting not only compositional parameters but also textural attributes.

The organoleptic quality of cheese is determined by complex changes that occur during ripening [10, 11]. The three reactions, which are primarily responsible for the development of texture and flavor, are proteolysis, glycolysis and lipolysis. Irudayaraj *et al.* (1999) investigated the use of FTIR spectroscopy to follow texture development in Cheddar cheese during ripening. They demonstrated that springiness could be successfully correlated with a number of bands in mid-infrared spectra. The development of cheese microflora during ripening is extremely important in the development of flavor and texture. Leifer *et al.* [12] demonstrated that FTIR spectroscopy could be used as a rapid and robust method for the qualitative analysis of cheese flora. They developed a model for the identification of *Lactococcus* sp. using a number of strains of *Lactcoccus lactis* ssp. *lactis* and *cermoris*. FTIR spectroscopy has also been used to evaluate the shelf-life period in which "freshness" is maintained in Crescenza cheese [13]. PCA was found to detect the decrease of Crescenza "freshness" and to define the critical day during shelf-life. Fagan *et al.* [14] determined the potential of FTIR spectroscopy coupled with PLS regression for the prediction of processed cheese instrumental texture and meltability attributes. The strongest models developed predicted hardness ($R^2 = 0.77$), springiness ($R^2 = 0.77$), cohesiveness ($R^2 = 0.81$) and Olson and Price meltability ($R^2 = 0.88$).

Fruit quality indices consist of internal quality, such as soluble solids content (SSC) and total acidity (TA), and external quality, such as size and weight [15]. Bureau *et al.* [16] studied the potential of FTIR spectroscopy to simultaneously determine the sugar and organic acid content of apricot fruit slurries. They identified the most suitable spectral region as 1500 to 900 cm^{-1} for model development ($R^2 \geq 0.74$ and root mean square error of prediction (RMSEP) $\leq$18%).

NIR spectroscopy has been widely employed for the prediction of the chemical composition of numerous products including, milk [17, 18], cheese [19–22], butter [23], meat [24–26], wheat [27, 28], and fruit [29].

For compositional analysis of products such as cheese NIR reflection has primarily been used. As early as 1982 NIR spectroscopy was applied to the determination of the fat, protein and moisture content of cheese [19]. More recently, NIR spectroscopy has also been applied to the prediction of the composition of processed cheese with much success (root mean square error of cross validation (RMSECV) = 0.45, 0.50, and 0.26% for fat, moisture and salt, respectively) [21]. Downey *et al.* [30] subsequently examined the potential of NIR spectroscopy to predict the sensory attributes and maturity of Cheddar cheese. The most successfully predicted parameters were maturity, crumbliness, rubberiness, chewiness, mouth-coating and massforming. A similar study on processed cheese in general found that it was possible to predict processed cheese sensory and instrumental texture parameters using NIR spectroscopy [31].

The texture of fruit is also a critical quality parameter. Valente *et al.* [32] employed a combination of visible and NIR spectroscopy and a nondestructive acoustic technique to estimate mango fruit firmness. The best calibration result was obtained using the second derivative spectra for two genetic algorithm selected spectral ranges (736–878 and 955–1022 nm) and had an associated RMSEP value of 3.28 ($R^2 = 0.82$), however, they were deemed not to be sufficiently good to allow automated fruit sorting.

Cayuela and Weiland [33] compared two commercial portable spectrometers for prediction of orange quality. The first spectrometer (Vis–NIR) used fiber optics to collect the reflected radiation and transmit it to three detectors (350–2500 nm). The second instrument (AOTF-NIR) had a reflectance post-dispersive optical configuration and InGaAs detector (1100–2300 nm). They predicted 11 parameters, that is, soluble solids content, acidity, titratable acidity, maturity index, flesh firmness, juice volume, fruit weight, rind weight, juice volume to fruit weight ratio, fruit color index and juice color index. They found the results were particularly good for the direct NIR prediction of soluble solids content, and maturity index (Vis–NIR spectrometer: RPD = 1.67–2.21 and AOTF-NIR spectrometer: 1.03 to 2.33).

A limited number of studies have employed Raman spectroscopy to characterize cheese during ripening. Some of the most common excitation wavelengths used today for Raman spectroscopy are 532, 633, 785 and 1064 nm. While the Raman scattering effect is more efficient at short wavelengths, the use of short wavelength lasers can induce fluorescence in the sample under test, which is stronger than the Raman scattering. NIR excitation does not give rise to such an effect and one of the most common NIR laser sources for Raman spectroscopy of foodstuffs has been the Nd: YAG (neodymium-doped yttrium aluminum garnet) solid state laser (1064 nm). However, when this laser is used water-based samples suffer from self-absorption of Raman scatter, as well as significant laser absorption [34]. Additional information on lasers used for Raman spectroscopy can be found elsewhere [34–36].

Rudzik *et al.* [37] have stated that Raman spectroscopy is an excellent tool for protein secondary structure determination. Fontecha *et al.* [38] used Raman spectroscopy with a 514.5 nm excitation laser to study the secondary structure of caseins from various sources, including ewes cheese. They assigned bands at 1609 and 1616 cm^{-1} to ring vibrations of aromatic amino acid side-chains, a band at 1655 cm^{-1} to the α-helix segments, and bands in the region of 1665 cm^{-1} to disordered and turn structures. They also stated that bands corresponding to the β-strand structures of casein were located in the region of 1680 cm^{-1} with β-sheet structures appearing at 1630 cm^{-1}. An increase in intensity at 1630 cm^{-1} occurred during cheese ripening as a result of increased β-sheet structures.

Raman spectroscopy has also been employed to understand how protein structure changes with processing to elucidate the impact on thermal processing on the functional properties of food [39].

Asfseth *et al.* [25] evaluated Raman spectroscopy for determining fatty acid composition and contents of the main constituents in a complex food model system. They developed a model system consisting of 70 different mixtures of protein, water, and oil blends. The PLS regression models predicted the total level of unsaturation in the form of the iodine value with an error of 2.8% of the total iodine value range. The same authors went on to apply Raman spectroscopy to the assessment of real food systems (salmon) [40]. The Raman spectra of intact salmon muscle, ground salmon and oil extracts were used to predict the degree of fatty acid unsaturation (iodine value) of the salmon ($n = 50$). The best model was developed using the oil extract spectra ($R = 0.87$, RMSECV = 2.5 g I_2/100 g fat) using only one latent variable.

3.3 Safety Applications

The area of photonics and food safety has grown over the last decade or so. Technologies have been developed which could be used to detect contaminants such as foreign objects, fecal matter, or microbial contamination. The potential to identify microbial populations and validate cleaning protocols has also been investigated.

The contamination of carcasses by fecal matter is the primary pathway for pathogens to contaminate such products. Fluorescence spectroscopy has been employed to determine excitation and emission characteristics of fecal matter derived from cows, deer, swine, chickens, and turkeys [41]. Spectra were recorded of fecal matter, animal meats, and swine, chickens', and turkeys' feedstuff. Excitation at approximately 410 to 420 nm yielded the highest level of fluorescence for both feces and feedstuffs. The emission maxima for feces were at 675 nm with the exception of chicken feces which occurred at 632 nm.

A number of studies have examined the potential of HSI to detect fecal contamination [42–46]. Park *et al.* [42] studied the performance of supervised classification algorithms and HSI to identify fecal and ingesta contaminants. The six different supervised classification algorithms examined were parallelepiped, minimum distance, Mahalanobis distance, maximum likelihood, spectral angle mapper and binary coding. The classification accuracies varied from 62.9 to 92.3%, depending on the classification method. The highest classification accuracy for identifying contaminants from corn-fed carcasses was 92.3% with a spectral angle mapper classifier. The overall mean classification accuracy for classifying fecal and ingesta contaminants was 90.3%. The same group also utilized a band-ratio image processing algorithm using 2-wavelengths (I_{565}/I_{517}) and 3-wavelength ($(I_{576} - I_{616})/(I_{529} - I_{616})$ models [43]. Figure 3.1 describes the hyperspectral image processing procedure and shows the results of the 3-wavelength models. The 2-wavelength model had a correct classification rate of 96.4% and 147 false positives, while the 3-wavelength models had a correct classification rate of 98.6% but 388 false positives. A similar approach was employed by Liu *et al.* [45] to detect fecal contamination on apples. They recorded hyperspectral images of two apple cultivars, and they used 2- (I_{725}/I_{811}) and 3-wavelength ($(I_{811} - I_{557})/(I_{725} - I_{557})$ band ratio equations to identify fecal contaminated skins, with the dual band ratio giving the best results.

Fourier transform Raman spectroscopy has been studied for the characterization and differentiation of six different microorganisms, including the pathogen *Escherichia coli* O157: H7 on whole apples [47]. They stated that the technology successfully discriminated between *E. coli* strains as well as accurately differentiating pathogens from non-pathogens. Sivakesava *et al.* [48] also attempted to identify and classify various species of microorganisms, in this case using FTIR and NIR spectroscopy. Using canonical variate analysis (CVA) and NIR spectra they achieved a 95.6% correct classification rate, while FTIR spectra achieved a lower correct classification rate of 93%, however, FTIR required a lower number of principal components. Harz *et al.* [49] employed micro-Raman spectroscopy to identify bacterial bulk material

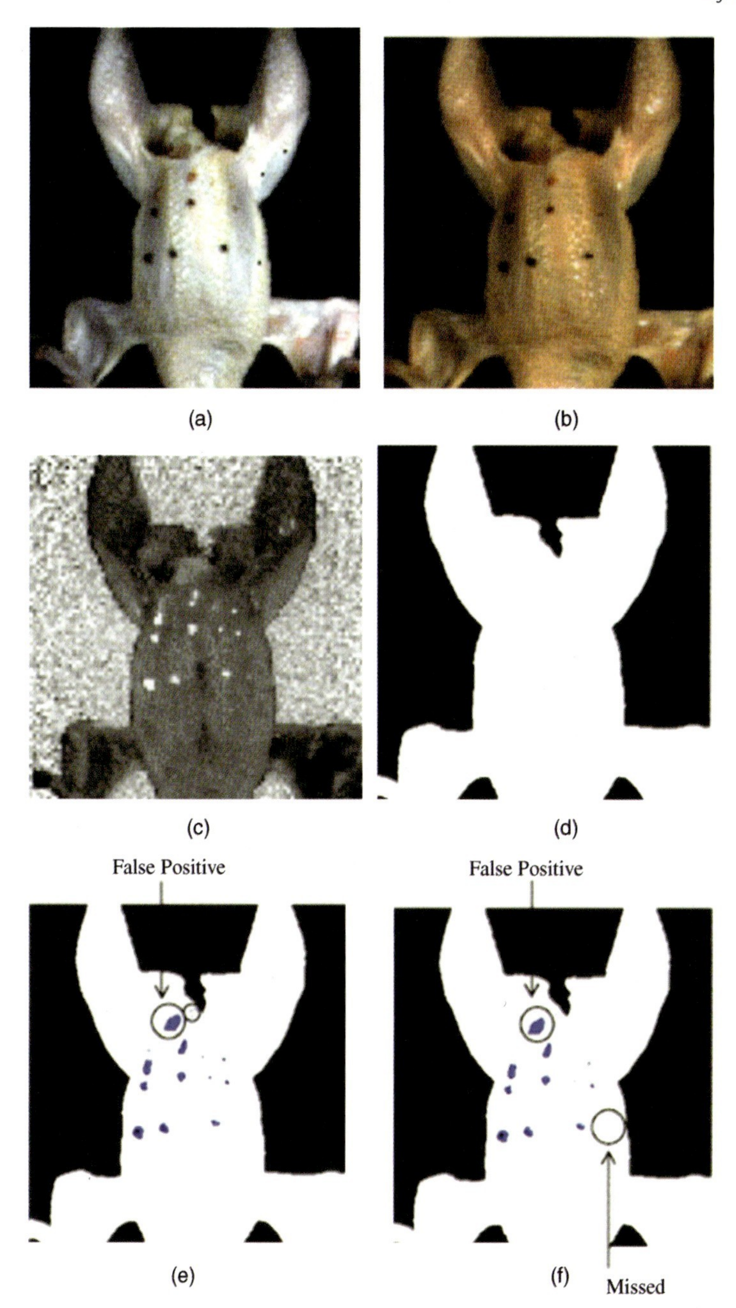

Figure 3.1 Procedure of hyperspectral image processing of a poultry carcass: (a) color composite; (b) calibrated color image; (c) ratio image (calibrated and smoothed); (d) background mask and band ratio results of three wavelengths (529, 576, and 616 nm) selected by single term regression model using the region of interest of the hyperspectral images at (e) threshold = 1.05; (f) threshold = 1.05 with filter. From [43].

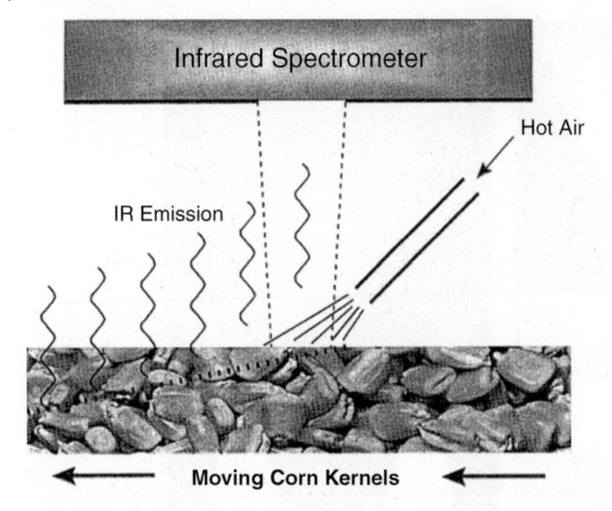

Figure 3.2 Schematic illustration of corn kernels moving past the detector in TIRS. [50].

and single cells of different species within the genus *Staphylococcus*. They found it was possible to make a distinction between strains in the bulk material when a support vector machine and hierarchical cluster analysis were employed. For single bacterial spectra they achieved an average recognition rate of 94.1% on the strain level and 97.6% on the species level.

Mycotoxins can pose a significant health risk to humans. Transient infrared (TIR) spectroscopy and FTIR photoacoustic spectroscopy have been employed to detect toxigenic fungi (*Aspergillus flavus*) in corn [50] (Figure 3.2). It was found that TIR spectroscopy effectively distinguished healthy corn from corn infected with *Aspergillus flavus* with a 85 or 95% success rate. Berardo *et al.* [51] challenged NIR spectroscopy to identify toxigenic fungi and their toxic metabolites produced in naturally and artificially contaminated products. They indicated that NIR could accurately predict the incidence of kernels infected by fungi, and by *Fusarium verticillioides* in particular. They were also able to predict the quantity of ergosterol and fumonisin B_1 in the meal. The best model predicted the percentage of global fungal infection and *Fusarium verticillioides* when data from the maize kernels ($R^2 = 0.75$ and standard error of cross validation (SECV) = 7.43) and maize meals ($R^2 = 0.79$ and SECV = 10.95).

3.4
Authenticity Application

Food manufacturers are required to demonstrate the authenticity of their products [2]. The European Union set out the rights of consumers and genuine food processors in terms of food adulteration and fraudulent or deceptive practices in food processing [52]. There are three EU schemes which are used to promote and protect the

Table 3.3 Examples of product names registered under protected designation of origin (PDO), protected geographical indication (PGI) and traditional specialty guaranteed (TSG) schemes.

Designation	Country	Type	Product Categrory
Scotch Lamb	United Kingdom	PGI	Class 1.1
Prosciutto di Parma	Italy	PDO	Class 1.2
Timoleague Brown Pudding	Ireland	PGI	Class 1.2
Edam Holland	Netherlands	PGI	Class 1.3
Mozzarella	Italy	TSG	Class 1.3
Emmental de Savoie	France	PGI	Class 1.3
Huile d'olive de Corse; Huile d'olive de Corse-Oliu di Corsica	France	PDO	Class 1.5
蠡县麻山药 (Lixian Ma Shan Yao)	China	PGI	Class 1.6
Yorkshire Forced Rhubarb	United Kingdom	PDO	Class 1.6
Brussels grondwitloof	Belgium	PGI	Class 1.6
Herefordshire cider/perry	United Kingdom	PGI	Class 1.8
Dortmunder Bier	Germany	PGI	Class 2.1
Pierekaczewnik	Poland	TSG	Class 2.3
Kalakukko	Finland	TSG	Class 2.3

Class 1.1 Fresh meat (and offal), Class 1.2 Meat products (cooked, salted, smoked), Class 1.3 Cheeses, Class 1.5 Oils and fats (butter, margarine, oil, etc.), Class 1.6 Fruit, vegetables and cereals fresh or processed, Class 1.8 other products of Annex I of the Treaty (spices etc.), Class 2.1 Beers, Class 2.3 Confectionery, bread, pastry, cakes, biscuits and other baker's wares.

names of quality agricultural products and foodstuffs. These are protected designation of origin (PDO), protected geographical indication (PGI) and traditional specialty guaranteed (TSG). Examples of products protected under these schemes are shown in Table 3.3. The authenticity of incoming raw materials is also of concern to food processors.

Detection of adulterated oils, for example, olive oil is critical both in terms of economic and health aspects [53]. A number of studies have been undertaken to examine the potential of a range of technologies to detect adulterated oils [53–57].

Lai [54] employed FTIR spectroscopy in conjunction with PCA and linear DA to determine the authenticity of vegetable oils. They found that samples clustered according to their origin (plant species) and it was also possible to discriminate between extra virgin and refined olive oils (correct classification rate: 93% calibration and 100% validation). Lai *et al.* [55] also examined the potential of FTIR spectroscopy to quantify the level of adulterants (refined olive oil and walnut oil) in extra virgin olive oil with some success (standard error of prediction (SEP) = 0.68–0.92 g $(100\,g)^{-1}$). The potential of Fourier transform Raman spectroscopy in such an application has also been reported [56]. They adulterated virgin olive oil with soybean, corn, and raw olive residue (olive pomace) oils at 1, 5, and 10%, respectively and also collected six genuine virgin olive oil samples. They achieved a 100% correct discrimination between genuine and adulterated samples and 91.3% correct classifications at different adulteration levels using principal component regression. Marigheto *et al.* [58] went on to compare the potential of FTIR and Raman spectroscopies to discriminate between oils of differing botanical origin, and for their ability to detect added adulterants. Ultimately they found that MIR, in combination with linear discriminant analysis, gave the best classification rates and adulteration detection levels were comparable with Raman spectroscopy. Determination of olive oil geographic origin using visible and NIR spectroscopy was also examined by Downey *et al.* [57]. They collected samples from three regions that is, Crete ($n = 18$), the Peloponese ($n = 28$), and other parts of Greece ($n = 19$), and spectral analysis was carried out in transflectance mode using a gold-plated accessory. The results indicated that the highest classification rate was achieved using raw spectra (400–2498 nm) and factorial DA, although it was noted that the samples were from a single harvest. A comparison between FTIR (MIR), FT-NIR, and FT-Raman spectroscopy for discrimination among 10 different edible oils and fats has also been carried out [59]. In agreement with other studies they found that FTIR spectroscopy was the most successful in discriminating between oils and fats (98%), followed by Fourier transform Raman spectroscopy (94%) and Fourier transform NIR spectroscopy (93%).

Cheese authenticity is an emerging research area, which is becoming increasingly important to the dairy sector [60]. Pillonel *et al.* [61] employed a small data set of Emmental cheeses from different regions, that is, Switzerland ($n = 6$), Allgau (Germany) ($n = 3$), Bretagne (France) ($n = 3$), Savoie (France) ($n = 3$), Vorarlberg (Austria) ($n = 3$) and Finland ($n = 2$) to compare NIR and FTIR spectroscopy for their potential to discriminate samples on the basis of geographic origins. They found that FTIR transmission spectra achieved 100% correct classification in conjunction with LDA when differentiating Swiss Emmental from the other samples pooled as one group, while NIR diffuse reflection spectroscopy allowed a classification by the six regions of cheese origin (100% correct classification). Karoui *et al.* [62] went on to combine FTIR and front-face fluorescence spectroscopy to discriminate between Emmental cheeses originating from different European countries: Austria ($n = 12$), Finland ($n = 10$), Germany ($n = 19$), France ($n = 57$), and Switzerland ($n = 65$). By pooling the first 20 principal components from both the FTIR and fluorescence spectral datasets, and subjecting them to factorial DA they achieved correct classifications of 76.7% of the validation spectra.

Aulrich and Molkentin [63] investigated the potential of NIR spectroscopy to discriminate between organically and conventionally produced milk on the basis of ω3-fatty acids predictions. They found it was possible to predict ω3-FA content and that, with due consideration to seasonal variations in the contents of the ω3-FA C18 : 3ω3 and C20 : 5ω3 in milk, that NIR spectroscopy could be used as screening tool in discriminating between organically and conventionally produced milk. The potential of FTIR spectroscopy to discriminate between milk samples on the basis of origin that is, cow, goat, or sheep has been investigated [64]. Blends of sheep and cow milk, goat and cow milk, and sheep and goat milk were prepared which contained between 0 and 100% of each milk in increments of 5%. A second set of samples was prepared containing a blend of sheep, cow, and goat milk, also in increments of 5%. The best result was achieved using kernel PLS regression for the first data set (error = 4 to 6%) while errors of 3.9 to 6.4% were achieved for the second dataset.

Over the last decade there has been a drive towards rapid nondestructive techniques for determining honey authenticity and geographic origin.

Goodacre *et al.* [65] employed Raman spectroscopy to discriminate between honey samples from different floral and geographical origins. Analysis was carried out on 51 samples, from 5 countries and from a number of botanical sources: acacia (7), chestnut (9), eucalyptus (4), heather (10), lime (4), rape (5), sunflower (4), citrus (2), lavender (2), rosemary (1), *Echium plantagineum* (1), orange (1), and fior di sulla (1). They stated that cluster analysis indicated that differences between the honeys were due to their botanical origin rather than their country of origin.

Sivakesava and Irudayaraj [66] employed FTIR spectroscopy as a screening tool for the determination of beet medium invert sugar adulteration in three different varieties of honey. They developed PLS regression models using first derivative spectra (1500–950 cm^{-1}) for determination of beet invert sugar in honey samples. The most successful model for discriminating adulterated honey samples was achieved with canonical variate analysis (classification rate 88–94%). However, care should be taken to ensure the experimental design does not facilitate classification on the basis of alteration of the solids content of the honey [67]. Therefore, honey samples and the adulterant solutions are usually adjusted to a standard solids (°Brix) level.

Detection of honey adulteration using FTIR spectroscopy has been examined [67]. Irish honey samples ($n = 221$) were adulterated using D-fructose and D-glucose solutions, while authentic samples ($n = 99$) were also collected. Samples and adulterant solutions were adjusted to 70°Brix. The classification model developed using PLS regression of first derivative spectra gave the best classification rate (93 or 99% if greater than 14% w/w adulterant). Kelly *et al.* [68] went on to examine the potential of FTIR spectroscopy to detect adulteration of Irish honeys by five adulterants. Authentic Irish honeys ($n = 580$) and honeys adulterated by fully inverted beet syrup ($n = 280$), high-fructose corn syrup ($n = 160$), partial invert cane syrup ($n = 120$), dextrose syrup ($n = 160$), and beet sucrose ($n = 120$) were adjusted to 70°Brix and analyzed. They were able to classify authentic honey and honey adulterated by beet sucrose, dextrose syrups, and partial invert corn syrup (correct classification rates: 96.2, 97.5, 95.8, 91.7% respectively). However, they were not able to clearly detect high-fructose corn syrup or fully inverted beet syrup adulteration. This group also examined the potential of NIR transflectance spectroscopy in such an application [69]. They used authentic honey ($n = 83$) and honeys adulterated with beet invert syrup ($n = 56$) and high fructose corn syrup ($n = 40$). In this scenario they found that NIR spectroscopy could be used as a rapid screening tool for detection of Irish honey adulterated with these compounds.

The work of Ruoff *et al.* [70, 71] investigated the potential of fluorescence and FTIR spectroscopy to classify honey according to its botanical and geographical origin. The samples analyzed were 11 unifloral (acacia, alpine rose, chestnut, dandelion, heather, lime, rape, fir honeydew, metcalfa honeydew, oak honeydew) and polyfloral honey types (fluorescence: $n = 371$; FTIR: $n = 411$).

Using fluorescence spectroscopy the discriminant models had error rates of less than 0.1% for polyfloral and chestnut honeys to 9.9% for fir honeydew honey, by using single spectral data sets. If two data sets were combined the error rates were from <0.1% (metcalfa honeydew, polyfloral, and chestnut honeys) to 7.5% (lime honey). For FTIR spectroscopy the error rates ranged from <0.1% (polyfloral, heather honeys, honeydew honeys from metcalfa, oak, and fir) to 8.3% (alpine rose honey). Fluorescence spectroscopy could only classify samples according to geographical origin within a group of honeys of the same botanical origin, however, linear DA of FTIR spectra of acacia, lime, dandelion, and honeydew honey samples from Germany and Switzerland resulted in an average correct classification rate of 85%. However, botanical origin had a more significant effect than geographical origin, hence, the results may in fact be due to the indirect effects of the botanical origin.

They have also tried to create a specific NIR spectral fingerprint for Corsican honey [60]. They collected NIR spectra from authentic Corsican samples, and mathematically processed them so as to develop a univariate specification based on fingerprint spectroscopy. The specification was then challenged by spectra of separate Corsican and non-Corsican honey samples over two production seasons. The best PLS discriminant models produced correct classification rates of 90.0% (Corsican), 90.3% (non-Corsican) and 90.4% (Corsican) and 86.3% (non-Corsican) for models validated via cross-validation and an independent test set, respectively.

Al-Jowder *et al.* [72] applied FTIR spectroscopy to the authenticity of fresh meats. Using PCA they were able to discriminate between chicken, pork and turkey meats. The basis for this discrimination was found to be the lipid (1740 cm^{-1}) and protein (1650 and 1550 cm^{-1}) bands with protein bands increasing in intensity for chicken, turkey and pork consecutively. They also differentiated between fresh samples and the frozen samples that had been thawed. McElhinney *et al.* [73] also investigated the potential of FTIR spectroscopy for quantitative analysis of meat mixtures. A PLS regression model was able to predict the lamb content of minced beef and lamb mixtures with an R value of 0.97, however, the study also developed models using NIR spectra, which were found to have an R value of 0.99. Al-Jowder *et al.* [74] went on to employ FTIR spectroscopy to discriminate between pure beef and beef containing 20% offal (heart, kidney, liver and tripe) in both raw and cooked samples. Ding and Xu [75] found that NIR spectroscopy could be used to discriminate between authentic beef hamburgers and those adulterated with 5–25% mutton, pork, skim milk powder, or wheat flour with an accuracy up to 92.7%. A comparative study has also been carried out to compare the potential of FTIR, NIR and visible spectroscopy to discriminate samples on the basis of origin (i.e., chicken, turkey, pork, beef, lamb) [76]. Difficulties arose when trying to discriminate between chicken and turkey. Therefore the authors developed models with both five meat classes and also with four, where chicken and turkey samples were combined as one group. For the five class model, incorporating FTIR spectra with visible and NIR spectra did not improve model accuracy. However, in the case of the four class model it did by improving the discrimination between chicken and turkey.

The use of fruit purees rather than whole fruits as a raw material is advantageous from an economic perspective, and in the case of some soft fruits a requirement [77]. However, determining the authenticity of fruit purees can be a challenge. FTIR spectroscopy has been used to discriminate between fruit purees using DA [77]. They could classify purees according to a number of factors that is, strawberry, raspberry, or apple (100% success); fresh or freeze-thawed (98.3% for strawberry, 75% for raspberry); sulfur dioxide addition (90% for apple); and Bramley versus non-Bramley apples (86%). Contal *et al.* [78] went on to examine if visible and NIR spectroscopy could be used for detecting strawberry and raspberry purees adulterated with apple. They prepared authentic strawberry ($n = 32$) and raspberry ($n = 32$) purees as well as strawberry and raspberry purees adulterated with 10, 20, 30, 50 and 75% apple ($n = 28$). Ultimately the developed model could detect apple adulteration at levels above 20% in strawberry and between 10 and 20% in raspberry. More recently, FT-NIR spectroscopy of cell wall components has been evaluated for determining the authenticity of fruit and the fruit content of products [79]. They collected NIR spectra of the alcohol-insoluble residues ($n = 92$) and hemicellulose ($n = 109$) fractions obtained from the samples. They found that the calibration models developed to predict alcohol-insoluble residues and hemicellulose correlated very well with the reference data ($R^2 = 0.98$).

A robust, rapid, and inexpensive method for quality assurance purposes is needed in the wine industry to monitor that wine parameters conform to specification, in order to ensure the quality of the final product delivered to the consumer [80]. FTIR spectroscopy has been used to classify 165 wines from three different geographical origins (Gaillac, Beaujolais and Touraine) which were made from the same grape variety, that is, Gamay. The spectra of dried samples were recorded in transmission mode. By employing factorial DA they were able to classify the samples according to geographic origin with a correct classification rate of 97% [81]. Bevin *et al.* [80] also employed FTIR spectroscopy for wine assessment. They analyzed three Australian wine varieties, Shiraz, Cabernet Sauvignon, and Merlot ($n = 161$). Spectra, which excluded the regions 1543 to 1717 cm^{-1}, 2971 to 3627 cm^{-1} and 2276–2431 cm^{-1}, were recorded in transmission mode. They found that for five of the six wineries they obtained a correct classification rate of 98%. Acevedo *et al.* [82] went on to examine if classification of wines according to their origin could be improved by using UV–Vis spectroscopy and support vector machines (SVM). They stated that SIMCA, k-neural network, and PLSR DA seemed to require more selective techniques/variables than SVM if a large variety of wines was to be discriminated according to the specific region in which they are produced. In terms of its simplicity the SVM was preferred over NN-MPL. Finally, Cozzolino *et al.* [83] compared and combined UV–Vis, NIR and FTIR spectroscopy to determine their efficacy as tools for the classification of commercial Sauvignon Blanc from Australia (seven regions) and New Zealand (four regions). Spectral analysis was carried out in transmission mode. The best results employed a model developed using PLS DA. Concatenation of the NIR and MIR data resulted in the highest overall correct classification rate (93%). While MIR was the next strongest (90%) followed by NIR (76%) and UV–Vis (67%).

3.5 Process Control

Development of technologies which facilitate the control and monitoring of critical control points during food processing is essential. This includes the monitoring and detection of microbial populations, which is critical not only from a food safety but also from a process monitoring perspective. Microbial populations are critical to flavor and texture development in products such as cheese. Oberreuter *et al.* [84] employed FTIR spectroscopy to rapidly monitor the population dynamics of bacteria in cheese during ripening. They found that only 1% of the coryneform bacteria could not be identified by FTIR spectroscopy. Gaus *et al.* [85] successfully employed UV-resonance Raman spectroscopy for the classification of lactic acid bacteria on a strain level; the observed clusters corresponding to the different strains when PCA and hierachical cluster analysis were used.

NIR technology has been successfully applied at laboratory and commercial scale for monitoring processes during dairy processing. In particular, the milk coagulation process during cheese production has received a great deal of attention. Cutting the coagulum either before or after the optimum point results in losses of curd and fat. An increase in cheese moisture also occurs if the gel is too firm when cut. Originally, the determination of the cutting time was established by the cheese maker. Although accurate this method is not feasible in closed commercial vats and, together with an increased desire for automation in the cheese industry, has led to the need for an on-line objective method for the monitoring of milk coagulation. Early methods, which utilized the changes in the optical properties of the milk, were reflection photometry [86] and absorbance [87]. Although the reflection photometry and absorbance methods were found to monitor coagulation they were rarely used. However, developments in fiber optics have overcome many of the problems associated with these techniques. Light in the NIR region can be transmitted through a fiber optic bundle and diffuse reflectance or transmission monitored. As the gel is formed the reflectance will increase while transmission will decrease. Payne *et al.* [88] developed a method based on changes in diffuse reflectance during milk coagulation. Reflectance was measured using a fiber optic probe, utilizing a photodiode light source at a wavelength of 940 nm. The time to the inflection point (t_{max}) was determined from the first derivative and was found to correlate well with Formograph cutting times. Linear prediction equations, which were considered to be of the form required for predicting cutting time, were also developed using t_{max}. This technology has been commercialized as the CoAguLite sensor (Reflectronics Inc, Lexington, KY). It could also be used in conjunction with other sensors, for example the FiberView dairy waste sensor system, which could be used to monitor waste streams in dairy facilities. This enables the location, occurrence or concentration of the discharge to be determined. It monitors solids concentration in dairy plant effluents in the range of 0 to 1% solids (or higher), and, due to its quick response to loss events, it allows operators to take corrective actions.

Syneresis is a critical phase in cheese manufacture, with the rate and extent of syneresis playing a fundamental role in determining the moisture, mineral and lactose content of drained curd, and hence that of the final cheese [89, 90]. Therefore, research is ongoing into the development of a syneresis control technology. A number of potentially noninvasive technologies have been investigated for such an application, including ultrasound and computer vision [91–94] and NIR sensing [95–97]. Initial studies focused on off-line optical sensing of whey samples [98]. An adaption of this technology led to the development of a sensor which could be installed in the wall of a cheese vat for on-line continuous monitoring of both coagulation and syneresis [99]. The sensor operated at 980 nm and was sensitive to casein micelle aggregation and curd firming during coagulation, and to changes in curd moisture and whey fat contents during syneresis. This sensor was also used to predict whey fat content (i.e., fat losses), curd yield and curd moisture content with SEPs of 2.37 g, 0.91 and 1.28%, respectively [100]. Further work used a wider spectral range (300–1100 nm) in conjunction with PLS regression to predict whey fat and curd moisture with RMSECV values of 0.094 and 4.066%, respectively [101]. Mateo *et al.* [102] developed another set of models which predicted the yield of whey ($R^2 = 0.83$, error = 6.13 g/100 g) using three terms, namely light backscatter, milk fat content and cutting intensity. These studies were carried out in laboratory scale cheese vats (7–11 l), therefore, further scaling up and development under commercial conditions of the technology will be required if it is to become viable at a commercial scale. Fluorescence spectroscopy has also been studied for its potential as a syneresis control tool [103]. In this preliminary evaluation it was found that tryptophan and riboflavin fluorescence had strong relationships with curd moisture and whey total solids content during syneresis, and that it was possible to develop simple one- and two-parameter models to predict curd moisture content, curd yield and whey total solids using parameters derived from the sensor profiles (error = 0.0005 to 0.394%; $R^2 = 0.963$ to 0.999).

NIR spectroscopy has also been investigated as a process control tool in yoghurt production. Cimander *et al.* [104] studied the potential of NIR spectroscopy to monitor yogurt fermentation in a 4.2-l laboratory scale vat. A sensor signal fusion approach was adopted with NIR (400 to 2500 nm), electronic nose, and standard bioreactor sensors installed as part of a multi-analyzer set-up. While the electronic nose followed changes in galactose, lactic acid, lactose and pH, the NIR sensor signal correlated well with the changes in the physical properties during fermentation. Therefore, the signals from the sensors were fused using a cascade artificial neural network (ANN). Results suggested that the accuracy of the neural network prediction was acceptable. This approach was further investigated by Navrátil *et al.* [105] under industrial conditions in a 1000-l vat. Signal responses from NIR and electronic nose sensors were subjected to PCA separately. The scores of the first principal component from each PCA were then used to make a trajectory plot for each fermentation batch. PLS regression of the NIR spectra was also used to predict pH and titratable acidity (expressed as Thorner degrees, °Th) during fermentation with reasonable success (SEPs of 0.17 and 6.6 °Th, respectively). MIR spectroscopy has also been employed to monitor the sorghum fermentation process [106]. FTIR spectroscopy was used to detect differences due to the effect of lactic bacteria on sorghum fermentation. It was found possible to differentiate between samples which used natural yogurt and *Lactobacillus fermentum* as inocula due to variations in protein and starch structures.

Optical sensors have also been developed to monitor meat emulsion stability [107–110]. Initial work focused on prediction of meat emulsion stability using reflection photometry [107]. The authors found that L^* values increased at the beginning of chopping, associated with reduced cooking losses, following 8 min of chopping there was a reduction in L^* and b^* values and an associated increase in cooking losses, which suggested the feasibility of an on-line optical sensor technology to predict the optimum endpoint of emulsification in the manufacture of finely comminuted meat products. These authors then recorded light backscatter intensity from beef emulsions manufactured with different fat/lean ratio and chopping duration using a dedicated fiber optic prototype [108]. They found several optically derived parameters to be significantly correlated with fat loss during cooking. In subsequent work they found normalized intensity decreased with increased chopping time as a result of emulsion homogenization, and with increased distance and that chopping time had a positive correlation with fat losses during cooking, which in turn had a negative correlation with normalized light intensity and loss of intensity. They suggest therefore that light extinction spectroscopy could provide information about emulsion stability [109].

3.6 Conclusions

NIR, MIR, fluorescence and Raman spectroscopic techniques are ideal PAT tools which could be implemented in a variety of quality, safety, authenticity and process control applications. Advances in equipment design, such as the move toward micro and miniature spectrometers will assist in the deployment of such technologies in the food industry. However, where studies have primarily been at laboratory scale further research is required to ensure appropriate scaling up and transfer of the technology to industry. The combination of spectroscopic and imaging technologies in emerging platforms may also allow enhanced identification of quality problems during food processing.

3.7 Glossary

Artificial neural network (ANN)
Attenuated total reflectance (ATR)
Canonical variate analysis (CVA)
Discriminant analysis (DA)
Fourier transform infrared (FTIR)
Hyperspectral imaging (HSI)
Mid-infrared (MIR)

Near-infrared (NIR)
Neodymium-doped yttrium aluminum garnet (Nd:YAG)
Partial least squares regression (PLSR)
Process analytical technology (PAT)
Protected designation of origin (PDO)
Protected geographical indication (PGI)
Principal components analysis (PCA)
Root mean square error of cross validation (RMSECV)
Root mean square error of prediction (RMSEP)
Soft independent modeling of class analogy (SIMCA)
Soluble solids content (SSC)
Standard error of prediction (SEP)
Traditional specialty guaranteed (TSG)
Transient infrared (TIR)
Total acidity (TA)

References

1 Balboni, M.L.(October 2003) Process analytical technology concepts and principles, *Pharm. Technol.*, 54–66.
2 Woodcock, T. *et al.* (2008) Application of Near and Mid-Infrared Spectroscopy to Determine Cheese Quality and Authenticity, *Food Bioproc. Technol.*, **1** (2), 117–129.
3 Martens, H. and Næs, T. (1989) *Multivariate Calibration*, Wiley, Chichester.
4 Martens, H. and Martens, M. (2001) *Multivariate Analysis of Quality: An Introduction*, John Wiley & Sons., Chichester.
5 Abbott, J.A. (1999) Quality measurement of fruits and vegetables, *Postharvest Biol. Technol.*, **15** (3), 207–225.
6 Vandevoort, F.R. (1992) Fourier-transform infrared-spectroscopy applied to food analysis, *Food Res. Int.*, **25** (5), 397–403.
7 Villé, H. *et al.* (1995) Determination of phospholipid content of intramuscular fat by Fourier Transform Infrared spectroscopy. *J. Meat Sci.*, **41**, 283–291.
8 Ripoche, A. and Guillard, A.S. (2001) Determination of fatty acid composition of pork fat by Fourier transform infrared spectroscopy. *Meat Sci.*, **58** (3), 299–304.
9 Lefier, D., Grappin, R., and Pochet, S. (1996) Determination of fat, protein, and lactose in raw milk by Fourier transform infrared spectroscopy and by analysis with a conventional filter-based milk analyzer. *J. AOAC Int.*, **79** (3), 711–717.
10 McSweeney, P.L.H. and Fox, P.F. (1993) Cheese: Methods of Chemical Analysis, In Cheese: Chemistry, Physics and microbiology, Vol 1, 2nd Ed (ed. P.F. Fox). Chapman & Hall, London, pp. 341–388.
11 Irudayaraj, J., Chen, M., and McMahon, D.J. (1999) Texture Development in Cheddar Cheese During Ripening, *Can. Agr. Eng.*, **41** (4), 253–258.
12 Lefier, D., Lamprell, H., and Mazerolles, G. (2000) Evolution of lactococcus strains during ripening in brie cheese using fourier transform infrared spectroscopy. *Lait*, **80** (2), 247–254.
13 Cattaneo, T.M.P. *et al.* (2005) Application of FT-NIR and FT-IR spectroscopy to study the shelf-life of Crescenza cheese. *Int. Dairy J.*, **15** (6–9), 693–700.
14 Fagan, C.C. *et al.* (2007) Evaluating mid-infrared spectroscopy as a new technique for predicting sensory texture attributes of processed cheese. *J. Dairy Sci.*, **90** (3), 1122–1132.
15 Liu, Y. *et al.* (2010) Nondestructive measurement of internal quality of

Nanfeng mandarin fruit by charge coupled device near infrared spectroscopy. *Comput. Electron. Agr.*, **71** (Supplement 1), S10–S14.

16 Bureau, S. *et al.* (2009) Application of ATR-FTIR for a rapid and simultaneous determination of sugars and organic acids in apricot fruit. *Food Chem.*, **115** (3), 1133–1140.

17 Carl, R.T. (1991) Quantification of the Fat-Content of Milk Using a Partial-Least-Squares Method of Data-Analysis in the near-Infrared. *Fresen. J. Anal. Chem.*, **339** (2), 70–71.

18 Kawamura, S. *et al.* (2007) Near-infrared spectroscopic sensing system for online monitoring of milk quality during milking. *Sens. Instrum. Food Qual. Safety*, **1** (1), 37–43.

19 Frank, J.F. and Birth, G.S. (1982) Application of near infrared reflectance spectroscopy to cheese analysis. *J. Dairy Sci.*, **65** (7), 1110–1116.

20 Molt, K. and Kohn, S. (1993) Near-IR spectroscopy. Chemometry for processed cheese products in quality control and process analysis. *Dtsch. Milchw.*, **44** (22), p. 1102, 1104–1107.

21 Blazquez, C. *et al.* (2004) Prediction of moisture, fat and inorganic salts in processed cheese by near infrared reflectance spectroscopy and multivariate data analysis. *JNIRS*, **12** (3), 149–157.

22 Rodriguezotero, J.L., Hermida, M., and Cepeda, A. (1995) Determination of fat, protein, and total solids in cheese by near-Infrared reflectance spectroscopy. *J. AOAC Int.*, **78** (3), 802–806.

23 Meagher, L.P. *et al.* (2007) At-line near-infrared spectroscopy for prediction of the solid fat content of milk fat from New Zealand butter. *J. Agr. Food Chem.*, **55** (8), 2791–2796.

24 Mlcek, J., Sustova, K., and Simeonovova, J. (2006) Application of FT NIR spectroscopy in the determination of basic chemical composition of pork and beef. *Czech J. Anim. Sci.*, **51** (8), 361–368.

25 Afseth, N.K. *et al.* (2005) Raman and near-infrared spectroscopy for quantification of fat composition in a complex food model system. *Appl. Spectrosc.*, **59** (11), 1324–1332.

26 Kim, Y.B. and Lee, M. (1997) Utilization on the near-infrared (NIR) for the chemical composition analysis of raw meat. *Korean J. Anim. Sci.*, **39** (1), 77–92.

27 Owens, B. *et al.* (2009) Prediction of wheat chemical and physical characteristics and nutritive value by near-infrared reflectance spectroscopy. *Br. Poult. Sci.*, **50** (1), 103–122.

28 Miralbes, C. (2003) Prediction chemical composition and alveograph parameters on wheat by near-infrared transmittance spectroscopy. *J. Agr. Food Chem.*, **51** (21), 6335–6339.

29 Peirs, A. *et al.* (2002) Comparison of Fourier transform and dispersive near-infrared reflectance spectroscopy for apple quality measurements. *Biosystems Eng.*, **81** (3), 305–311.

30 Downey, G. *et al.* (2005) Prediction of maturity and sensory attributes of Cheddar cheese using near-infrared spectroscopy. *Int. Dairy J.*, **15** (6–9), 701–709.

31 Blazquez, C. *et al.* (2006) Modelling of sensory and instrumental texture parameters in processed cheese by near infrared reflectance spectroscopy. *J. Dairy Res.*, **73** (1), 58–69.

32 Valente, M. *et al.* (2009) Multivariate calibration of mango firmness using vis/NIR spectroscopy and acoustic impulse method. *J. Food Eng.*, **94** (1), 7–13.

33 Cayuela, J.A. and Weiland, C. (2010) Intact orange quality prediction with two portable NIR spectrometers. *Postharvest Biol. Technol.*, **58** (2), 113–120.

34 Wilson, H.M.M. *et al.* (1996) Fourier transform Raman spectroscopy: A comparison of Nd:YAG and lower-frequency sources. *Vib. Spectrosc.*, **10** (2), 89–104.

35 Pan, M.-W., Benner, R.E., and Smith, L.M. (2002) Continuous lasers for raman spectrometry, in *Handbook of Vibrational Spectroscopy* (eds J. Chalmers and P. Griffiths), John Wiley & Sons Ltd., Chichester, pp. 490–506.

36 Saleh, B.E.A. and Teich, M.C. (2007) *Fundamentals of Photonics*, John Wiley & Sons Ltd., Hoboken.

37 Rudzik, L. *et al.* (2007) Raman spectroscopic investigations on the

secondary structure of proteins in cheese. *Dtsch. Milchw.*, **58** (6), 196–198.

38 Fontecha, J., Bellanato, J., and Juarez, M. (1993) Infrared and Raman-Spectroscopic Study of Casein in Cheese - Effect of Freezing and Frozen Storage. *J. Dairy Sci.*, **76** (11), 3303–3309.

39 Kilil, R. and Irudayaraj, J. (2007) Applications of Raman spectroscopy for food quality measurement, in *Nondestructive Testing of Food Quality* (eds J. Irudayaraj and C. Reh), Blackwell Publishing, Oxford, pp. 143–163.

40 Afseth, N.K., Wold, J.P., and Segtnan, V.H. (2006) The potential of Raman spectroscopy for characterisation of the fatty acid unsaturation of salmon. *Anal. Chim. Acta*, **572** (1), 85–92.

41 Kim, M.S., Lefcourt, A.M., and Chen, Y.R. (2003) Optimal fluorescence excitation and emission bands for detection of fecal contamination. *J. Food Prot.*, **66** (7), 1198–1207.

42 Park, B. *et al.* (2006) Performance of supervised classification algorithms of hyperspectral imagery for identifying fecal and ingesta contaminants. *Trans. ASABE*, **49** (6), 2017–2024.

43 Park, B. *et al.* (2006) Performance of hyperspectral imaging system for poultry surface fecal contaminant detection. *J. Food Eng.*, **75** (3), 340–348.

44 Heitschmidt, G.W. *et al.* (2007) Improved hyperspectral imaging system for fecal detection on poultry carcasses. *Trans. ASABE*, **50** (4), 1427–1432.

45 Liu, Y.L. *et al.* (2007) Development of simple algorithms for the detection of fecal contaminants on apples from visible/near infrared hyperspectral reflectance imaging. *J. Food Eng.*, **81** (2), 412–418.

46 Park, B. *et al.* (2007) Fisher linear discriminant analysis for improving fecal detection accuracy with hyperspectral images. *Trans. ASABE*, **50** (6), 2275–2283.

47 Yang, H. and Irudayaraj, J. (2003) Rapid detection of foodborne microorganisms on food surface using Fourier transform Raman spectroscopy. *J. Mol. Struct.*, **646** (1–3), 35–43.

48 Sivakesava, S., Irudayaraj, J., and DebRoy, C. (2004) Differentiation of microorganisms by FTIR-ATR and NIR spectroscopy. *Trans. ASABE*, **47** (3), 951–957.

49 Harz, M. *et al.* (2005) Micro-Raman spectroscopic identification of bacterial cells of the genus Staphylococcus and dependence on their cultivation conditions. *Analyst*, **130** (11), 1543–1550.

50 Gordon, S.H. *et al.* (1999) Transient infrared spectroscopy for detection of toxigenic fungus in corn: Potential for on-line evaluation. *J. Agr. Food Chem.*, **47** (12), 5267–5272.

51 Berardo, N. *et al.* (2005) Rapid detection of kernel rots and mycotoxins in maize by near-infrared reflectance spectroscopy. *J. Agr. Food Chem.*, **53** (21), 8128–8134.

52 European Commission (2002) Article 8, Regulation (EC) No. 178/2002D. Laying down the general principles and requirements of food law, establishing the European Food Safety Authority and laying down procedures in matters of food safety.

53 Lichan, E. (1994) Developments in the Detection of Adulteration of Olive Oil. *Trends Food Sci. Tech.*, **5** (1), 3–11.

54 Lai, Y.W., Kemsley, E.K., and Wilson, R.H. (1994) Potential of Fourier Transform-Infrared Spectroscopy for the Authentication of Vegetable-Oils. *J. Agr. Food Chem.*, **42** (5), 1154–1159.

55 Lai, Y.W., Kemsley, E.K., and Wilson, R.H. (1995) Quantitative-Analysis of Potential Adulterants of Extra Virgin Olive Oil Using Infrared-Spectroscopy. *Food Chem.*, **53** (1), 95–98.

56 Baeten, V. *et al.* (1996) Detection of virgin olive oil adulteration by Fourier transform Raman spectroscopy. *J. Agr. Food Chem.*, **44** (8), 2225–2230.

57 Downey, G., McIntyre, P., and Davies, A.N. (2003) Geographic classification of extra virgin olive oils from the eastern Mediterranean by chemometric analysis of visible and near-infrared spectroscopic data. *Appl. Spectrosc.*, **57** (2), 158–163.

58 Marigheto, N.A. *et al.* (1998) A comparison of mid-infrared and Raman spectroscopies for the authentication of edible oils. *J. Am. Oil Chem. Soc.*, **75** (8), 987–992.

59 Yang, H., Irudayaraj, J., and Paradkar, M.M. (2005) Discriminant analysis of edible oils and fats by FTIR, FT-NIR and

FT-Raman spectroscopy. *Food Chem.*, **93** (1), 25–32.

60 Woodcock, T., Downey, G., and O'Donnell, C.P. (2009) Near infrared spectral fingerprinting for confirmation of claimed PDO provenance of honey. *Food Chem.*, **114** (2), 742–746.

61 Pillonel, L. *et al.* (2003) Analytical methods for the determination of the geographic origin of Emmental cheese: mid- and near-infrared spectroscopy. *Eur. Food Res. Technol.*, **216** (2), 174–178.

62 Karoui, R. *et al.* (2004) Determining the geographic origin of Emmental cheeses produced during winter and summer using a technique based on the concatenation of MIR and fluorescence spectroscopic data. *Eur. Food Res. Technol.*, **219** (2), 184–189.

63 Aulrich, K. and Molkentin, J. (2009) Potential of Near infrared Spectroscopy for differentiation of organically and conventionally produced milk. *Landbauforsch. Volk.*, **59** (4), 301–307.

64 Nicolaou, N., Xu, Y., and Goodacre, R. (2010) Fourier transform infrared spectroscopy and multivariate analysis for the detection and quantification of different milk species. *J. Dairy Sci.*, **93** (12), 5651–5660.

65 Goodacre, R., Radovic, B.S., and Anklam, E. (2002) Progress toward the rapid nondestructive assessment of the floral origin of European honey using dispersive Raman spectroscopy. *Appl. Spectrosc.*, **56** (4), 521–527.

66 Sivakesava, S. and Irudayaraj, J. (2001) Detection of inverted beet sugar adulteration of honey by FTIR spectroscopy. *J. Sci. Food Agr.*, **81** (8), 683–690.

67 Kelly, J.F.D., Downey, G., and Fouratier, V. (2004) Initial study of honey adulteration by sugar solutions using midinfrared (MIR) spectroscopy and chemometrics. *J. Agr. Food Chem.*, **52** (1), 33–39.

68 Kelly, J.D., Petisco, C., and Downey, G. (2006) Application of Fourier transform midinfrared spectroscopy to the discrimination between Irish artisanal honey and such honey adulterated with various sugar syrups. *J. Agr. Food Chem.*, **54** (17), 6166–6171.

69 Kelly, J.D., Petisco, C., and Downey, G. (2006) Potential of near infrared transflectance spectroscopy to detect adulteration of Irish honey by beet invert syrup and high fructose corn syrup. *JNIRS*, **14** (2), 139–146.

70 Ruoff, K. *et al.* (2006) Authentication of the botanical and geographical origin of honey by front-face fluorescence spectroscopy. *J. Agr. Food Chem.*, **54** (18), 6858–6866.

71 Ruoff, K. *et al.* (2006) Authentication of the botanical and geographical origin of honey by mid-infrared spectroscopy. *J. Agr. Food Chem.*, **54** (18), 6873–6880.

72 Al-Jowder, O., Kemsley, E.K., and Wilson, R.H. (1997) Mid-infrared spectroscopy and authenticity problems in selected meats: A feasibility study. *Food Chem.*, **59** (2), 195–201.

73 McElhinney, J., Downey, G., and O'Donnell, C. (1999) Quantitation of lamb content in mixtures with raw minced beef using visible, near and mid-infrared spectroscopy. *J. Food Sci.*, **64** (4), 587–591.

74 Al-Jowder, O., Kemsley, E.K., and Wilson, R.H. (2002) Detection of adulteration in cooked meat products by mid-infrared spectroscopy. *J. Agr. Food Chem.*, **50** (6), 1325–1329.

75 Ding, H.B. and Xu, R.J. (2000) Near-infrared spectroscopic technique for detection of beef hamburger adulteration. *J. Agr. Food Chem.*, **48** (6), 2193–2198.

76 Downey, G., McElhinney, J., and Fearn, T. (2000) Species identification in selected raw homogenized meats by reflectance spectroscopy in the mid-infrared, near-infrared, and visible ranges. *Appl. Spectrosc.*, **54** (6), 894–899.

77 Defernez, M., Kemsley, E.K., and Wilson, R.H. (1995) Use of Infrared-Spectroscopy and Chemometrics for the Authentication of Fruit Purees. *J. Agr. Food Chem.*, **43** (1), 109–113.

78 Contal, L., Leon, V., and Downey, G. (2002) Detection and quantification of apple adulteration in strawberry and raspberry purees using visible and near infrared spectroscopy. *JNIRS*, **10** (4), 289–299.

79 Kurz, C. *et al.* (2010) Evaluation of fruit authenticity and determination of the fruit content of fruit products using FT-NIR spectroscopy of cell wall components. *Food Chem.*, **119** (2), 806–812.

80 Bevin, C.J. *et al.* (2006) Development of a rapid "fingerprinting" system for wine authenticity by mid-infrared spectroscopy. *J. Agr. Food Chem.*, **54** (26), 9713–9718.

81 Picque, D., Cattenoz, T., and Corrieu, G. (2001) Classification of red wines analysed by middle infrared spectroscopy of dry extract according to their geographical origin. *J. Int. Sci. Vigne Vin*, **35** (3), 165–170.

82 Acevedo, F.J. *et al.* (2007) Classification of wines produced in specific regions by UV-Visible spectroscopy combined with support vector machines. *J. Agr. Food Chem.*, **55** (17), 6842–6849.

83 Cozzolino, D. *et al.* (2011) Can spectroscopy geographically classify Sauvignon Blanc wines from Australia and New Zealand? *Food Chem.*, **126** (2), 673–678.

84 Oberreuter, H. *et al.* (2003) Fourier-transform infrared (FT-IR) spectroscopy is a promising tool for monitoring the population dynamics of microorganisms in food stuff. *Eur. Food Res. Technol.*, **216** (5), 434–439.

85 Gaus, K. *et al.* (2006) Classification of lactic acid bacteria with UV-resonance Raman spectroscopy. *Biopolymers*, **82** (4), 286–290.

86 Hardy, J. and Fanni, J. (1981) Application of Reflection Photometry to the Measurement of Milk Coagulation. *J. Food Sci.*, **46** (6), 1956–1957.

87 McMahon, D.J., Brown, R.J., and Ernstrom, C.A. (1984) Enzymic coagulation of milk casein micelles. *J. Dairy Sci.*, **67** (4), 745–748.

88 Payne, F.A., Hicks, C.L., and Shen, P.S. (1993) Predicting optimal cutting time of coagulating milk using diffuse reflectance. *J. Dairy Sci.*, **76** (1), 48–61.

89 Pearse, M.J. and Mackinlay, A.G. (1989) Biochemical aspects of syneresis: A review. *J. Dairy Sci.*, **72** (6), 1401–1407.

90 Lawrence, R.C. and Gilles, J. (1980) The assessment of the potential quality of young Cheddar cheese. *New Zeal. J. Dairy Sci. Technol.*, **15** (1), 1–12.

91 Taifi, N. *et al.* (2006) Characterization of the syneresis and the firmness of the milk gel using an ultrasonic technique. *Meas. Sci. Technol.*, **17** (2), 281–287.

92 Tellier, C. *et al.* (1993) Evolution of water proton muclear magnetic relaxation during milk coagulation and syneresis: structural implications. *J. Agr. Food Chem.*, **41** (12), 2259–2266.

93 Everard, C.D. *et al.* (2007) Computer vision and color measurement techniques for inline monitoring of cheese curd syneresis. *J. Dairy Sci.*, **90** (7), 3162–3170.

94 Fagan, C.C. *et al.* (2008) Application of image texture analysis for online determination of curd moisture and whey solids in a laboratory-scale stirred cheese vat. *J. Food Sci.*, **73** (6), E250–E258.

95 Fagan, C.C. *et al.* (2009) Visible-near infrared spectroscopy sensor for predicting curd and whey composition during cheese processing. *Sens.Instrum. Food Qual. Safety*, **3**, 62–69.

96 Fagan, C.C. *et al.* (2007) Novel online sensor technology for continuous monitoring of milk coagulation and whey separation in cheesernaking. *J. Agr. Food Chem.*, **55**, 8836–8844.

97 Castillo, M., Payne, F., and Shea, A. (2005) Development of a combined sensor technology for monitoring coagulation and syneresis operations in cheese making. *J. Dairy Sci.*, **88**, 142–142.

98 Castillo, M. *et al.* (2005) Optical sensor technology for measuring whey fat concentration in cheese making. *J. Food Eng.*, **71** (4), 354–360.

99 Fagan, C.C. *et al.* (2007) Novel online sensor technology for continuous monitoring of milk coagulation and whey separation in cheese making. *J. Agr. Food Chem.*, **22**, 8836–8844.

100 Fagan, C.C. *et al.* (2008) On-line prediction of cheese making indices using backscatter of near infrared light. *Int. Dairy J.*, **18** (2), 120–128.

101 Fagan, C. *et al.* (2009) Visible-near infrared spectroscopy sensor for predicting curd and whey composition during cheese processing. *Sens. Instrum.Food Qual. Safety*, **3** (1), 62–69.

102 Mateo, M.J. *et al.* (2009) Influence of curd cutting programme and stirring speed on the prediction of syneresis indices in cheese-making using NIR light backscatter. *Lwt-Food Sci. Technol.*, **42** (5), 950–955.

103 Fagan, C.C. *et al.* (2011) Preliminary Evaluation of Endogenous Milk Fluorophores as Tracer Molecules for Curd Syneresis. *J. Dairy Sci.*, **94** (11), 5350–5358.

104 Cimander, C., Carlsson, M., and Mandenius, C.-F. (2002) Sensor fusion for on-line monitoring of yoghurt fermentation. *J. Biotechnol.*, **99** (3), 237–248.

105 Navratil, M., Cimander, C., and Mandenius, C.F. (2004) On-line multisensor monitoring of yogurt and Filmjolk fermentations on production scale. *J. Agr. Food Chem.*, **52** (3), 415–420.

106 Correia, I. *et al.* (2005) Sorghum fermentation followed by spectroscopic techniques. *Food Chem.*, **90** (4), 853–859.

107 Alvarez, D. *et al.* (2007) Prediction of meat emulsion stability using reflection photometry. *J. Food Eng.*, **82** (3), 310–315.

108 Alvarez, D. *et al.* (2009) A novel fiber optic sensor to monitor beef meat emulsion stability using visible light scattering. *Meat Sci.*, **81** (3), 456–466.

109 Alvarez, D. *et al.* (2010) Application of light extinction to determine stability of beef emulsions. *J. Food Eng.*, **96** (2), 309–315.

110 Alvarez, D. *et al.* (2010) Prediction of beef meat emulsion quality with apparent light backscatter extinction. *Food Res. Int.*, **43** (5), 1260–1266.

111 Gonzalez-Martin, M.I. *et al.* (2011) Prediction of sensory attributes of cheese by near-infrared spectroscopy. *Food Chem.*, **127** (1), 256–263.

112 Hennessy, S., Downey, G., and O'Donnell, C. (2008) Multivariate Analysis of Attenuated Total Reflection-Fourier Transform Infrared Spectroscopic Data to Confirm the Origin of Honeys. *Appl. Spectrosc.*, **62** (10), 1115–1123.

113 Tgersen, G. *et al.* (1999) On-line NIR analysis of fat, water and protein in industrial scale ground meat batches. *Meat Sci.*, **51** (1), 97–102.

114 Karoui, R. *et al.* (2004) Fluorescence and infrared spectroscopies: a tool for the determination of the geographic origin of Emmental cheeses manufactured during summer. *Lait*, **84** (4), 359–374.

115 Blasco, J. *et al.* (2007) Citrus sorting by identification of the most common defects using multispectral computer vision. *J. Food Eng.*, **83** (3), 384–393.

116 Nakariyakul, S. and Casasent, D.P. (2008) Hyperspectral waveband selection for contaminant detection on poultry carcasses. *Opt. Eng.*, **47** (8).

117 Karoui, R. and De Baerdemaeker, J. (2007) A review of the analytical methods coupled with chemometric tools for the determination of the quality and identity of dairy products. *Food Chem.*, **102** (3), 621–640.

4
Application of Optical Methods for Quality and Process Control of Topically Applied Actives in Cosmetics and Dermatology

Juergen Lademann, Martina C. Meinke, Maxim E. Darvin, and Joachim W. Fluhr

4.1
Introduction

The skin is not only the largest organ of the human body, but also its barrier to the sometimes harsh environment. It provides protection against dehydration and penetration of pollutants and microorganisms. The skin barrier is mainly formed by the uppermost skin layer, the stratum corneum. It consists of dead cornified cells, which are surrounded by bilamellar lipid layers and is easily accessible for non-invasive assessment [1–3]. The skin is composed of the epidermis and the dermis. The stratum corneum is the uppermost part of the epidermis.

In dermatology and cosmetics a multitude of active substances are topically applied. These substances are intended to target the epidermis and eventually the dermis. To exert their effect they have to penetrate through the stratum corneum. As healthy skin has a strong barrier, only very small amounts of topically applied substances reach the living cells [4]. Maibach and Feldmann [5] demonstrated that, in the case of steroids, only approximately 0.1% of the topically applied substances reach the target structures in the living tissue. In contrast, an impaired epidermal barrier allows the penetration of larger amounts of drugs, thus reaching their target structures in deeper parts of the skin.

To develop and optimize cosmetic products and drugs the penetration of topically applied substances into and through the skin barrier must be analyzed. Diffusion cell experiments are among the most important *in vitro* techniques employed in such cases [6]. During these experiments, membrane samples of artificial or excised human and animal skin are placed into a diffusion cell. In most cases the tissue membranes are in contact with a receptor fluid so that the topically applied substances which penetrate the skin can be detected in the receptor fluid. A general drawback of the *in vitro* techniques is that they reflect the *in vivo* conditions only to a limited extent. Penetration investigations require measurements over an extended period of time. For ethical reasons it is not possible to take large numbers of biopsies from volunteers or patients for penetration studies.

Handbook of Biophotonics. Vol.3: Photonics in Pharmaceutics, Bioanalysis and Environmental Research, First Edition.
Edited by Jürgen Popp, Valery V. Tuchin, Arthur Chiou, and Stefan Heinemann

In recent years, optical methods have been widely established for analyzing the penetration of topically applied substances *in vivo*, and investigating structural tissue changes during the application of medicinal and cosmetic products [7]. In contrast to the internal organs, the skin is easily accessible to optical investigation methods.

The optical and spectroscopic analytical methods that are currently used in dermatology and cosmetic testing can be divided into two major groups: The first group comprises microscopic techniques suited to investigation of cellular structures noninvasively up to a depth of approximately 150 μm; this is limited by the light penetration into the skin and the scattering properties of the skin [8, 9]. This allows one to analyze the effect of topically applied substances and their action at the cellular level. The second group comprises spectroscopic methods, which are mainly based on the detection of topically applied substances in the skin. Again, microscopic techniques are used, for example Raman microscopy, which is capable of analyzing at an axial resolution of approximately 5 μm the distribution of specific substances in various layers of the human skin, and specifically in the epidermis [10]. Furthermore, this group includes remission measurements, which are widely employed for the noninvasive detection of topically applied substances in human skin over an extended period of time. In this chapter, both groups of analytical methods are discussed and explained, based on practical examples.

4.2 Laser Scanning Microscopy

4.2.1 Fluorescence Measurements

Three main laser scanning microscopy techniques are currently used for analyzing the cellular structures on human skin [11]. The first technique is based on fluorescence measurements with fluorescent dyes, for example fluorescein, being injected into the skin as a contrast medium [12]. By means of UV excitation, for example, at 488 nm (the wavelength of an argon laser system), cellular structures can be visualized as the dye accumulates in the lipid layers surrounding the cellular structures. In Figure 4.1 various cellular structures which were analyzed by fluorescence measurements are shown. Figure 4.1a shows the corneocytes of the stratum corneum. Distinct differences in the structures of the living cells of the stratum basale are visible in Figure 4.1b. In deeper skin layers even the papillary structure can be visualized (Figure 4.1c). This method is suited to detection of changes in the cellular structure. Figure 4.2a, for example, shows the stratum corneum of dry skin. Dry skin loses its regular honeycomb-like arrangement of corneocytes [13]. The skin surface exhibits a relief-like structure. Following the application of a moisturizing cream, the barrier characteristics of the concerned skin area were improved. Figure 4.2b shows that the regular corneocyte arrangement has regenerated.

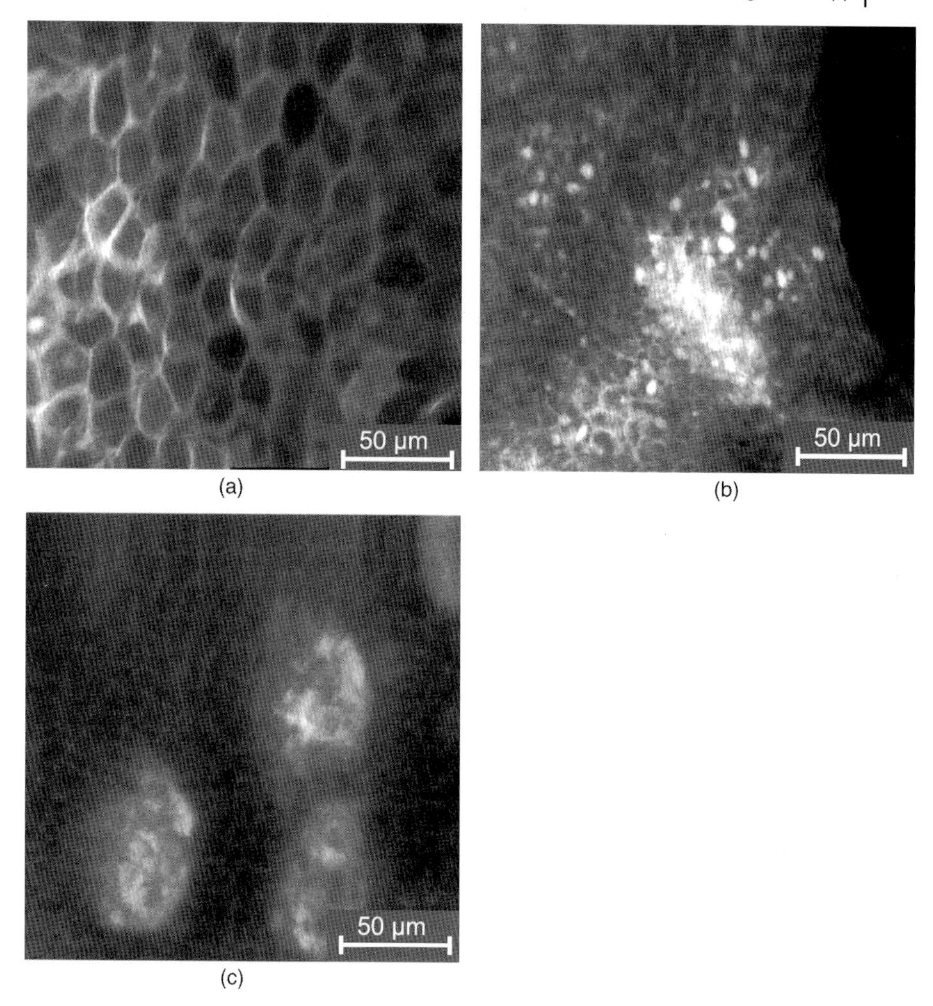

Figure 4.1 Cellular structure analyzed by fluorescence measurements. Corneocytes of the stratum (a), stratum basale (b) and papillary structure (c).

The latest studies have demonstrated that investigations to determine the barrier properties by *in vivo* laser scanning microscopy are a convenient alternative for analysis of the state of the skin compared to transepidermal water loss (TEWL) measurements, which have been the gold standard, so far [14]. This is because TEWL measurements are strongly influenced by external factors, such as ambient temperature and air humidity. In addition, these measurements are disturbed by topically applied substances themselves. To what extent topically applied substances, like steroids or ointments, accelerate wound healing can be analyzed by *in vivo* laser scanning microscopy while the epidermal barrier (stratum corneum) is regenerating [15]. Figure 4.3 represents the histological section of the stratum corneum,

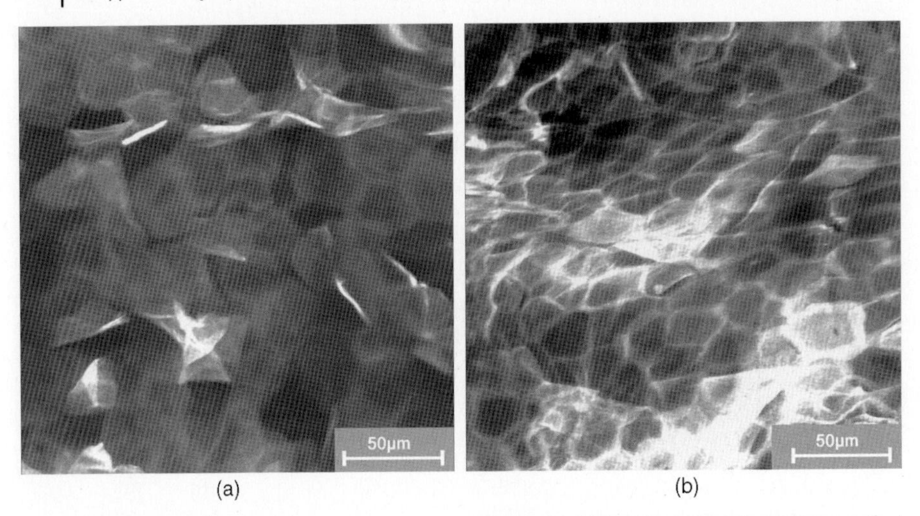

Figure 4.2 Stratum corneum of dry skin (a) and regular corneocyte arrangement after regeneration (b).

the epidermis and the dermis. The thickness of the barrier can be determined by focusing the laser upon the skin surface and adjusting it into the skin until the boundary between the stratum corneum and the stratum granulosum becomes visible [16]. This situation is illustrated in Figure 4.4. Figure 4.4a shows the skin surface with the characteristic structures of the corneocytes of the stratum corneum, whereas Figure 4.4b depicts the boundary between the stratum corneum and the stratum granulosum. Thus the transition allows the differentiation of cellular structures. The laser focus adjustment between the skin surface and the transition from the stratum corneum to the stratum granulosum corresponds to the thickness of the stratum corneum. This *in vivo* technique permits noninvasive detection of the time-dependent regeneration of an impaired epidermal barrier [17].

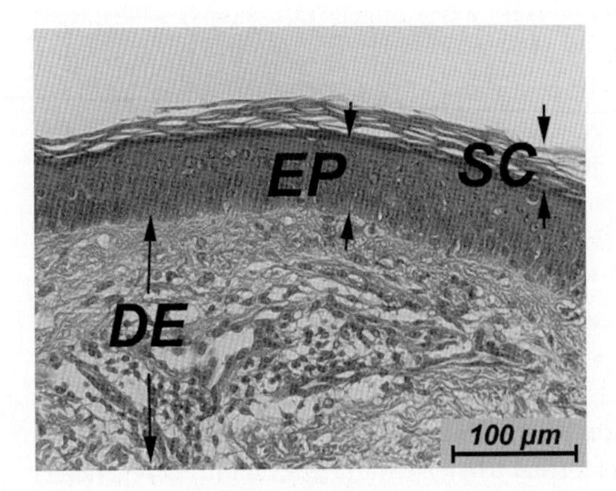

Figure 4.3 Histological section of the stratum corneum, the epidermis and the dermis.

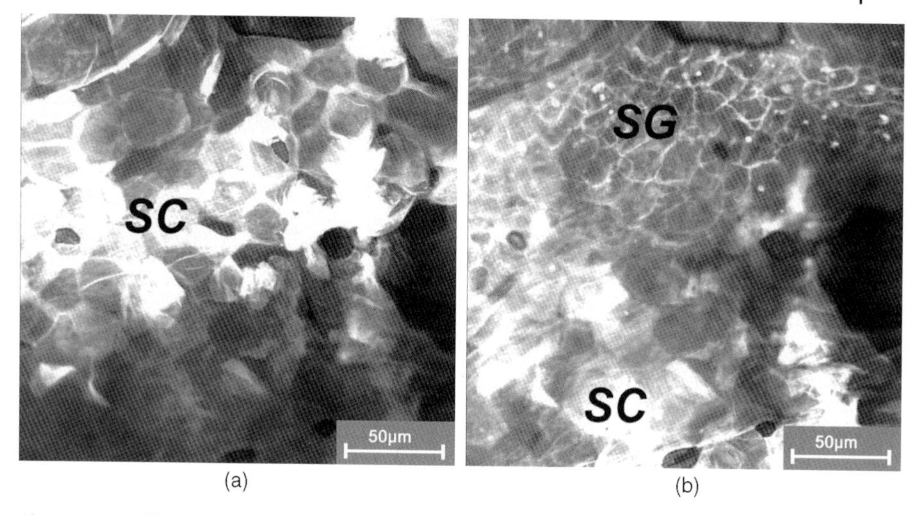

Figure 4.4 Differentiation in the cellular structures: corneocytes of the stratum corneum (a) and the boundary between the stratum corneum and the stratum granulosum (b).

Comparing the wound healing process of untreated skin with that of skin treated with ointments or steroids allows quantification of the repair capacity of topically applied products [18, 19].

If the *in vivo* laser scanning microscopy is used in fluorescence mode it is possible not only to evaluate topically applied substances in cosmetics and dermatology for their effects on cellular structures, but also to directly detect the penetration of these substances, provided these have been labeled with a fluorescent dye [20, 21]. Figure 4.5 illustrates the distribution of pigments used in sunscreens on the skin surface in a furrow. The pigments were labeled with a fluorescent dye which made them detectable when located in different skin structures. In Figure 4.5a the pigments are homogeneously distributed on the skin surface. The skin surface without pigments would appear black. If the laser focus is adjusted only a few microns deeper into the stratum corneum, a thin pigment film becomes visible, which covers the wall of the skin furrow homogeneously (Figure 4.5b). Furthermore, Figure 4.5b shows that the pigments are exclusively located on the skin surface. At the sites where the pigments were visible in Figure 4.5a, the fluorescence signal is no longer detectable in Figure 4.5b because the focus is located below the first layers of corneocytes. Moving the focus on the bottom of the furrow (Figure 4.5c), the bottom is covered by a pigment film. Using this investigation method one can test whether substances, such as pigments in sunscreens, are homogeneously distributed on the skin and form a homogeneous protective layer on structured skin areas.

The penetration kinetics of substances for topical application, as used in drugs and cosmetics, can be visualized by *in vivo* laser scanning microscopy in the fluorescence mode. A schematic drawing of this penetration pathway is shown in Figure 4.6. According to the bricks and mortar model [22] the stratum corneum is represented here as a wall consisting of bricks (corneocytes) and mortar (lipids).

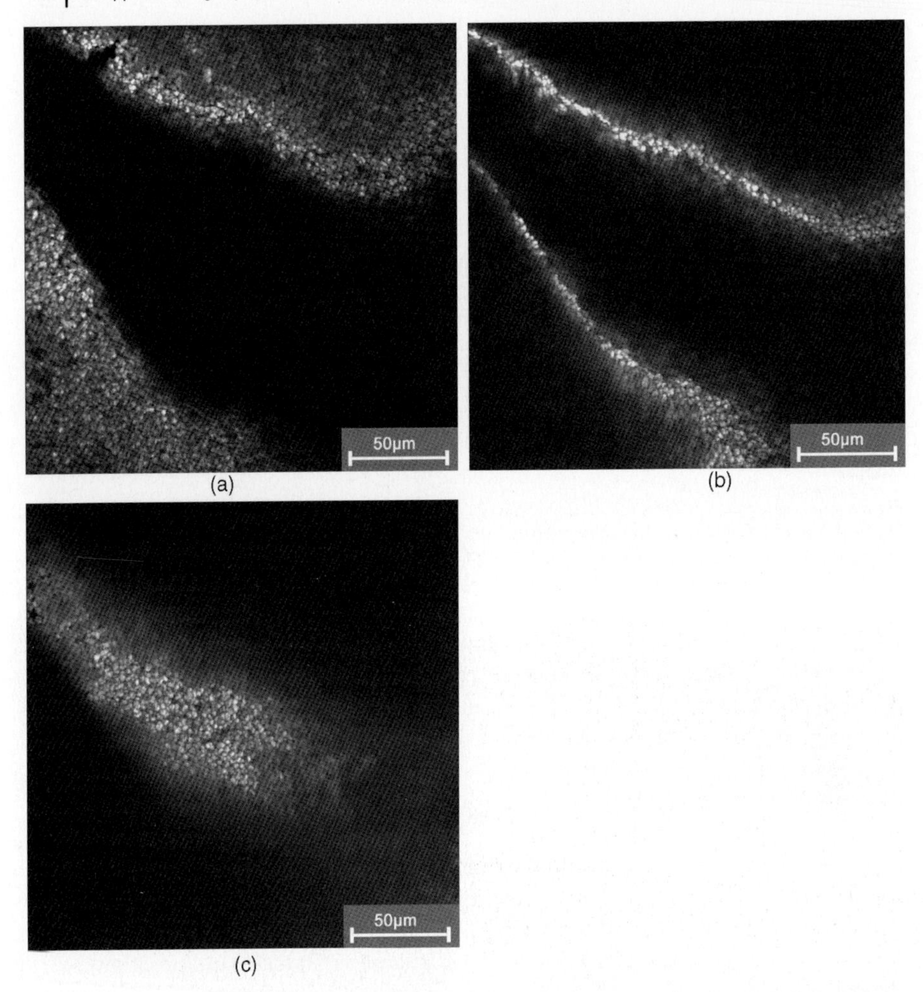

Figure 4.5 Penetration of fluorescein-labeled substance (white spots on the figure) into the skin measured with laser scanning microscopy on the skin surface near the furrow at 5 μm depth (a), 15 μm depth (b) and 25 μm depth (c).

Under real conditions, the corneocytes (black in this figure) are surrounded by lipid layers (white structures). If a substance is applied topically and penetrates into the stratum corneum, it reaches increasingly deeper corneocyte cell layers with time (Figure 4.6a–c). As the corneocytes are very thin structures, permeable to laser light, the various cell layers of the corneocytes can be detected if they are reached by fluorescence labeled substances. This situation is depicted in Figure 4.7 under *in vivo* conditions. Figure 4.7a shows the distribution of the substance 5 min after topical application. The fluorescence is detectable only within the first layer of corneocytes. After 20 min a fluorescence signal is also visible in the spaces of deeper cell layers (Figure 4.7b). This means that the topically applied substance has penetrated into

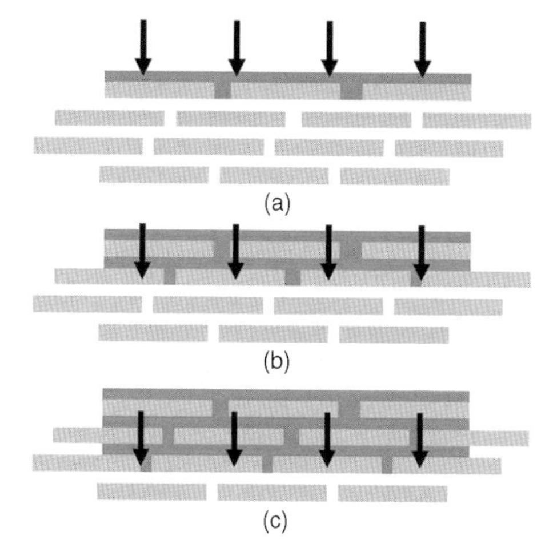

Figure 4.6 Schematic penetration pathway based on the classical bricks and mortar model.

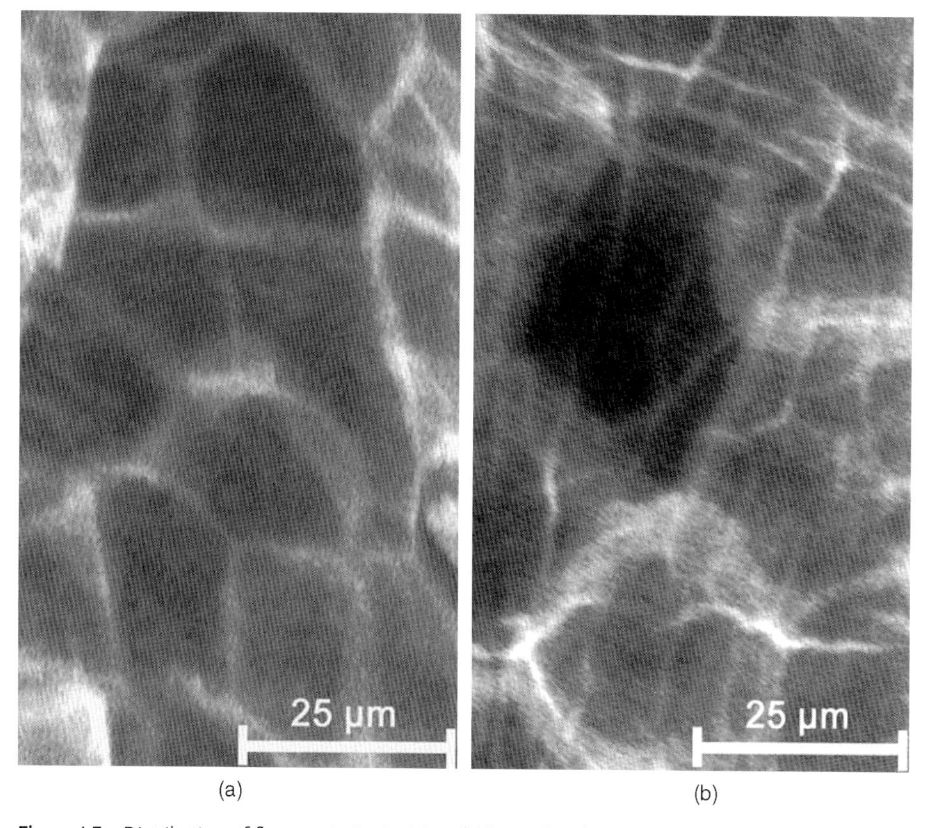

Figure 4.7 Distribution of fluorescein 5 min (a) and 20 min (b) after its topical application.

deeper skin layers within 20 min. In general, the intercellular penetration, that is, the penetration within the lipid layers around the corneocytes, can be distinguished from follicular penetration inside the hair follicles [23, 24]. While the intercellular penetration had long been supposed to be the only penetration pathway, it is now known that follicular penetration is also an important pathway, allowing topically applied substances to pass the skin barrier [25–27].

The hair follicles represent an interesting target structure [28, 29]. Surrounded by a dense network of blood capillaries, they also accommodate dendritic and stem cells [30]. The penetration of a fluorescence-labeled substance into the hair follicles is shown in Figure 4.8. The substance is located in the orifice of the follicle, before it penetrates deeper into the hair follicle, as seen in Figure 4.8a–c.

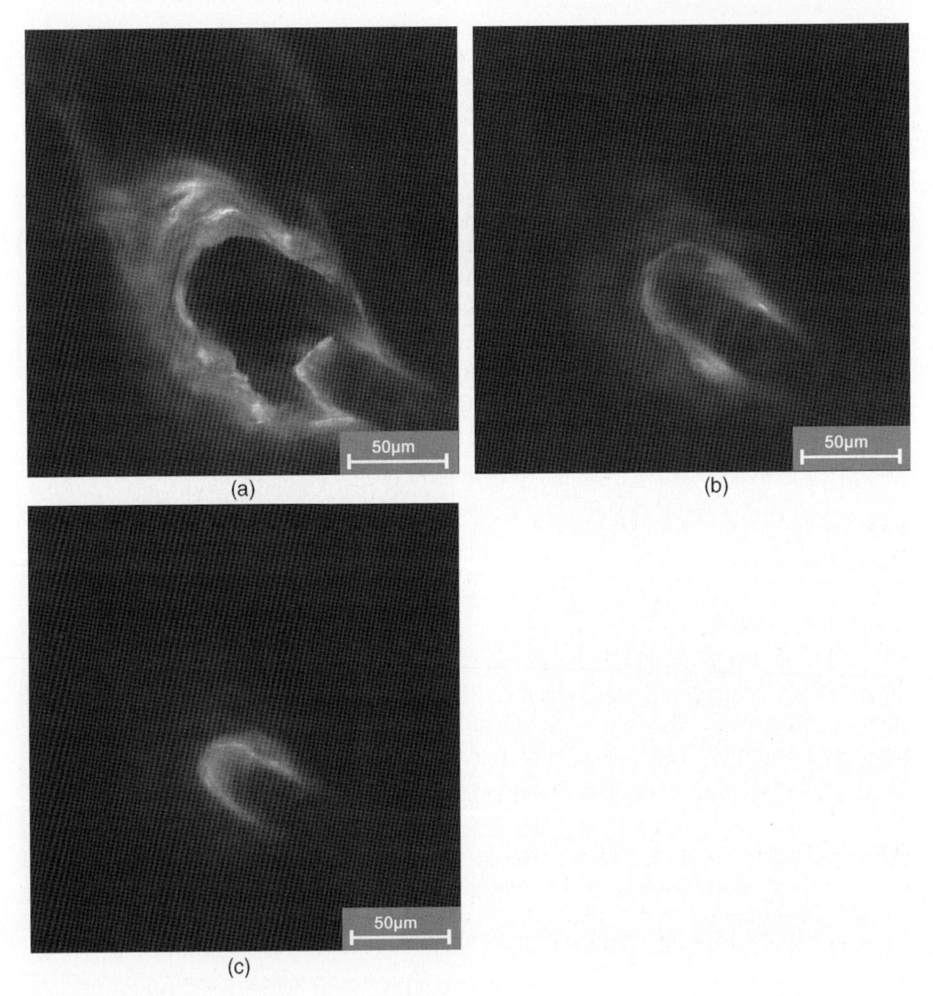

Figure 4.8 The penetration of a fluorescence-labeled substance into the hair follicle.

4.2.2
Remission Measurements

The presented examples demonstrate that investigations using *in vivo* laser scanning microscopy in the fluorescence mode are suitable to analyze changes in cellular structures and the kinetics of fluorescence-labeled topically applied substances at the same time. The disadvantage of this technique is that it requires a fluorescence marker. As long as this fluorescence marker is applied topically, which is necessary to differentiate between dry and "normal" healthy skin, the use of such markers does not entail any limitations. Most topically applied cosmetic products and drugs do not exhibit an auto-fluorescence signal, thus impeding direct detection of their penetration and distribution in fluorescence mode. The analysis of cellular changes in the epidermis and in the dermis is limited, because this analysis requires the injection or topical application of a fluorescence dye. Whereas this method can be used on healthy skin, its application is strongly restricted in the case of skin lesions or skin cancer to dyes approved for *in vivo* application in humans. These cases require analytic methods which are not based on fluorescence markers. To overcome the above-mentioned disadvantages of laser scanning microscopy using the fluorescence mode, reflectance measurements can be performed to visualize the morphology due to differences in the optical properties of the cellular structures [31]. As the optical properties of these skin layers show only slight differences, the contrast is less than that obtained by fluorescence measurements. Nevertheless, using this technique permits verification of the hypothesis about cellular changes during cosmetic or dermatological therapy [32, 33]. The reflectance measurements are performed at higher wavelengths, above 700 nm, leading to a higher penetration depth of the light. Therefore, visualization up to the dermis becomes feasible. Modern instruments offer a dermatoscopic overview of the skin allowing selection of the optimal skin area.

4.2.3
Multiphoton Measurements

In recent years two-photon microscopy has gained considerable importance [34, 35]. This technique uses ultrashort femtosecond pulses to excite by two-photon processes the autofluorescence of tissue. This makes it possible to analyze structures without any dyes. The results obtained by this method are at least similar to *in vivo* laser scanning microscopy, with fluorescence markers being unnecessary. Typical images obtained by *in vivo* two-photon microscopy are shown in Figure 4.9. Due to the initial wavelength being above 700 nm, the penetration depth of the light reaches the dermis discussed for single photon LSM in reflectance mode using red light. With multiphoton irradiation collagen structures can produce second harmonic generation (SHG) signals and can be visualized with an enhanced contrast [36].

Moreover, the short pulses used in two-photon microscopy lend themselves to the measurement of fluorescence life time generation. Some topically applied substances exhibit life time fluorescences which are distinctly different from those of the

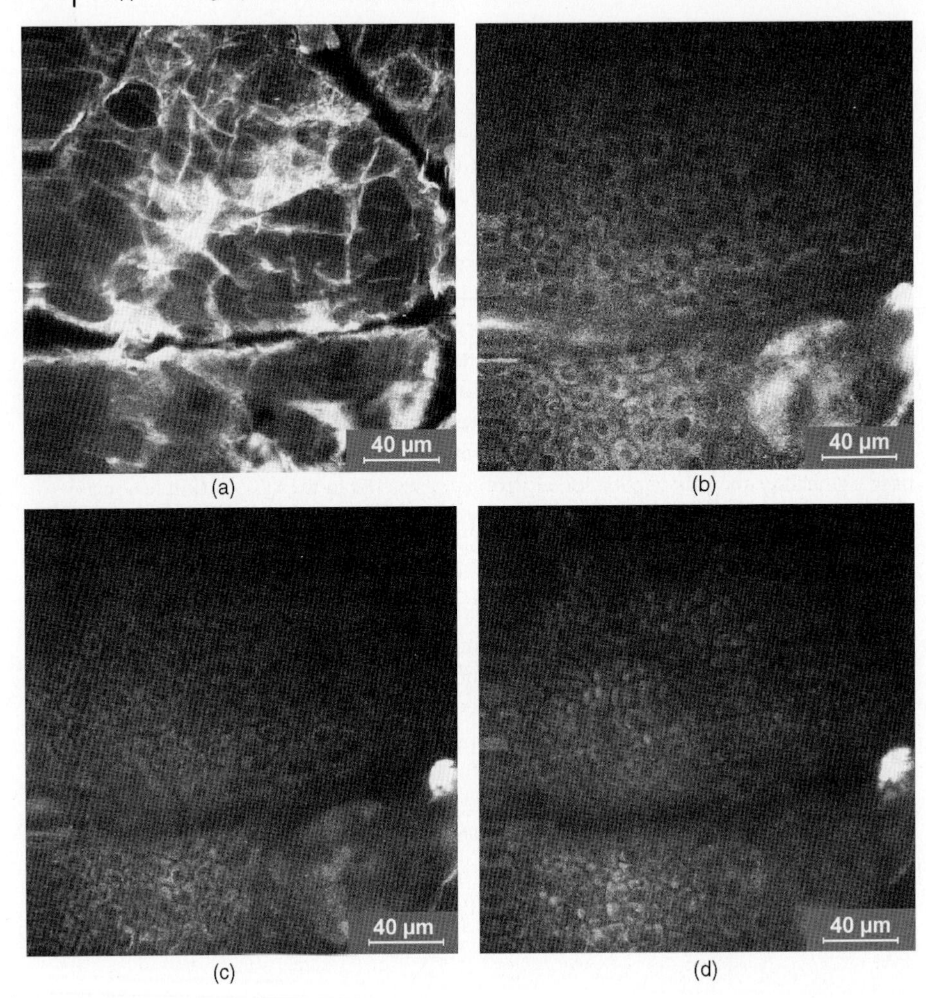

Figure 4.9 Typical images of human skin obtained by *in vivo* two-photon microscopy. Stratum corneum (a), stratum granulosum (b), stratum spinosum (c), stratum basale (d).

autofluorescence of the tissue. In such cases, the penetration of these substances into human skin can also be investigated *in vivo* using fluorescence life time imaging (FLIM) [37].

The first experiments to set up a system, which combines two-photon microscopy imaging of cellular structures with CARS (coherent anti-Stokes Raman scattering) microscopy [38] for the detection of substances in the tissue, have been performed. This opens new possibilities for analyzing the penetration of topically applied substances and their effects on the cellular structures of the skin. Using such a system, the therapeutic effect on the structure of the skin lipids and the distribution of the inflammatory cells in psoriasis patients could be successfully analyzed. This development is still at an early stage, but will become increasingly important in future dermatological and cosmetic research. The drawback of this system is its high cost. However, for many investigations it is not necessary to image cellular structures at different depths of the skin, at the same time detecting in these skin layers the topically applied substances.

4.3 Raman Spectroscopic Measurements

Microscopic methods like Raman microscopy, permitting substances to be detected in the skin at high spatial resolution, are well suited to analysis of penetration processes [39]. Raman microscopes can be used to detect the distribution of moisture in the different cell layers of human skin [10, 40]. These investigations are suitable for both characterizing wound healing in dermatology and evaluating cosmetic products, such as moisturizing creams. Various other substances can also be detected in skin by laser scanning Raman microscopy. These substances include carotenoids, which are the main constituents of the antioxidants in human skin, but other topically applied substances also show clearly detectable Raman bands [41]. Figure 4.10 represents a typical example of the carotenoid distribution within the stratum corneum after disinfectant application. These studies were performed with an *in vivo* Raman micro-spectroscope (River Diagnostics Ltd., Rotterdam). Figure 4.10 shows the natural stratum corneum distribution of carotenoids prior to

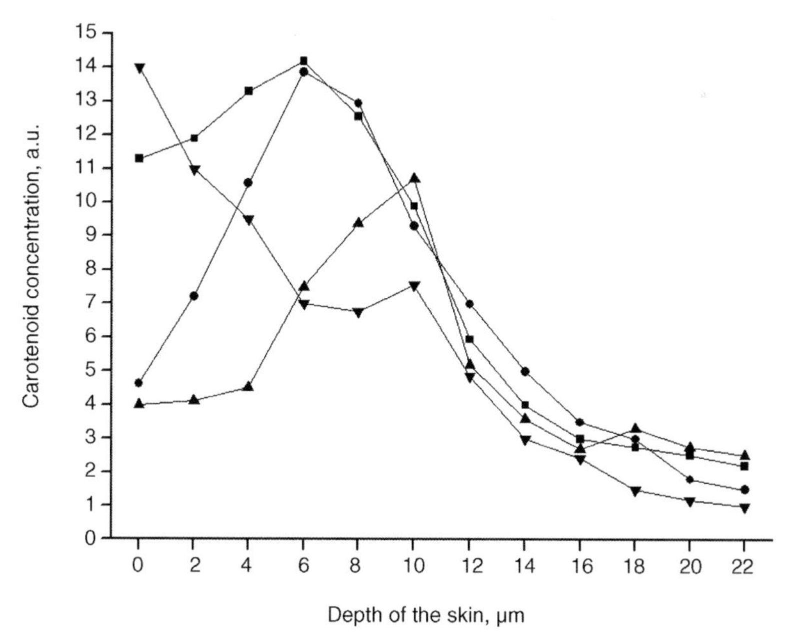

Figure 4.10 Distribution of carotenoids within the stratum corneum after disinfectant application measured with confocal Raman microscopy at different time points.

contact with the disinfectant (baseline; $t = 0$ min). After disinfectant application the carotenoid concentration declines specifically in the upper cell layers of the stratum corneum, whereas deeper layers remain unaffected ($t = 20$ min). At later time points, the penetrating disinfectant leads also to a reduction in the carotenoid concentration in deeper stratum corneum layers ($t = 40$ min). After 60 min a regeneration process is initiated showing increasing carotenoid concentrations in the upper cell layers. This is because the carotenoids are delivered continuously with the sweat and sebaceous lipids onto the skin surface, where they spread and penetrate into the skin like topically applied substances [42]. Consequently, the carotenoids re-penetrate into the upper cell layers after 60 min, while the carotenoid concentration in the deeper layers of the stratum corneum still remains unchanged. By carotenoid delivery with sweat and sebaceous lipids onto the skin surface and repenetration into the skin, the original profile of the carotenoid distribution in the stratum corneum is gradually restored (data almost similar to the $t = 0$ value; not shown in the present figure). In this case the penetration of the disinfectant could be indirectly detected through the decreased carotenoid concentration. In many cases it is also possible to detect substances directly by Raman spectroscopy and analyze their distribution at different depths of the skin. Figure 4.11 represents a typical distribution of the water concentration in the stratum corneum. This distribution can be changed by a disturbed skin barrier or by topically applied drugs and cosmetic products.

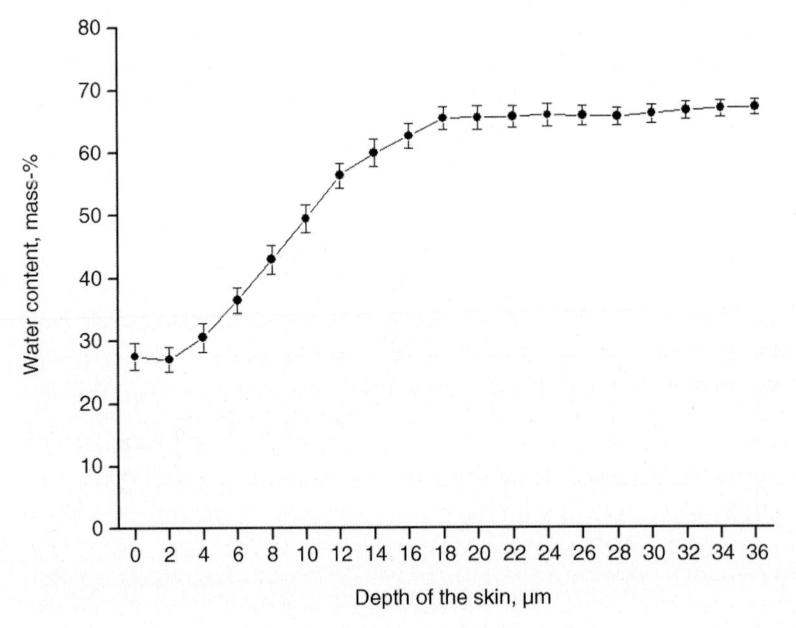

Figure 4.11 A typical distribution of water concentration in the stratum corneum measured with confocal Raman microscopy.

Using Raman microscopy it is possible to investigate the distribution of specific substances, either topically applied or constituents of our body, at different depths of the skin.

4.4 Resonance Raman Spectroscopy

There are also techniques which can detect substances in human skin without spatial resolution exclusively by remission methods. This was demonstrated by a measuring system for the *in vivo* detection of carotenoids in human skin. Carotenoids absorb in the blue–green range of the spectrum [43, 44]. Thus, due to their resonant excitation, an argon laser operated at 488 and 514.5 nm is well suited. Based on the differences in absorption spectra for beta-carotene and lycopene under excitation at 488 and 514.5 nm, and, as a result, on the differences in Raman scattering efficiencies, the measuring system is capable of detecting the carotenoid concentration in human skin selectively and sensitively [45]. The resonance Raman signal of the carotenoids measured at 1525 cm^{-1} is acquired through a detection channel and delivered to a spectrometer attached to an evaluation unit. The measuring skin volume is determined by the penetration depth of the laser radiation at 488 and 514 nm into the skin, which is approximately 150 μm, and the laser spot diameter on the skin surface. To avoid any disturbances due to inhomogeneities and the microstructure of the skin surface, the laser beam diameter upon the skin surface was approximately 6.5 mm. To reduce the dermal fluorescence background, which could influence the accuracy of Raman measurements, the photo-bleaching effect is used [46]. With the help of this system the storage of carotenoids contained in cosmetic products, such as anti-aging creams, can be followed for an extended period of time [47] without the use of expensive laser scanning microscopy.

4.5 Conclusions

The skin is the organ of our body ideally suited to noninvasive analysis by laser scanning microscopy for quality and process control of topically applied drugs and cosmetic products, as well as for diagnostic and therapy control. For this purpose a wide range of optical methods is available, extending from simple remission measurements to sophisticated combined investigations by two-photon CARS microscopy. Which method is best suited for specific investigations depends decisively on the research objective. In any case, *in vivo* investigations have already contributed to considerably deepening our knowledge of penetration and storage of topically applied substances in human skin. As a result, follicular penetration was recognized and verified as a new penetration pathway besides intercellular penetration. The use of optical techniques in the development and optimization of drugs and cosmetic products for topical application is likely to improve their efficacy.

With the ongoing miniaturization of the optical systems, and the related price reduction, these techniques will become increasingly established in clinical research, routine diagnostics and therapy control of topically applied substances.

References

1 Elias, P.M. and Steinhoff, M. (2008) Outside-to-inside (and now back to outside) pathogenic mechanisms in atopic dermatitis. *J. Invest. Dermatol.*, **128**, 1067–1070.

2 Darlenski, R., Sassning, S., Tsankov, N., and Fluhr, J.W. (2009) Non-invasive in vivo methods for investigation of the skin barrier physical properties. *Eur. J. Pharm. Biopharm.*, **72**, 295–303.

3 Elias, P.M. (1983) Epidermal lipids, barrier function, and desquamation. *J. Invest. Dermatol.*, **80** (Suppl), 44s–49s.

4 Rougier, A., Rallis, M., Krien, P., and Lotte, C. (1990) In vivo percutaneous absorption: a key role for stratum corneum/vehicle partitioning. *Arch. Dermatol. Res.*, **282**, 498–505.

5 Feldmann, R.J. and Maibach, H.I. (1967) Regional variation in percutaneous penetration of 14C cortisol in man. *J. Invest. Dermatol.*, **48**, 181–183.

6 Trauer, S., Lademann, J., Knorr, F., Richter, H., Liebsch, M., Rozycki, C., Balizs, G., Buttemeyer, R., Linscheid, M., and Patzelt, A. (2010) Development of an in vitro modified skin absorption test for the investigation of the follicular penetration pathway of caffeine. *Skin Pharmacol. Physiol.*, **23**, 320–327.

7 Ansari, R.R., Gonchukov, S.A., and Lademann, J. (2003) Laser methods in medicine and biology. *Laser Phys.*, **13**.

8 Gotter, B., Faubel, W., and Neubert, R.H. (2008) Optical methods for measurements of skin penetration. *Skin Pharmacol. Physiol.*, **21**, 156–165.

9 Bashkatov, A.N., Genina, E.A., Kochubey, V.I., and Tuchin, V.V. (2005) Optical properties of human skin, subcutaneous and mucous tissues in the wavelength range from 400 to 2000nm. *J. Phys. D. Appl. Phys.*, **38**, 2543–2555.

10 Caspers, P.J., Lucassen, G.W., Carter, E.A., Bruining, H.A., and Puppels, G.J. (2001) In vivo confocal Raman microspectroscopy of the skin: noninvasive determination of molecular concentration profiles. *J. Invest. Dermatol.*, **116**, 434–442.

11 Meyer, L., Otberg, N., Richter, H., Sterry, W., and Lademann, J. (2006) New prospects in dermatology: fiber-based confocal scanning laser microscopy. *Laser Phys.*, **16**, 758–764.

12 Jacobi, U., Toll, R., Audring, H., Sterry, W., and Lademann, J. (2005) The porcine snout--an in vitro model for human lips? *Exp. Dermatol.*, **14**, 96–102.

13 Byrne, A.J. (2010) Bioengineering and subjective approaches to the clinical evaluation of dry skin. *Int. J. Cosmet. Sci.*, **32**, 410–421.

14 Patzelt, A., Sterry, W., and Lademann, J. (2010) In vivo measurements of skin barrier: comparison of different methods and advantages of laser scanning microscopy. *Laser Phys. Lett.*, **7**, 843–852.

15 Lange-Asschenfeldt, B., Alborova, A., Kruger-Corcoran, D., Patzelt, A., Richter, H., Sterry, W., Kramer, A., Stockfleth, E., and Lademann, J. (2009) Effects of a topically applied wound ointment on epidermal wound healing studied by in vivo fluorescence laser scanning microscopy analysis. *J. Biomed. Opt.*, **14**.

16 Lademann, J., Richter, H., Astner, S., Patzelt, A., Knorr, F., Sterry, W., and Antoniou, C. (2008) Determination of the thickness and structure of the skin barrier by in vivo laser scanning microscopy. *Laser Phys. Lett.*, **5**, 311–315.

17 Rieger, T., Teichmann, A., Richter, H., Sterry, W., and Lademann, J. (2007) Application of in-vivo laser scanning microscopy for evaluation of barrier creams. *Laser Phys. Lett.*, **4**, 72–76.

18 Schatzlein, A. and Cevc, G. (1998) Non-uniform cellular packing of the stratum corneum and permeability barrier function of intact skin: a high-resolution

confocal laser scanning microscopy study using highly deformable vesicles (Transfersomes). *Br. J. Dermatol.*, **138**, 583–592.

19 Teichmann, A., Sadeyh Pour Soleh, H., Schanzer, S., Richter, H., Schwarz, A., and Lademann, J. (2006) Evaluation of the efficacy of skin care products by laser scanning microscopy. *Laser Phys. Lett.*, **3** (10), 507–509.

20 Alvarez-Roman, R., Naik, A., Kalia, Y.N., Fessi, H., and Guy, R.H. (2004) Visualization of skin penetration using confocal laser scanning microscopy. *Eur. J. Pharm. Biopharm.*, **58**, 301–316.

21 Martschick, A., Teichmann, A., Richter, H., Schanzer, S., Antoniou, C., Sterry, W., and Lademann, J. (2007) Analysis of the penetration profiles of topically applied substances by laser scanning microscopy. *Laser Phys. Lett.*, **4**, 395–398.

22 Elias, P.M. (2005) Stratum corneum defensive functions: an integrated view. *J. Invest. Dermatol.*, **125**, 183–200.

23 Otberg, N., Richter, H., Schaefer, H., Blume-Peytavi, U., Sterry, W., and Lademann, J. (2004) Variations of hair follicle size and distribution in different body sites. *J. Invest. Dermatol.*, **122**, 14–19.

24 Teichmann, A., Jacobi, U., Ossadnik, M., Richter, H., Koch, S., Sterry, W., and Lademann, J. (2005) Differential stripping: determination of the amount of topically applied substances penetrated into the hair follicles. *J. Invest. Dermatol.*, **125**, 264–269.

25 Otberg, N., Richter, H., Knuttel, A., Schaefer, H., Sterry, W., and Lademann, J. (2004) Laser spectroscopic methods for the characterization of open and closed follicles. *Laser Phys. Lett.*, **1**, 46–49.

26 Otberg, N., Richter, H., Schaefer, H., Blume-Peytavi, U., Sterry, W., and Lademann, J. (2003) Visualization of topically applied fluorescent dyes in hair follicles by laser scanning microscopy. *Laser Phys.*, **13**, 761–764.

27 Jung, S., Otberg, N., Thiede, G., Richter, H., Sterry, W., Panzner, S., and Lademann, J. (2006) Innovative liposomes as a transfollicular drug delivery system: penetration into porcine hair follicles. *J. Invest. Dermatol.*, **126**, 1728–1732.

28 Vogt, A., Hadam, S., Heiderhoff, M., Audring, H., Lademann, J., Sterry, W., and Blume-Peytavi, U. (2007) Morphometry of human terminal and vellus hair follicles. *Exp. Dermatol.*, **16**, 946–950.

29 Krause, K. and Foitzik, K. (2006) Biology of the hair follicle: the basics. *Semin. Cutan. Med. Surg.*, **25**, 2–10.

30 Oshima, H., Rochat, A., Kedzia, C., Kobayashi, K., and Barrandon, Y. (2001) Morphogenesis and renewal of hair follicles from adult multipotent stem cells. *Cell*, **104**, 233–245.

31 Meyer, L.E., Otberg, N., Sterry, W., and Lademann, J. (2006) In vivo confocal scanning laser microscopy: comparison of the reflectance and fluorescence mode by imaging human skin. *J. Biomed. Opt.*, **11**, 044012-.

32 Dietterle, S., Lademann, J., Rowert-Huber, H.J., Stockfleth, E., Antoniou, C., Sterry, W., and Astner, S. (2008) In-vivo diagnosis and non-inasive monitoring of imiquimod 5% cream for non-melanoma skin cancer using confocal laser scanning microscopy. *Laser Phys. Lett.*, **5**, 752–759.

33 Ulrich, M., Roewert-Huber, J., Gonzalez, S., Rius-Diaz, F., Stockfleth, E., and Kanitakis, J. (2011) Peritumoral clefting in basal cell carcinoma: correlation of in vivo reflectance confocal microscopy and routine histology. *J. Cutan. Pathol.*, **38**, 190–195.

34 Koehler, M.J., Vogel, T., Elsner, P., Konig, K., Buckle, R., and Kaatz, M. (2010) In vivo measurement of the human epidermal thickness in different localizations by multiphoton laser tomography. *Skin Res. Technol.*, **16**, 259–264.

35 Konig, K. (2008) Clinical multiphoton tomography. *J. Biophoton.*, **1**, 13–23.

36 Koehler, M.J., Hahn, S., Preller, A., Elsner, P., Ziemer, M., Bauer, A., Konig, K., Buckle, R., Fluhr, J.W., and Kaatz, M. (2008) Morphological skin ageing criteria by multiphoton laser scanning tomography: non-invasive in vivo scoring of the dermal fibre network. *Exp. Dermatol.*, **17**, 519–523.

37 Sanchez, W.Y., Prow, T.W., Sanchez, W.H., Grice, J.E., and Roberts, M.S. (2010) Analysis of the metabolic deterioration of ex vivo skin from ischemic necrosis

through the imaging of intracellular NAD (P)H by multiphoton tomography and fluorescence lifetime imaging microscopy. *J. Biomed. Opt.*, **15**, 046008-.

38 Evans, C.L. and Xie, X.S. (2008) Coherent anti-stokes Raman scattering microscopy: chemical imaging for biology and medicine. *Annu. Rev. Anal. Chem. (Palo Alto CA.)*, **1**, 883–909.

39 Caspers, P.J., Williams, A.C., Carter, E.A., Edwards, H.G., Barry, B.W., Bruining, H.A., and Puppels, G.J. (2002) Monitoring the penetration enhancer dimethyl sulfoxide in human stratum corneum in vivo by confocal Raman spectroscopy. *Pharm. Res.*, **19**, 1577–1580.

40 Caspers, P.J., Lucassen, G.W., and Puppels, G.J. (2003) Combined in vivo confocal Raman spectroscopy and confocal microscopy of human skin. *Biophys. J.*, **85**, 572–580.

41 Darvin, M.E., Fluhr, J.W., Caspers, P., van der, P.A., Richter, H., Patzelt, A., Sterry, W., and Lademann, J. (2009) In vivo distribution of carotenoids in different anatomical locations of human skin: comparative assessment with two different Raman spectroscopy methods. *Exp. Dermatol.*, **18**, 1060–1063.

42 Lademann, J., Caspers, P.J., van der Pol, A., Richter, H., Patzelt, A., Zastrow, L., Darvin, M., Sterry, W., and Fluhr, J.W. (2009) In vivo Raman spectroscopy detects increased epidermal antioxidative potential with topically applied carotenoids. *Laser Phys. Lett.*, **6**, 76–79.

43 Darvin, M.E., Gersonde, I., Albrecht, H., Meinke, M., Sterry, W., and Lademann, J. (2006) Non-invasive in vivo detection of the carotenoid antioxidant substance lycopene in the human skin using the resonance Raman spectroscopy. *Laser Phys. Lett.*, **3**, 460–463.

44 Darvin, M.E., Gersonde, I., Ey, S., Brandt, N.N., Albrecht, H., Gonchukov, S.A., Sterry, W., and Lademann, J. (2004) Noninvasive detection of beta-carotene and lycopene in human skin using Raman spectroscopy. *Laser Phys.*, **14**, 231–233.

45 Darvin, M.E., Gersonde, I., Meinke, M., Sterry, W., and Lademann, J. (2005) Non-invasive in vivo determination of the carotenoids beta-carotene and lycopene concentrations in the human skin using the Raman spectroscopic method. *J. Phys. D Appl. Phys.*, **38**, 2696–2700.

46 Darvin, M.E., Brandt, N.N., and Lademann, J. (2010) Photobleaching as a method of increasing the accuracy in measuring carotenoid concentration in human skin by Raman spectroscopy. *Opt. Spectrosc.*, **109**, 205–210.

47 Darvin, M.E., Sterry, W., and Lademann, J. (2010) Resonance Raman spectroscopy as an effective tool for the determination of antioxidative stability of cosmetic formulations. *J. Biophotonics*, **3**, 82–88.

Part Two
On-Site Analysis

Handbook of Biophotonics. Vol.3: Photonics in Pharmaceutics, Bioanalysis and Environmental Research, First Edition.
Edited by Jürgen Popp, Valery V. Tuchin, Arthur Chiou, and Stefan Heinemann

5
Agricultural Applications: Animal Epidemics and Plant Pathogen Detection

Robert Möller

5.1
Introduction

Livestock and field crops are always threatened by various natural threats. While farmers always had to accept and adapt to bad weather conditions, like freezing temperatures, hail, drought and others, threatening especially their crops, they can actively fight other threats like pests and pathogens. However, a timely implementation of eradication and containment strategies to limit the effects and spread of pest and pathogens requires a clear detection and identification of the biological threat. While most pests can more or less easily be identified, because most of them can be seen by the naked eye or with a low magnification, the identification of pathogens is much harder. As pathogens are microorganisms, a conventional direct identification is only possible using the classical basic steps of isolation, cultivation and optical detection. These identification strategies are basically the same for all microorganisms and are based on the principles of microbial detection established by Louis Pasteur and Robert Koch. Especially for bacterial or fungal pathogens, the cultivation steps are comparably simple; a big drawback of this strategy is the time that is needed. Often many days or even weeks are lost till a clear identification of a pathogen can be achieved and defined countermeasures can be taken. So a rapid detection and identification of the pathogen would also be of high relevance for agricultural applications.

As agriculture is an important infrastructure the fast and accurate detection and identification of pathogens should be of the highest interest. A natural, accidental or deliberate introduction of pathogens into the farming industry in the western world could have devastating economic, social and environmental effects. This was demonstrated in 2001 by the reappearance of foot-and-mouth disease (FMD) in the UK. The outbreak resulted in multi-billion dollar losses associated not only with agriculture, but also a wide range of activities including the pharmaceutical and tourist industries [1]. The outbreak could only be controlled by slaughtering millions of animals, most of which were not infected, to quickly achieve eradication of the virus and to achieve the FMD-free status of the UK [2]. The main problem during this FMD outbreak was that the clinical screening was time-consuming and

Handbook of Biophotonics. Vol.3: Photonics in Pharmaceutics, Bioanalysis and Environmental Research, First Edition.
Edited by Jürgen Popp, Valery V. Tuchin, Arthur Chiou, and Stefan Heinemann

labor-intensive. Because of the scale of the outbreak national reference laboratories using the standardized testing procedures were unable to test the hundreds and thousands of individual animals on suspected infected premises. During a large scale outbreak of a highly infectious disease the system of national reference laboratories quickly reaches its capacity limits. After the 2001 FMD outbreak the use of rapid diagnostic assays was recommended by two major reports [3, 4]. The driving force for the establishment of novel testing devices for on-site diagnosis has been largely influenced by the desire to reduce the time to perform diagnostic tests, so that objective data can be used to support the decision-making process during the outbreak [5].

The development of on-site diagnostics will be discussed in the following, mostly using the example of FMD and *Phytophthora* detection. FMD is one of the most significant animal diseases affecting trade. Although rarely fatal in adult animals, the appearance of FMD in a disease-free country results in severe trade restrictions and agricultural losses [6]. The disease is caused by the foot-and-mouth disease virus (FMDV), leading to vesicles on the foot, mouth, tongue, and teats of cloven-hooved animals, and is one of the most contagious disease agents known. FMD is classified as a reportable disease by the Office International des Epizooties (OIE) [6].

As a second example, the detection of the notifiable plant pathogen *Phytophthora* is chosen. We have chosen to focus on two specific examples because there are over 500 plant pathogens alone that can cause major disease losses [7]. *Phytophthora* is a genus of plant-damaging oomycetes. Certain species can cause enormous economic losses on crops worldwide, as well as environmental damage in natural ecosystems. The most prominent *Phytophthora* species is probably *P. infestans* the infective agent of late blight or potato blight. The most famous occurrence of *P. infestans* was in the middle of the 19th century, when it caused the Great Irish Famine, but it is still a problem today, causing estimated damage worth 6 billion $ a year worldwide. As the spread and large scale outbreak of *Phytophthora infestans* can be at least partially controlled by fungicides, recently, other members of the genus have raised interest in fast diagnostic devices. *Phytophthora ramorum,* the causal agent of the sudden oak death [8], and the more recently described pathogen *P. kernoviae,* the cause of dieback and leaf blight on a broad range of plant species [9], are two species of the genus that are under surveillance. To prevent the introduction and spread of the pathogens emergency phytosanitary measures were enforced in the EU in 2002 [5]. Specific and sensitive detection devices are necessary to avoid the planting of contaminated material, as it is difficult to control diseases caused by the two *Phytophthora* species, because of the lack of efficient products for chemical treatment under field conditions [10].

5.2
Diagnosis Under Field Conditions

As already described, the effective disease management and implementation of plant and animal health legislation is reliant upon rapid and accurate disease diagnosis, based upon recognition of symptoms in the field and identification of the causal agent [11]. For important notifiable diseases, samples from suspected cases are normally sent to regional or national reference laboratories, where validated assays are deployed and results can be reported to national competent authorities [5]. However, as time is a critical factor during a disease outbreak, valuable time is lost by transporting the samples. Furthermore, during a large scale outbreak these laboratories using routine diagnostic procedures have only a limited capacity to handle a large number of samples in a short amount of time. Fast on-site testing might be the solution to these problems, but specific requirements have to be fulfilled to allow testing away from dedicated laboratory facilities. These assays are commonly referred to as "point-of-care tests" (POCT), "pen-side", "portable", "on-site", "field tests" or "point of decision" tests [5]. To allow the detection and identification of pathogens away from a centralized laboratory infrastructure under field conditions the applied assays and the devices used have to be easy to handle and robust.

The diagnosis of a plant or animal disease can be relative simple when typical, definitive symptoms are evident. However, symptoms are not always unique and can be confused with other diseases [7]. Especially important for the detection of a pathogen is the taking of the sample. This should be normally be done by trained personnel (farmers, veterinarians, or inspectors) who can recognize the signs of a disease.

The first tests that would allow detection of plant viruses on site were introduced 30 years ago. These tests were based on chloroplast agglutination [12] and latex agglutination [13, 14]. Both tests relied on an immunological reaction that leads to the agglutination of chloroplasts or latex after a short period. Although extremely simple, these tests were never widely accepted, being not sufficiently robust for routine use in the field. It was also difficult to distinguish a positive agglutination from a false-positive clumping [15]. By replacing the latex with sensitized *Staphylococcus aureas* agglutination tests were improved and are still in use for the detection of bacterial pathogens, especially in laboratories, but they can also be used in the field. The problems with the agglutination test illustrate the problems and requirements for on-site testing. As already mentioned, ease of handling, a good discrimination of positive and negative results, a high sensitivity and specificity, and a robustness of the test are the main requirements for field tests.

5.3 Immunological Based-Techniques

All immunological tests use the specific antigen/antibody interaction for the detection of pathogens or substances. The underlying principle is basically an adaptation of the enzyme linked immunoabsorbent assay (ELISA). The interaction of antibody and antigen is visualized by an enzyme-mediated color change reaction. This method has been adapted also for on-site diagnostics.

5.3.1
Flow Through Format

In this format the capture antibodies for the detection are either bound to a membrane or to modified filter plugs. The so-called Alert test (Neogen Corporation) used a membrane with three spots (Figure 5.1a). The modified membrane is housed in a plastic container on a hydroscopic "cork". When the extracted sample is added to the membrane it flows over the membrane and the antigen can bind to the antibody. An enzyme-tagged antibody is then added, which binds to the captured antigen. Finally a substrate (4 chloronaphthol) for the enzyme is flowed over the disc. In the presence of the captured enzyme a blue precipitate is formed. If no antigen is present no enzyme is bound and no color change detected. In a final step the membrane is rinsed to stabilize the color reaction. The test has three spots or discs, the smallest disc is loaded with the antigen, serving as positive control. One of the larger discs carries no antibody (negative control) while the final disc is pre-treated with a specific antibody [15]. Alert kits are available for several plant pathogens including *Phytophthora* and have been demonstrated for monitoring diseases and as a valuable tool for the management of fungicide applications [16, 17]. Even so, these tests are relatively sophisticated and are quite laborious to perform. Because of these constraints the test has been applied only in high value applications.

By assembling three specifically modified filter plugs in the tip of a syringe the so-called AffiniTip (Hydros Inc.) is formed (Figure 5.1b). One filter plug holds the antigen (positive control), one contains no antibody, and the third is modified with a target specific antibody. The test is performed by attaching the tip to a syringe and all the necessary solutions are drawn over the filters. All necessary chemicals are packed in a small disposable container. A drawback of the system is the relatively long test time and some skill is also needed in handling the syringe accurately [15].

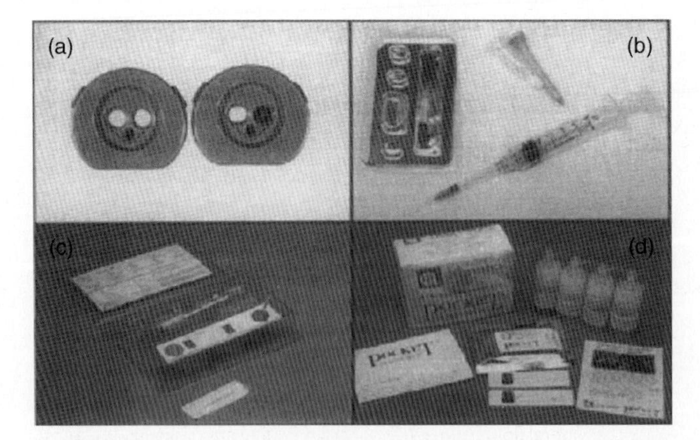

Figure 5.1 Four methods of immunological testing, flow through format Alert (a) and AffiniTip (b) and two lateral flow assays from DeTechtor (c) and Pocket Diagnostic (d) [15].

5.3.2
Lateral Flow Assays

The previously described test schemes still rely on the use of an enzyme for the detection of the antigen; this makes the assay time consuming and complicates the assay. The enzyme can be replaced by modified colloidal gold, latex or silica particles. These particles are sensitized by modification with a specific antibody. If these labels accumulate they can be easily visualized.

A typical lateral flow assay or lateral flow device (LFD) (Figure 5.1c and d) consists of a membrane on which a specific capture antibody is immobilized on a line. On a second line an antibody is immobilized that directly binds the sensitized particle label, thus serving as a positive control. As the sample is loaded to the release pad the antigen binds to the antibodies on the labeled particles and the fluid is drawn over the membrane. In the presence of the antigen the labels accumulate on both lines and become visible (Figure 5.2). If no antigen is present the particles only bind on the second line.

LFDs were initially developed during the mid 1980s for clinical applications and were first commercialized for home pregnancy testing. The test was first introduced by Unipath in 1988. Since then the test scheme has been adapted to a variety of clinical and nonclinical applications. The devices are rapid, inexpensive, disposable, and easy to use, and test results can easily be interpreted by a nonspecialist. Ferris and coworkers have developed and validated LFDs (Figure 5.3) for the detection of all seven serotypes of FMDV [18].

Even though the analytical sensitivity appeared to be lower than RT-PCR, the ability to sample multiple animals within a herd, with this simple and cost-efficient on-site detection, increases the confidence in the results (at herd level) [5]. Field trials with LFDs were taken during the FMD outbreak in Southern England in 2007. These tests showed that results could be achieved in as little as 10 min [19].

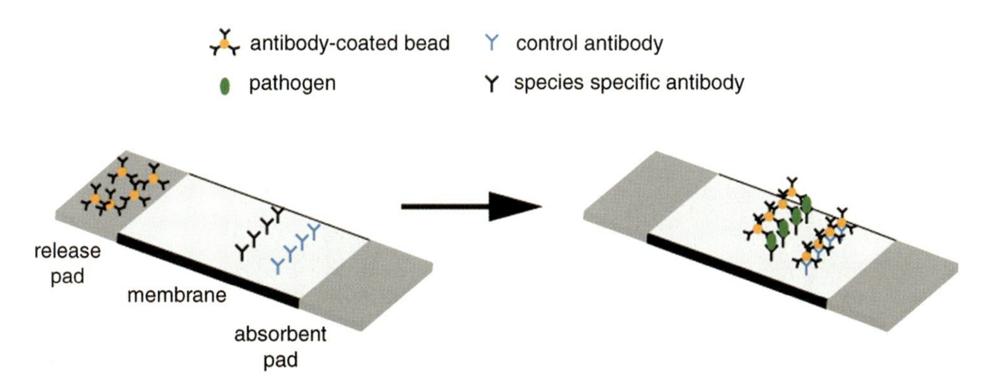

Figure 5.2 Scheme of a one-step lateral flow assay. If the sample is loaded onto the release pad the fluid flows over the membrane releasing the modified beads. If the specific pathogen is present the beads bind to the pathogen and are accumulated at the specific lines.

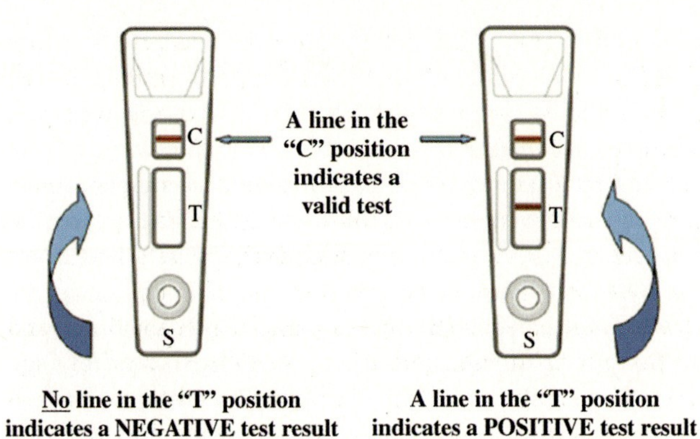

Figure 5.3 Depiction of negative and positive results on a lateral flow assay [18].

By using LFDs the presence of *Phytophthora* can also be detected in symptomatic plant material [11]. However, the detection was not species specific, so positive samples had to be sent to a laboratory for specific molecular testing for *P. ramorum* and *P. kernoviae*. When compared to species-specific methods (both PCR-based and cultural methods) the LFDs showed a high diagnostic sensitivity, indicating their suitability as a pre-screening method in the field [20]. LFDs have also been used for the rapid detection of DNA for the identification of pathogens. Tomlinson and coworkers combined an isothermal amplification of extracted DNA with detection on an LFD to identify *P. ramorum* and *P. kernoviae* [21].

5.4
Nucleic Acid-Based Testing

Although the previously described method of lateral flow assay seems well suited for on-site application, its application is limited. Especially when relying on the immunological reaction, low concentrations of pathogens are often not detectable. Furthermore, specific and stable antibodies are needed and are not always easily available for all pathogens. To enable a more sensitive and more specific, down to the strain level, detection and identification of pathogens, nucleic acid-based methods are used. The most common method is probably PCR. However, it seems technologically quite challenging to develop portable devices for this highly sensitive and specific testing method.

The major bottleneck for the establishment of on-site PCR systems seems to be the extensive sample preparation that is often necessary to isolate DNA out of a complex sample [22]. Nevertheless, there have already been successful on-site PCR tests using existing equipment. This has been done for the FMDV [23–25] as well as for *Phytophthora* [26] detection, and for *P. ramorum* assays have been successfully used in the field [27]. For these tests Cepheid SmartCyclers (Cepheid, Sunnyvale, CA, USA) have been used. Other portable devices that enable pathogen detection via real time PCR are the Enigma FL (Enigma Dagnostics, Port Down, GB) and the Bioseeq

Vet (Smith Detection, Watford, GB). These systems are automated, starting with sample preparation, followed by the PCR and then the readout of the results. Despite the fact, that the systems allow a fast and specific detection, their broad application is mainly limited by their high equipment and per assay costs.

The use of highly precise optical components for the detection and instrumentation for precise temperature control are probably the two expensive components in portable real-time PCR devices. In order to reduce equipment costs isothermal amplification methods have been investigated for their possible use in on-site applications. Especially, loop-mediated isothermal amplification (LAMP) [28] and nucleic acid sequence-based amplification (NASBA) [29, 30] have been of high interest. These isothermal amplification strategies seem to be well suited for field settings as only a stable reaction temperature is needed, instead of a cycling of different reaction temperatures. RT-LAMP and LAMP assays have been developed for the detection of FMDV [31] and *P. ramorum* [32]. Isothermal amplification strategies can also be combined with a simple detection of the PCR product on a simple lateral flow device [21]. This also eliminates costly optical components for the detection, enabling the development of simple, robust and cost-efficient devices for on-site testing.

5.5 Emerging Technologies

Compared to medical applications, the development of systems suitable for on-site diagnosis for agricultural applications is just at the beginning. This is probably due to an uncertainty over whether a viable market exists for the on-site detection of plant and livestock diseases. However, the requirements for on-site diagnostics are very similar to those for medical applications, surprisingly, many detection platforms have only been developed for medical applications. In order to identify possible developments, it is necessary to search for the latest developments in the medical point-of-care field.

One of the biggest trends in recent years in the field of pathogen diagnostics for point-of-care applications is the development of so-called lab-on-a-chip (LOC) or micro total analysis systems (μTAS). The objective of these devices is to integrate as many as possible, or even all, steps necessary for the analysis on a small device. The main steps that can be identified for these systems are sampling, sample preparation, isolation of the compound of interest, amplification, detection, and readout [33]. Different concepts for these devices and the surrounding technologies are reviewed elsewhere [34–37].

Although the main goal is the integration of the entire analysis process in one small device, extensive work has been done on integrated preanalytics, as sample preparation is still the key bottleneck for on-site analytics. By using a laser irradiation magnetic bead system (LIMBS) Lee and coworkers have realized a portable sample preparation system [38]. Instead of magnetic particles, Au nanorods can be used to employ an optothermal effect for the lysis of bacteria cells. The sample can then be used for a real-time PCR without removing the nanorods [39]. The LIMBS principle has been transferred to a polymer-based CD, with preloaded reagents, employing microchannels and centrifugal forces to guide the fluids [40]. Using this system an analysis of viral or bacterial pathogens was demonstrated in 12 min.

There have also been a couple of successful demonstrations of true sample-to-answer systems integrating all necessary steps in one small device. For the analysis of whole blood a chip has been realized by Easley and colleagues that purifies DNA by a solid phase extraction, followed by a PCR amplification, and detection by microchip electrophoresis [41]. Another system used real-time PCR for the detection of the avian flu virus H5N1 [42]. Sample preparation, RT-PCR and fluorescence detection were done in droplets on a Teflon-coated chip (Figure 5.4). The droplet itself functions as a solid phase extractor and real-time thermocycler. Using superparamagnetic beads, viral RNA is isolated, purified and concentrated. The droplet is then moved clockwise over the chip for thermocycling. The real-time detection is achieved by using SYBR Green and placing the chip on an integrated optical detection system [43]. Interestingly, the authors have also developed a pocket-size real-time PCR system [44]. By doing so, they have created a true point-of-care system, which could serve as a platform for a variety of applications.

Besides the analysis of genetic material, other sample-to-answer systems rely on immunological detection schemes. By combining open channel electrophoresis and laser induced fluorescence detection the identification of swine influenza virus has been shown [45]. By using antibody-coated microbeads and a microfluidic device marine fish iridovirus has been detected. When the virus is present in the sample the particles conjugate and are trapped in a filter, followed by washing and fluorescence detection [46]. This relatively simple approach shortened the analysis time to 30 min from over 3 h for the conventional ELISA, and also showed improved sensitivity. By using a micr flow cytometer and antibody-coated magnetic beads Yang and coworkers have relized a complete analysis system with multiplexing potential [47].

Even though these LOC and μTAS approaches are all quite interesting, it is still uncertain if they can be used in routine diagnostic applications. Many of the described systems are highly sophisticated and expensive. In order to be used as

Figure 5.4 Example of a sample-to-answer system for the detection of avian flu virus H5N1 [42].

standard analytical tools these devices have to be produced reproducibly at low cost for the consumables (chips, reagents, etc.) and the detection units.

A totally different approach for the fast on-site diagnosis of foot-and-mouth disease has been demonstrated by using infrared thermography (Figure 5.5)[6]. Using this technique it was possible to rapidly identify infected cattle before vesicular lesion could be observed. Even though the systems allow fast diagnosis of whether the cattle have fever or not, and allow the fast screening of large numbers of animals, the method relies only on relatively unspecific symptoms but could be used as a pre-screening method during a FMD outbreak.

5.6 Conclusion

The developments of recent years, from LFDs to portable PCR platforms and isothermal amplification techniques, leave no doubt that an on-site diagnosis of livestock or plant disease is possible and will probably revolutionize pathogen diagnostics. However, many of these developments still lack proper sample preparation, making the entire diagnostic process still quite laborious. The advances in the development of true integrated devices and sample-to-answer systems over the last decade show the possibilities of this technology. So far these systems have not been developed for agricultural application, but an adaption will come as these devices are established first as routine diagnostic tools in medical applications.

The use of on-site analytics in agriculture will have significant impact, not only on the detection of pathogens, but also on other sectors of agriculture and horticulture. Faster decision making will allow the reduced and targeted use of pesticides, better risk prediction, and reduced waste in supply chains, benefiting both farmers and consumers alike [5]. Furthermore, the implementation of decentralized diagnostics will probably be most beneficial to developing countries, as they lack established

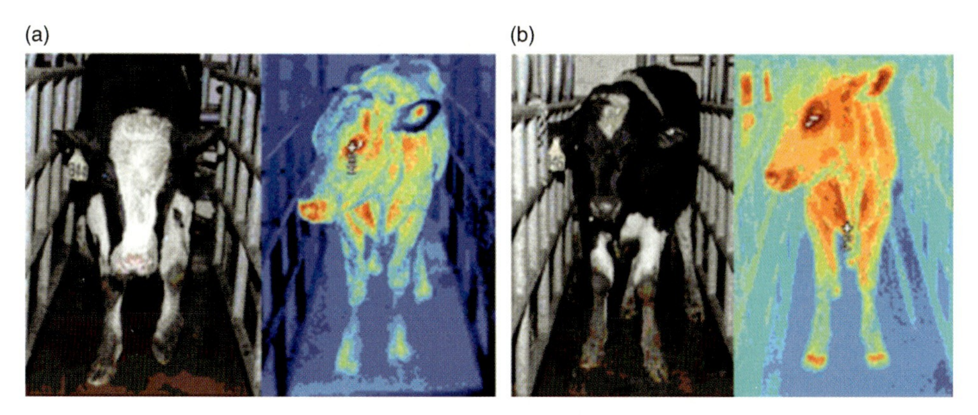

Figure 5.5 Using infrared thermography for the detection of FMD. Digital and infrared images of cattle without (a) or with (b) fever and viremia at 24 h post challenge, before vesicular lesions were observed. [6]

structures for centralized testing. Comparable to the mobile phone industry, this might lead to a development where these countries leapfrog the installation of costly centralized laboratory structures, enforcing the decentralized and mobile detection technology.

Even though the principal strategies have been successfully demonstrated and the advantages of on-site testing are clear, much work has to been done to transfer the described results into routine diagnostics.

References

1 Thompson, D. *et al.* (2002) Economic costs of the foot-and-mouth disease outbreak in the United Kingdom in 2001. *Rev. Sci. Tech.*, **21**, 675–687.

2 Davies, G. (2002) The foot-and-mouth disease (FMD) epidemic in the United Kingdom 2001. *Comp. Immunol. Microbiol. Infect. Dis.*, **25**, 331–343.

3 Anderson, I. (2002) Foot and Mouth Disease 2001: Lessons to be Learned, Inquiry Report, The Stationary Office: London.

4 Society, R. (2002) *Infectious Diseases in Livestock*, Royal Society, London.

5 King, D.P. *et al.* (2010) Plant and veterinary disease diagnosis: a generic approach to the development of field tools for rapid decision making? *Bull. OEPP/EPPO Bull.*, **40** (40), 34–39.

6 Rainwater-Lovett, K. *et al.* (2009) Detection of foot-and-mouth disease virus infected cattle using infrared thermography. *Vet. J.*, **180** (3), 317–324.

7 Schaad, N.W. *et al.* (2003) Advances in molecular-based diagnostics in meeting crop biosecurity and phytosanitary issues. *Annu. Rev. Phytopathol.*, **41**, 305–324.

8 Werres, S. *et al.* (2001) Phytophthora ramorum sp. nov. a new pathogen on Rhododendron and Viburnum. *Mycol. Res.*, **105**, 1155–1165.

9 Brasier, C.M. *et al.* (2005) Phytophthora kernoviae sp. nov., an invasive pathogen causing bleeding stem lesions on forest trees and foliar necrosis of ornamentals in the UK. *Mycol. Res.*, **109** (Pt 8), 853–859.

10 Lopez, M.M. *et al.* (2003) Innovative tools for detection of plant pathogenic viruses and bacteria. *Int. Microbiol.*, **6** (4), 233–243.

11 Lane, C.R. *et al.* (2007) Evaluation of rapid diagnostic field test kit for the identification of Phytophthora species, including P. ramorum and P. kernoviae at the point of inspection. *Plant Pathology*, **56**, 828–835.

12 Seaby, D.A. and Caughey, C. (1977) Mechanization of the slide agglutination test. *Potato Res.*, **20**, 343–344.

13 Talley, J. *et al.* (1980) A simple kit for detection of plant viruses by latex serological tests. *Plant Pathol.*, **29**, 77–79.

14 Torrance, L. and Jones, R.A.C. (1981) Recent development in serological methods suited for the use in routine testing for plant viruses. *Plant Pathol.*, **30**, 1–24.

15 Danks, C. and Barker, I. (2000) On-site detection of plant pathogens using lateral-flow devices. *Bull. OEPP/EPPO Bull.*, **30**, 421–426.

16 Miller, S.A. *et al.* (1988) Application of rapid, field-useable immunoassays for the diagnosis and monitoringof fungal pathogens in plants. *Proc. Br. Crop. Protect. Conf.*, **2**, 795–803.

17 Holmes, S.J. (1996) The validation and commercial development of immunological diagnostic technology for non-medical applications. *Proc. BCPC Symp.: Diagn. Crop. Prod*, **1**, 333–342.

18 Ferris, N.P. *et al.* (2009) Development and laboratory validation of a lateral flow device for the detection of foot-and-mouth disease virus in clinical samples. *J. Virol. Methods*, **155** (1), 10–17.

19 Ryan, E. *et al.* (2008) Clinical and laboratory investigations of the outbreaks of foot-and-mouth disease in southern

England in 2007. *Vet. Rec.*, **163** (5), 139–147.
20 Kox, L.F. *et al.* (2007) Diagnostic values and utility of immunological, morphological, and molecular Methods for in planta detection of Phytophthora ramorum. *Phytopathology*, **97** (9), 1119–1129.
21 Tomlinson, J.A., Dickinson, M.J., and Boonham, N., (2010) Rapid detection of Phytophthora ramorum and P. kernoviae by two-minute DNA extraction followed by isothermal amplification and amplicon detection by generic lateral flow device. *Phytopathology.*, **100** (2), 143–149.
22 King, D.P. *et al.* (2008) Prospects for rapid diagnosis of foot-and-mouth disease in the field using reverse transcriptase-PCR. *Vet. Rec.*, **162** (10), 315–316.
23 Callahan, J.D. *et al.* (2002) Use of a portable real-time reverse transcriptase-polymerase chain reaction assay for rapid detection of foot-and-mouth disease virus. *J. Am. Vet. Med. Assoc.*, **220** (11), 1636–1642.
24 Donaldson, A.I., Hearps, A., and Alexandersen, S. (2001) Evaluation of a portable, "real-time" PCR machine for FMD diagnosis. *Vet. Rec.*, **149** (14), 430.
25 Hearps, A., Zhang, Z., and Alexandersen, S. (2002) Evaluation of the portable Cepheid SmartCycler real-time PCR machine for the rapid diagnosis of foot-and-mouth disease. *Vet. Rec.*, **150** (20), 625–628.
26 Tomlinson, J.A. *et al.* (2005) On-site DNA extraction and real-time PCR for detection of Phytophthora ramorum in the field. *Appl. Environ. Microbiol.*, **71** (11), 6702–6710.
27 Hughes, K.J.D. *et al.* (2006) On-site real time PCR detection of Phytophthora ramorum causing dieback of Parrotia persica in the UK. *Plant Pathol.*, **55**, 813.
28 Notomi, T. *et al.* (2000) Loop-mediated isothermal amplification of DNA. *Nucleic Acids Res.*, **28** (12), E63.
29 Compton, J. (1991) Nucleic acid sequence-based amplification. *Nature*, **350** (6313), 91–92.
30 Lau, L.T. *et al.* (2008) Detection of foot-and-mouth disease virus by nucleic acid sequence-based amplification (NASBA). *Vet. Microbiol.*, **126** (1–3), 101–110.
31 Dukes, J.P., King, D.P., and Alexandersen, S. (2006) Novel reverse transcription Loop-Mediated Isothermal Amplification (RT-Lamp) for rapid detection of Foot-and-Mouth disease virus. *Arch. Virol.*, **151** (6), 1093–1106.
32 Tomlinson, J.A., Barker, I., and Boonham, N. (2007) Faster, simpler, more-specific methods for improved molecular detection of Phytophthora ramorum in the field. *Appl. Environ. Microbiol.*, **73** (12), 4040–4047.
33 Schulze, H. *et al.* (2009) Multiplexed optical pathogen detection with lab-on-a-chip devices. *J. Biophoton.*, **2** (4), 199–211.
34 Liu, W.T. and Zhu, L. (2005) Environmental microbiology-on-a-chip and its future impacts. *Trends Biotechnol.*, **23** (4), 174–179.
35 Ligler, F.S. (2009) Perspective on optical biosensors and integrated sensor systems. *Anal. Chem.*, **81** (2), 519–526.
36 Auroux, P.A. *et al.* (2004) Miniaturised nucleic acid analysis. *Lab. Chip.*, **4** (6), 534–546.
37 Myers, F.B. and Lee, L.P. (2008) Innovations in optical microfluidic technologies for point-of-care diagnostics. *Lab. Chip.*, **8** (12), 2015–2031.
38 Lee, J.G. *et al.* (2006) Microchip-based one step DNA extraction and real-time PCR in one chamber for rapid pathogen identification. *Lab. Chip.*, **6** (7), 886–895.
39 Cheong, K.H. *et al.* (2008) Gold nanoparticles for one step DNA extraction and real-time PCR of pathogens in a single chamber. *Lab Chip*, **8** (5), 810–813.
40 Cho, Y.K. *et al.* (2007.) One-step pathogen specific DNA extraction from whole blood on a centrifugal microfluidic device. *Lab Chip*, **7** (5), 565–573.
41 Easley, C.J. *et al.* (2006) A fully integrated microfluidic genetic analysis system with sample-in-answer-out capability. *Proc. Natl. Acad. Sci. USA*, **103** (51), 19272–19277.
42 Pipper, J. *et al.* (2007) Catching bird flu in a droplet. *Nat. Med.*, **13** (10), 1259–1263.
43 Pipper, J. *et al.* (2008) Clockwork PCR including sample preparation. *Angew. Chem. Int. Ed. Engl.*, **47** (21), 3900–3904.

44 Neuzil, P. *et al.* (2010) Rapid detection of viral RNA by a pocket-size real-time PCR system. *Lab Chip,* **10** (19), 2632–2634.

45 Reichmuth, D.S. *et al.* (2008) Rapid microchip-based electrophoretic immunoassays for the detection of swine influenza virus. *Lab Chip,* **8** (8), 1319–1324.

46 Liu, W.T. *et al.* (2005) Microfluidic device as a new platform for immunofluorescent detection of viruses. *Lab Chip,* **5** (11), 1327–1330.

47 Yang, S.Y. *et al.* (2008) Micro flow cytometry utilizing a magnetic bead-based immunoassay for rapid virus detection. *Biosens. Bioelectron.,* **24** (4), 861–868.

6
On-Site Analysis

Günther Proll and Günter Gauglitz

6.1
Introduction

Chemical contamination, allergens, and pathogenic compounds are ubiquitous in the environment. They can be found in contaminated ground water, in sediments, and in soil. Accordingly, this contamination can be absorbed by plants and transferred through the food chain to animals and humans. Furthermore, contact between humans is also responsible for inter-human transfer of such toxic substances. In the last decade, nanoparticles have come increasingly into focus regarding their influence on environmental and health problems. Because of their ability to act as carriers for any pathogenic compounds, due to their adsorptive effects, their influence on health care (in addition to their own potential toxicity) is now also considered.

State-of-the-art monitoring of environmental conditions and pollution in the food chain makes use of a large variety of classical analytical instruments [1]. Hyphenated techniques allow detection and identification of such molecules as allergens or pathogens and, in principal, of all chemical contaminants, even in complex matrices, in mixtures, sometimes at low concentrations (especially after enrichment procedures), and with necessary specificity [2–4]. Sampling, sample preparation and analysis are time-consuming and usually require a well equipped laboratory and qualified personnel. These facts cause problems for monitoring with respect to time, expenditure, and costs [5]. On the other hand, in the past decade the EU Commission has published quite a few directives which expect Communities and companies producing food to permanently monitor water quality [6–8] and/or foodstuff purity [9, 10]. Also, the US Environmental Protection Agency (EPA) enforces analytical monitoring of water to reduce cases of illnesses and deaths due to potential fecal contamination and waterborne pathogen exposure [11]. A similar requirement is enforced in many European member states, for example, Germany [12]. Recent trends in new developments of sensors for measuring priority pollutants are discussed using a list of these substances and comparing some methods [13]. Sensors and biosensors can be applied to support the EU Water Framework Directives and the related Marine Strategy Framework Directives [14].

Handbook of Biophotonics. Vol.3: Photonics in Pharmaceutics, Bioanalysis and Environmental Research, First Edition.
Edited by Jürgen Popp, Valery V. Tuchin, Arthur Chiou, and Stefan Heinemann

For this reason, new possibilities of fast, cost-effective and reliable analyses have been considered and developed within recent years. Photonics provides perfect tools for these upcoming tasks. Presently, a large variety of optical detection methods are being applied in bio- and chemo-sensors, some are even at a high level of development. Further improvement and production strategies are expected as photonics is increasingly used to provide miniaturized improved instrumentation for analytical purposes. These optical methods are especially suitable for miniaturization and parallelized detection of multi-analyte samples in cost-effective and fast monitoring of contaminants in water, food, and the environment. Miniaturization allows reduction of the sample volume and the number of agents. Thus, these improvements in instrumentation reduce cost and waste, and support portability.

Presently, the urgent problem is the requirement that, especially when looking for bacteria in many applications, just a few molecules have to be detected in small sample volumes. Instrumentation has not yet overcome the limitations of LODs (limits of detection) in such application areas, although quite a few approaches to enrichment strategies have been published and applied down to $10^{-19}\,mol\,l^{-1}$. To give an idea of the problem, drinking water regulations require the verification of the absence of *E. coli* and coliforms in a 100 mL water sample. Thus, 1 cfu (colony forming unit) has to be detected. Accordingly, the EPA defined standard methods for enrichment of salmonella in biosolids [15] many years ago. Best results of "enrichment" have been achieved using PCR (polymerase chain reaction). However, this approach is time-consuming and regarded as unsuitable for on-site analysis. Therefore, in recent years some techniques of enrichment have been developed, allowing automated pre-concentration and combination with detection platforms. However, most of them are not fully automated and cannot always achieve high recovery rates, but at least they avoid cultivation steps. Thus, different concentration–elution methods using filtration, membranes or liquid–liquid extraction [16, 17] have been compared. Other concepts of enrichment use cross-flow microfiltration [18], by which a hollow fiber membrane module allows enrichment factors of more than 100. Another approach is immunomagnetic separation (IMS) [19] which takes advantage of coupling nanoparticles to antibodies or affinity ligands and by these means isolating bacterial cells, even in complex samples. Epoxy-based monoliths used for the affinity capturing of bacteria have also been reported [20].

In recent publications, such techniques are discussed for various applications [21]. In many papers enrichment using PCR is proposed. Combined enrichment and real-time PCR is described to obtain quantitative data for salmonellae [22]. Sandwich ELISA (enzyme-linked immunosorbent assay) in combination with PCR amplification has been developed for enrichment and detection of *legionella pneumophila* [23].

However, such approaches can amplify not only the bacteria that are being looked for, but also other components in the sample. Therefore, cross-flow microfiltration combined with immunomagnetic separation as pre-concentration steps and subsequent PCR amplification with DNA microarray detection are considered to come close to the required limits of detection [24]. Thereby, the stipulated limit of 0 bacteria is rather problematic from an analytical point of view. Regulation defining a limit of 1 cell in 100 ml is preferable as a LOD.

The concentration requirements are rather restrictive. However, there are quite a few areas in which monitoring, quantification and identification of contaminants in various samples, such as soil, water and food, are of high interest, even at higher concentrations. A rather expensive market research study covering advantages in label-free detection technologies gives a technical insight into developments at different companies, a description of platform technologies, patents and key application areas [25]. New methods to identify trace amounts of infectious pathogens rapidly, accurately and with high sensitivity are in constant demand in order to prevent epidemics and loss of lives, and have been reviewed recently [26]. This publication states that there is an urgent need for sensitive, specific, accurate, user-friendly diagnostic tests. Photonics shows promising perspectives for future development and application.

6.2 Substances to be Monitored

Contaminants of the emerging compound classes (listed in Table 6.1) are suspected to cause adverse effects in humans and wildlife. For instance, pentabromobiphenylether, 4-nonylphenol, C10-C13 chloroalkanes and the di(2-ethylhexyl)phthalate (DEHP) have been listed as priority hazardous substances in the field of water policy by EC Water Directive 2000/60/EC [7] and the final EU decision No. 2455/2001/EC. Active hormonal substances (natural hormones are active at levels of ng l^{-1}) such as estrogens, anti-inflammatory cortico-steroids and anabolic androgens are being widely used in human and veterinary medicine [27].

Table 6.1 List of emerging compound classes.

Emerging Compound Classes	Examples
biocides and plant protection products (e.g., pesticides)	atrazine, chlorpyrifos, diuron, endosulfan, isoproturon, simazin, trifluralin
pharmaceuticals	
veterinary and human antibiotics	sulfonamides, tetracyclines
analgesics, anti-inflammatory drugs	codeine, ibuprofen, diclofenac
psychiatric drugs and so on	diazepam
β-blockers	metoprolol, propanolol, timolol
X-ray contrasts	iopromide, iopamidol, diatrizoate
Surfactants and surfactant metabolites	nonylphenol, octylphenol, alkylphenol carboxylates
flame retardants	polybrominated diphenyl ethers (PBDEs), tetrabromo bisphenol A,
industrial additives and agents	chelating agents (EDTA), aromatic sulfonates
gasoline additives	dialkyl ethers, methyl-*t*-butyl ether (MTBE)
hormones and EDCs	estradiol, estrone, ethinylestradiol, bisphenol A, phthalic acid, nonylphenol, atrazine
natural toxins	saxitoxin, zearalenone, microcystins

It is important to notice that the mentioned emerging compounds can accumulate in different matrices in the environment (e.g., water, sediment, soil, etc.) with varying persistence and bioaccumulation. For many of these emerging contaminants, there are no data for eco-toxicological potential and risk assessments available. Therefore, it is difficult to predict what health effects they may have on humans, terrestrial and aquatic organisms, and ecosystems.

6.3 Optical Methods for Monitoring

There are a large number of different detection methods. These can be classified according to, for example, electrochemical, mass-sensitive, thermal, or optical methods. In this chapter, especially optical methods will be considered. There are quite a few recent review articles and book chapters which cover trends in optical detection techniques, considering various aspects. Present optical biosensors and perspective developments are presented and discussed in some chapters of a book by Ligler [28]. Beginning with fiber optic-based biosensors, a large number of evanescent field-based set-ups are discussed. The special principle of surface plasmon resonance (SPR), which is also commercialized, is described in detail. A special chapter covers plasmonic amplification, as has been realized in SERS (surface enhanced Raman spectroscopy) arrangements. In addition, the principle of fluorescence life time measurements and the use of planar wave guides for fluorescence sensors are covered. Electrochemiluminescence as a sensitive tool is reviewed. Cavity ring-down, cantilever, nanoparticles and fluorescence-based intracellular biosensing are handled, demonstrating future trends.

Fluorescence is the main approach for detecting biomolecular interactions. Presently, most assays are based on this method. Therefore many review articles have been published, and books or book chapters have been written [29]. The field of chemical and biochemical sensors based on fluorescence detection techniques has been classified, former development and present status as well as trends are given [30]. Optical chemical sensors (OCS) have experienced increasing interest: Their advantages of being non-invasive and allowing on-line detection are presented in a review article on multiple fluorescent sensing and imaging [31]. The state of the art in terms of spectroscopic methods, materials, and selected examples for dual and triple sensors is discussed, together with a look into the future. The different methods using direct optical detection have been reviewed [32, 33], and commercially available instrumentation as well as typical applications are presented [34].

It is possible to distinguish between direct measurements of pollutants without reagents, which can be the case for bacteria, virus, or self-fluorescent poly-aromatic hydrocarbons, and monitoring contamination by interaction between a reagent and a pollutant. These methods can be further divided into experiments based on interaction partners where at least one of the partners is labeled, and other methods using non-labeled partners. In addition, different techniques provide optimum approaches if the interaction between analyte and reagent takes place in the homogeneous phase

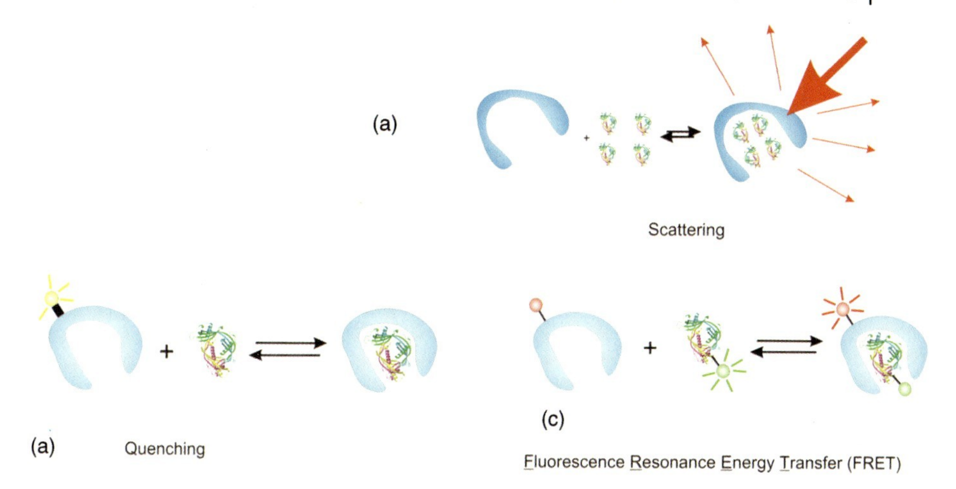

Figure 6.1 Interaction processes between analyte and receptor. (a) Interaction results in modified structure or increased size, allows measurement with scattering methods, (b) interaction can cause quenching, (c) fluorescence resonance energy transfer (FRET), fluorescence emission wavelength is different.

or in the heterogeneous phase of a transduction element, as used in bio- or chemosensors. These different principle experimental set-ups are surveyed in Figures 6.1 and 6.2.

With regard to the homogeneous phase, it is possible to distinguish between interactions between non-labeled compounds and those between partners where one or both are labeled. Accordingly, in Figure 6.1 different detection methods of these interactions are used. In the case of (a), methods of light scattering can quantify the amount of interaction of non-labeled partners. For (b), quenching effects of the labeled molecule during interaction can be considered, but indirect detection using the transduction element is preferable. Fluorescence resonance energy transfer (FRET) is state of the art in many applications (c). Diffusion to the transducer has always to be taken into account in all the other possibilities of monitoring interaction, as is demonstrated in Figure 6.2, where detection at the interface to the transduction element (heterogeneous phase detection) presents a variety of principles. Even the simplest and most preferred approach of direct measurements (a), with the “analyte” as a ligand to a receptor being immobilized on the transducer surface (also considering additional biopolymer layers against non-specificity) is influenced by this diffusion process. The signal is usually only high enough for the quantification of such an interaction if the measured partner is large or heavy enough (b) or the analyte is self-fluorescent (for case (a)). If the receptor is labeled in assay (b), the measurement is also successful. This indirect measurement is more frequently applied in competitive or binding inhibition tests (c), where the receptor in the homogeneous phase is not necessarily labeled by a fluorophore (e). For this approach, a derivative of the analyte has to be immobilized at the transducer as a recognition element. Case (d) demonstrates the same assay for a system where the receptor is labeled and

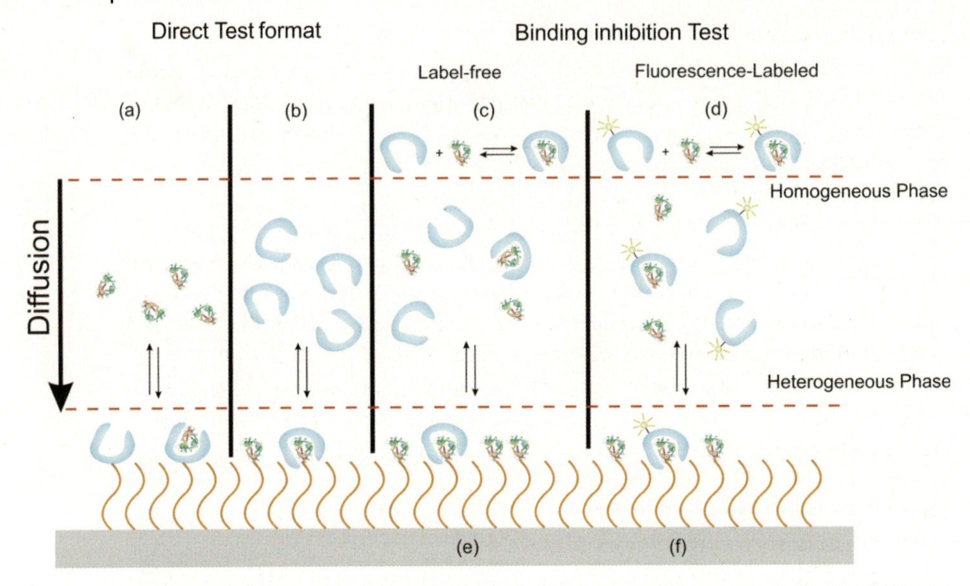

Figure 6.2 Interface with biomolecular layer and nonspecific binding and diffusion layer responsible for mass transport limitation (a) direct assay with immobiliised receptors attracting small analytes, (b) rather large analytes interacting with small receptors immobilized at the transducer, (c) derivative of analyte immobilized to biopolymer layer, in the equilibrium of the homogeneous phase non-blocked receptors can be detected in a second equilibrium in the heterogeneous phase (e), this type of interaction can also be monitored if the receptor is labeled ((d) and (f)).

interaction is monitored at the transducer interface (f). For these different approaches a variety of optical detection principles are suitable, either based on fluorescence detection or using direct optical detection principles. New methods based on Raman enhanced spectroscopy provide, in most of these approaches, not only quantitative information about the amount of interaction but also allow characterization of the interacting specimen.

For monitoring in the homogeneous phase, typically at least one of the partners requires a label, such as a fluorophore. However, if the pollutant can cause light scattering or if the interaction-complex between contaminant and reagent molecules causes drastic changes in either size or form, changes in scattering properties can be used for detection. Accordingly, light scattering [35, 36] is a typical detection principle used in flow cytometry, which allows detection and characterization of cells or other particles suspended in a stream of fluid. This method is frequently used in environmental microbiology and biotechnology [37], or even in the food industry, as reviewed recently [38]. In the latter case, mostly labeled particles are measured, also taking advantage of fluorescence-activated cell-sorting (FACS). A modification of the light scatter-based method was introduced at Purdue university with BARDOT (bacteria rapid detection using optical scattering technology) [39] as a label-free system looking at unique patterns found by laser light scattering from bacteria colonies. Pathogens can be detected and classified. An instrument to measure nanoparticles with respect to size, concentration, zeta potential, and aggregation, as well as population in liquids in a multi-parameter approach is provided by Nanosight [40]. The real-time measurement provides information about the kinetics of protein aggregation, the development of viral vaccines, nano-toxicology and biomarker detection, and is also useful for pollutant detection. Applied to pathogens, light scattering is reported as a fast and sensitive detection tool [41].

In environmental control, electrochemical and optical detection methods are currently popular. Recent application of flow-based methods competes with chromatographic methods. They have been compared recently [42]. Flow methodology has been modified and multiparameter flow-through optosensors are reviewed and current trends of such detection techniques are critically discussed.

Fluorescence intensity or quenching effects of an analyte detecting reagents such as labeled antibodies can be determined in dependence on analyte concentration. If the diffusion property of the labeled reagent is changed significantly through interaction with a large analyte, fluorescence correlation spectroscopy becomes measurable. Also, detection in the homogeneous phase is possible if both partners of the interaction process are labeled and fluorescence resonance energy transfer takes place. FRET requires a labeled molecule as donor and another labeled molecule as acceptor, that is, the fluorescence of the donor must be within the wavelength range of the absorption of the acceptor molecule. As soon as these partners become very close, that is, if the distance between them is less than approximately 10 nm, energy from the donor is transferred to the acceptor which starts to fluoresce at the long wavelength side [43, 44]. In general, Förster resonance energy transfer-based analytical techniques have found great interest in bioanalysis because they allow detection of protein–protein interactions and conformational changes of biomolecules. This can be done at the nanometer scale, both *in vitro* and *in vivo* in cells, tissues, and organisms. FRET measurement principles and procedures have been continuously improved in recent years. Nevertheless, in some cases only qualitative or semi-quantitative information can be obtained. Recent applications are given and their advantages and limitations are discussed in[45]. Special applications allow the monitoring of intra- and intermolecular reactions occurring in microfluidic reactors performing highly efficient and miniaturized biological assays for the analysis of biological entities such as cells, proteins and nucleic acids [46]. Even multiplexed analytics using quantum dots are used to improve signals [47]. A review of bioluminescence resonance energy transfer in bioanalysis is found in [48].

Measurements in the heterogeneous phase provide an even larger number of methods for detection of the final interaction state, and even the kinetics of the interaction process. Some of these methods even allow the characterization of bacteria. An ELISA providing colorimetric read-out is one of the basic methods to detect pathogens. Immunographic assay strips supply a simple method of multiplex detection on a somewhat semi-quantitative basis [49].

For many years chemiluminescence or bioluminescence have been used as methods to detect low amounts of molecules [50]. A chemical or biochemical reaction leads to electronically excited products which decay, emit photons, or pass their energy to other species which will emit light. Some advantages in comparison to normal fluorescence can be discussed [51]. In principle, all the chemiluminescence-based techniques are fundamentally limited by system properties, such as reaction yield and quantum yield. However, new technology has emerged which uses metal surface plasmons to amplify chemiluminescence signals. A tutorial review gives a survey on the next-generation of chemiluminescent-based technologies [52]. In the detection of salmonella, legionella and other pathogenic bacteria such detection principles, in combination with flow-through techniques have been successfully applied [53, 54].

A large number of different fluorescence methods exist. They can be used in different types of assays for detection, including heterogeneous phase assays. They demonstrate low limits of detection. However, labeling can influence the interaction process and reduce bioactivity. For this reason, in recent years, direct optical detection techniques have gained increasing interest. Therefore, methods like Raman measurements and reflectance-based monitoring will be discussed next. Figure 6.3 shows reflectance at a thin layer. Radiation is reflected at each interface between media of different refractive index [55]. Using a thin layer the reflected beams at the two interfaces can superimpose (grey lines, Figure 6.3a). This can be considered as micro-reflectometry [56]. If radiation is coupled into such a thin layer then total reflection occurs beyond a certain critical angle of incidence and radiation is guided (see Figure 6.3b) within the fiber or the waveguide. In addition, the guided radiation's electrical field vector couples to the outside of the waveguide creating an exponential decay field vector which is influenced by the refractive index next to this interface. Accordingly, as demonstrated in Figure 6.4, this so-called evanescent field can either excite fluorophores close to the interface or read-out changes in refractive indices close to this grating. A simple application of this effect has been demonstrated using long waveguides with long interaction lengths and measuring finally the reduction of

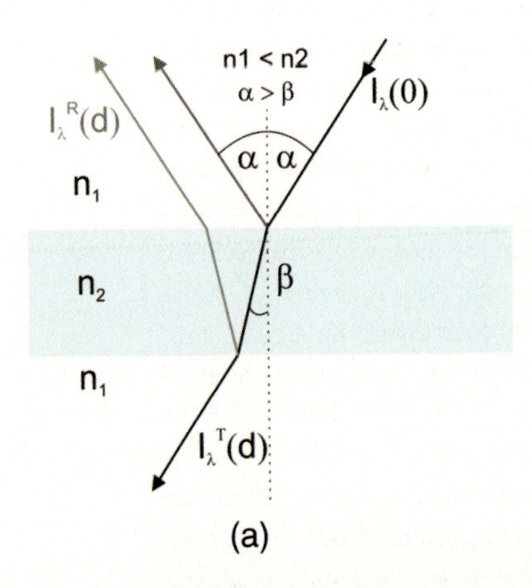

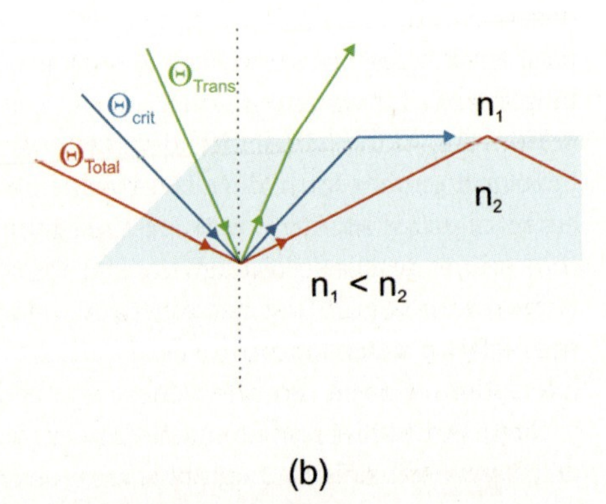

Figure 6.3 (a) Reflectance at a thin layer, refraction into layer at smaller angle to optical axis for incidence from medium at lower optical density to one at higher optical density. Multiple reflections at interfaces (grey lines) forming superimposed reflected partial beams. (b) reflectance at interface between medium of higher optical density and one with lower optical density can result in total reflectance and guided waves.

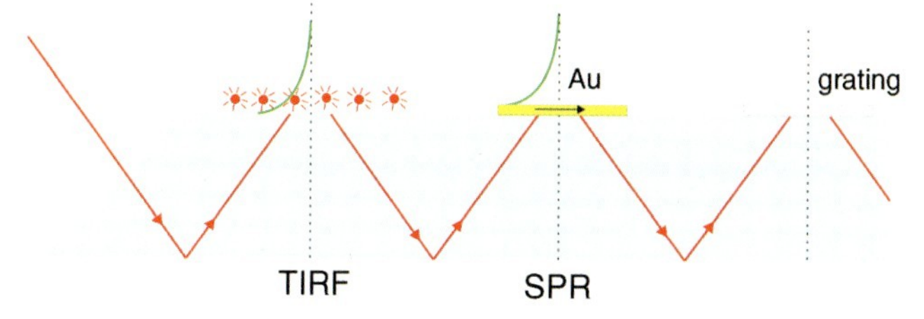

TIRF: Total Internal Reflection Fluorescence

SPR: surface plasmon resonance

Figure 6.4 A guided wave couples to an evanescent field at the outside of the waveguide which decays exponentially. This can be absorbed in part and excite fluorophores in close contact to the waveguide or excite surface plasmons in a metal coated on the waveguide (SPR) or couple out at a grating in the interface of the waveguide in dependence on changes at this grating.

guided radiation intensity [57], or using integrated optics to improve stability [58]. Modern waveguide based set-ups use a variety of read-outs, such as surface plasmon resonance (SPR), grating coupler, resonant mirror, Young interferometers, and Mach-Zehnder chips. They and some more can be summarized as micro-refractive principles and are discussed in [59].

In the heterogeneous phase chemical modification of the surface is essential. This layer has to reduce (better suppress) non-specific binding, allow high loading (as many recognition sites as possible) and to be very stable (for potential regeneration steps). A recent review covers the surface functionalization methods regarding the recognition elements, the immobilization techniques, self-assembled monolayers and covalent coupling [60]. In addition, microfluidic devices are presented for different materials and various optical detection techniques for label-free assays are discussed; evanescent field devices are listed, resonant cavities and especially surface plasmon devices discussed. The special polyethylene glycol layer, a well established biopolymer coating of transducer surfaces against non-specific binding, was characterized and discussed with respect to its feasibility [61]. The new approach to biosensing, using scaffolds in complex matrices has been recently presented [62]. An essential part of any device for monitoring pathogenic compounds is the quality of the flow cell. Nowadays, these devices are miniaturized to fit to many detection principles, to reduce the amount of reagent, and to optimize flow conditions combined with defined diffusion processes to the transducer surface. Such miniaturized tools have been reviewed recently [63].

Many approaches to the monitoring of pollutants in the environment are based on microarray technology. Even protein microarrays are suggested, which have been reviewed some years ago [64]. These arrays can use direct optical detection techniques [65], which is becoming of increasing interest. Interferometric reflectance imaging sensors allow sensitive protein and DNA detection in real time and even on a high through-put basis [66]. However, as mentioned, most approaches are based on fluorescence detection [67]. In principle, DNA sequences supply good chances for selective detection of pollutants with such read-out techniques [68]. Such DNA microarray technology is especially considered for microbial pathogen detection [69]. Some of these different detection platforms [70] are named, a large number of references are given, and the applicability to detection of pathogenic bacteria is discussed [71].

Fluorescence is one of the most used methods in analytics. Because of the capability to measure low concentrations, fluorescence is the basis of many microarrays. In the general review on DNA and protein mircroarrays [67] fluorescent labeling of nucleic acids and protein strains is discussed. Fluorescence labeling of nanoparticles is classified according to polymeric nanoparticles, silica particles, and quantum dots. The latter have a semiconductor core of different particle sizes (CdSe, CdTe e.g., 1–10 nm in diameter). The core is covered by a transparent layer, for example, ZnS, to increase quantum effects and stability. The surface is functionalized with biocompatible material which can be used for proteins and makes them water soluble. The advantage is that a single wavelength of excitation can be used to offer various emission wavelengths, just dependent on the particle size [72]. Optical, magnetic and electrochemical methods of nanoparticle-based biosensors are compared in an overview of the progress, the limitations and future challenges of such devices for detection of pathogenic bacteria [73]. The synthesis of various types of nanoparticles supplies a large variety of properties with gold magnetic or quantum dot nanoparticles. A review gives a large number of recent applications in environmental analysis, it discusses different detection principles and mentions key trends as well as future perspectives [74]. The use of such nanoparticles, even in combination with new recognition elements like scaffolds, results in interesting advantages for nano-structured biosensors [75]. Crystalline europium-doped gadolinium oxide nanoparticles also offer a large Stokes shift and long fluorescence lifetime (1 ms) [76].

The read-out of microarrays can consist of either a scanner or an imaging system. Commercially available readers are compared on the internet [77].

Excitation of fluorescence close to a waveguide's interface is considered as total internal fluorescence reflectance (TIRF) [78]. This method is frequently used to excite fluorescence of appropriate arrays. Another approach is to measure fluorescence anisotropy, time resolved fluorescence, or fluorescence lifetime imaging (FLIM) [79]. FRET is also used [45].

Using interaction at a heterogeneous interface, or at the interface between a surface with recognition elements and the analyte in solution, direct optical detection becomes possible in addition to the detection of fluorophores. This can be done either in a single-channel system or in an array of detection elements, called a micro-array, with a corresponding parallel read-out of all interaction spots. Under these conditions, all read-out methods for microarrays can usually also be used for single-channel readout. For the read-out of fluorescence, a large variety of fluorophores labels is available commercially. The excitation even ranges up to the far-red wavelength range, which improves the quality of the fluorescence signal with respect to noisy scattering. Besides the measurement of fluorescence intensity, especially time-resolved fluorescence is preferable and also the increase in sensitivity by nanoparticle-enhanced signals using the effect of plasmons. In all cases, the photo-stability of the label and quenching effects can reduce the efficiency of excitation and signal-to-noise ratio. The gain using such plasmon enhancing effects can be demonstrated in antigen–antibody interaction studies [80]. It even allows measurements using microscopy and multispot approaches [81]. The urgent need to determine infectious pathogens rapidly, accurately, and at low concentration is argued in detail using modern approaches in surface modification and fluorescent nanoparticles in a general review [82].

It had been mentioned that labeling can reduce bioactivity. Therefore, in recent years direct optical detection methods have gained increasing interest, keeping in mind the problems with matrices and the greater influence of non-specific binding and limitations of limits of detection. Nevertheless, quite a few good arguments exist for these direct methods. They have been classified above as either micro-refractometric or micro-reflectometric. Figure 6.4 demonstrated the existence of an evanescent field which is dependent on the refractive index of a solution or a layer close to the interface of the transducer. Binding of analytes or receptor molecules to the recognition elements in this layer causes changes in the refractive index. The monitoring evanescent field couples to the field of the guided wave within the waveguide or fiber. There exist a large number of optical principles to read-out changes of the propagating wave, thus allowing calibration of a pollutant's concentration with changes in the effective refractive index [83–85]. A huge number of publications deal with such evanescent field-based biosensors. The most discussed methods are grating couplers [86–88], resonant mirrors [89–91] and SPR [84, 92–94]. The effect of refractometric measurement can be combined with interference. Such is the case in Mach–Zehnder chips and in its modified form, the Young interferometer. Both split the waveguide into two arms, one of which is undisturbed (reference arm) and the other is in contact with the analytical process. Thus, a phase shift between these two arms takes place and can be very sensitively observed. Calculations prove that limits of detection far beyond the normal could be expected. However, even in Mach–Zehnder rearrangements measurable concentrations are less limited by the optics than by microfluidics and the biomolecular interaction process. Measurements of pesticides have been published [95, 96]. If the arms are not re-unified, but a free radiation optical arrangement is chosen without rejoining the waveguide arms then this is called a Young interferometer [97]. Such set-ups are rather costly at present and lack mechanical stability.

SPR was commercialized many years ago [98]. Therefore, it is the most commonly used evanescent field technique. Normally, a 50 nm thick gold layer is coated on a prism (half-cylinder); radiation is incident on the gold film via this prism. Electron density fluctuations can be excited at total internal reflection conditions at a suitable angle and/or wavelength of incident radiation. Then, reflected radiation is reduced in intensity at the resonance condition. This resonance condition depends on the refractive index close to the gold film in the solution inside the sample cell. Resonance is either observed at a shifting angle of reflection or wavelength. Applications to bio- and chemo-sensing were reviewed many years ago [99]. The analytical approach using SPR to measure contaminants like pesticides, mycotoxins, or surfactants in food is discussed with respect to immobilization strategies (avoiding non-specific binding) and assay formats [100]. At present, an urgent aim is to multiplex SPR [101]. To advance food-safety towards high-throughput control, development of SPR is pushed forward in the areas of medical, pharmaceutical, biotechnological, and agri-food industries [102].

Recently, some applications of SPR in simultaneous multianalyte antibiotic detection in milk samples have been published for three important antibiotic families (fluoroquinolones, sulfonamides and phenicols) with LODs below 1 $\mu g\,l^{-1}$ [103]. To avoid contamination of food the rapid detection of foodborne pathogens is of vital importance. The developed biosensor array chip is able to specifically detect the presence of two known pathogens. This biosensor array is integrated into a self-contained PDMS microfluidic chip [104]. The detection of microcystins (MCs) in drinking water has also been developed. Among several, the selected assay format is based on a competitive inhibition assay, in which microcystin-LR (MCLR) is covalently immobilized onto the surface of a SPR chip functionalized with a self-assembled monolayer [105]. By these means SPR is even combined with fluorescence [106, 107] which improves detection.

In all the discussed optical detection techniques the product of refractive index and the physical thickness of the considered interaction layer ("optical thickness") is measured. In the case of evanescent field techniques the change in refractive index has the dominant influence on the signal. Unfortunately, this refractive index is rather dependent on temperature. For this reason temperature control is essential and requires well referenced measurements together with thermostating. This fact is also valid, in part, for reflectometric methods. Thus, ellipsometry is also temperature dependent. Multiple reflections at interfaces of a layer system cause superpositions of the partial radiation beams resulting in an interference pattern. Both polarization states of the electromagnetic radiation are used. The two modes (two states of polarization) are differently influenced by the effective refractive index. Thus, in spectral ellipsometry, the phase shift and change in amplitude of the two modes allow information about physical thickness and refractive index [108, 109]. Such instrumentation has been very expensive, however, process control of wafers has widened the applications and reduced instrument prices.

Ellipsometry can be further simplified just by using non-polarized light. This method is called reflectometric interference spectroscopy (RIfS) [110, 111]. The principle is demonstrated in Figure 6.5. Changes in optical thickness shift the interference spectrum. Since time-resolved measurements are no problem, binding processes can be observed. Glass or transparent polymer layers can be used, even ITO-coated glass plates are suitable. Thus, the material is cheap. Even measurement at one wavelength (or better at four) is possible [112, 113]. Temperature dependence is negligible in contrast to evanescent field techniques. Increase in temperature will reduce the refractive index, in contrast to the increasing physical thickness. Thus, both effects nearly compensate and allow measurements without thermostating. The optical robustness and simple set-up allow easy parallelization, 700 spots per square centimeter have been achieved [114]. The principle has been used to measure pesticides [115] and endocrine disruptors [116] in food.

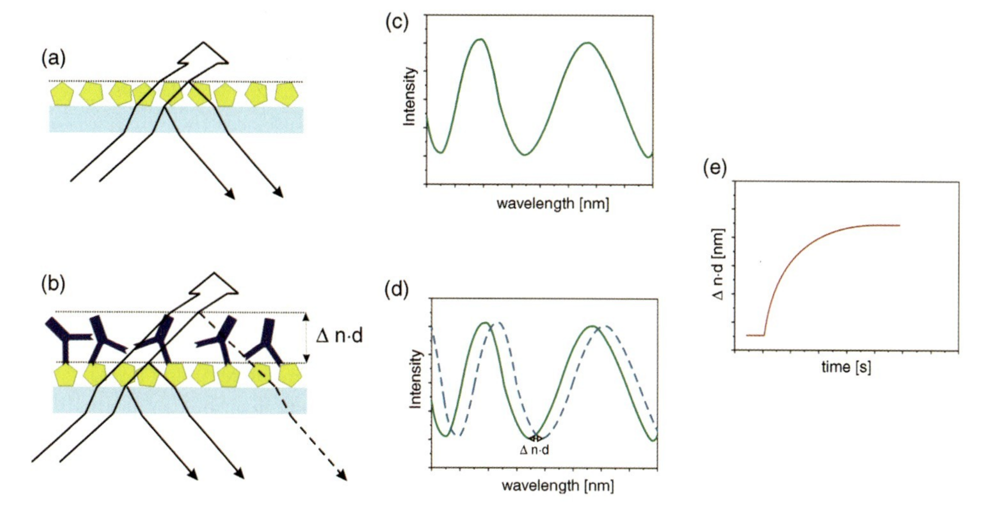

Figure 6.5 Multiple reflection takes place at interfaces of layers in dependence on physical thickness, wavelength, refractive index and angle of incidence. The result is an interference pattern (c). Changes in either refractive index or physical thickness cause a shift in interference pattern. Time-resolved measurement can be correlated to a binding process (e).

Since Raman spectroscopy has drastically developed towards simple and robust instrumentation, it is considered to be suitable for on-line analysis [117]. Metallic surfaces or even metallic nanoparticles can increase the signal, resulting in surface enhanced Raman scattering/spectroscopy (SERS) [118]. In Raman spectroscopy vibrational and rotational excitations of molecules are studied. It is an inelastic scattering process. The Raman scattering is excited by a laser in the visible or near-infrared range. It is complementary to infrared spectroscopy. Spontaneous Raman scattering is typically very weak. Accordingly, the main difficulty of Raman spectroscopy is in separation of the weak inelastically scattered light from the intense Rayleigh scattered light. Use of lasers has overcome some of the problems. A number of advanced types of Raman spectroscopy, including surface-enhanced Raman (SERS), resonance Raman, tip-enhanced Raman (TERS), polarized Raman, and hyper Raman have advanced Raman to a powerful analytical tool. Modern instrumentation even offers hand-held devices, or at least systems usable out of lab [119, 120]. A recent special issue in the *Journal of Analytical and Bioanalytical Chemistry* lists in the editorial the advantages, developments and possibilities [121]. Within this volume some applications are given.

The capabilities of SERS as a technique for applications related to environmental analysis and monitoring are presented in a recent review [122]. This method has proven to give good results in the structural characterization of soils, ultrasensitive detection of pollutants and heavy metal ions, and the analysis of plants, tissues and microorganisms. In addition, a critical approach with respect to the drawbacks and difficulties associated with the various experimental configurations and enhancing substrates is introduced. Raman spectroscopy is a direct technique to detect pathogenic bacteria in food without prior enrichment. This is demonstrated for three different milk-contaminating bacteria; *Escherichia coli, Brucella melitensis,* and *Bacillus thuringensis* [123]. Vibrational spectroscopy is feasible for the validation of bacteria in microfluidic devices, as demonstrated by the use of SERS [124].

Recent biosensor developments use photonic crystals. These optical components are formed by structured transparent semiconductors, glasses, or polymers with dielectric structures with a periodicity of refractive index which influence refraction and interference of photons. The dimensions of these structures have the dimensions of wavelengths and are three-dimensional. This is in contrast to interference filters or gratings [125, 126]. They also can be used for Bragg reflection [127]. A review of the recent progress and novel applications of photonic fibers is given for photonic crystal fibers [128]. Different types are discussed and even the terahertz guidance is discussed. An overview of such systems can also be found for various applications to sensing [129]. Even though cantilevers are easily damaged and difficult to set up for field applications, recently, an online portable micro-cantilever biosensor for salmonella has been described [130].

6.4 Assays

For specificity, an analytical method can either use an intrinsic property of the analyte (e.g., IR fingerprint) or specific recognition elements capable of discriminating between the analyte molecule and other compounds In the latter, the specific interaction of a recognition element with an analyte is then transferred into a detectable physical signal change by the transducer element. Together with the developments in the area of biophotonics for on-site environmental analysis, the need for automated assays has lead to the development of new test formats. Usually, these methods combine the principles of various immunoassays or chemical sensing approaches performed on sensor surfaces, taking advantage of microfluidics and different fluorescence or label-free transduction technologies. Current developments in microfluidics and microarray technology are the basis for multiplexed quantitative measurements in the area of environmental on-site analysis [67, 133, 134].

Each assay type or test format, together with the applied detection principle, provides individual properties suited for various applications, as indicated in Section 6.5. The predominant test formats used for environmental analysis are enzyme linked immunosorbent assay (ELISA, see Figure 6.6), replacement assay, sandwich assay, binding inhibition assay, and the direct test format. The selection of the test format, in combination with the transduction principle, has to be driven by the analytical needs of each individual application. In addition, the properties of the recognition element (e.g., antibody, aptamer, scaffold, molecular imprinted polymer, etc.) have to meet the requirements for a given test format. This is not only for the affinity/avidity or selectivity, but is also a question of chemical stability and the diffusion relations in flow injection analysis (FIA) systems [135].

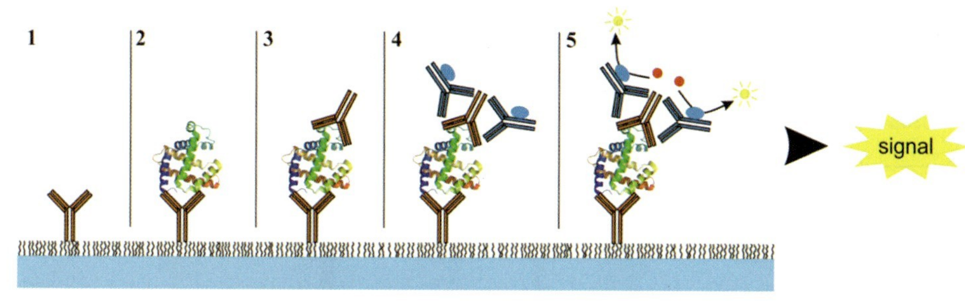

Figure 6.6 ELISA (enzyme linked immunosorbent assay).

The ELISA was developed more than 40 years ago and is one of the best known test formats with a wide range of applications and variations [136]. Because of the stop-flow conditions, an ELISA needs washing steps between the incubation phases. As shown in Figure 6.6 for the most common type of ELISA assay, the first incubation phase starts with a capture antibody-coated surface, which is incubated with the sample. After washing, a detection antibody is applied, followed by an incubation phase and further washing. The next steps produce the read-out signal via a secondary antibody directed towards the detection antibody, and linked with an enzyme as a chromogenic reporter system. There are many commercially available ELISA tests available for a wide range of analytes. However, their usability as on-site tests is limited due to the washing steps, which have to be performed manually or via complex automated liquid handling. Therefore, the developments in the area of optical (bio)sensors included flow injection analysis (FIA) [137–140] systems or microfluidics.

One of the test formats used for environmental immunoassays is the replacement assay (Figure 6.7a). For this testformat the sensor surface is modified with an antibody carrying a labeled analyte analog. During the incubation step the analyte in the sample replaces the bound labeled analog. Replacement assays can handle large and small analytes and should be carried out with an antibody offering a not too high dissociation rate constant for the analyte analog. Usually the replacement assay is combined with fluorescence read out. Limitations of this test format are the stability of the immobilized antibodies during regeneration procedures, which is necessary for a quasi-continuous monitoring.

For the sandwich assay (Figure 6.7b) a capture antibody (or other recognition element) is immobilized onto a sensor surface (heterogeneous phase). The sample can be either incubated with this surface, followed by subsequent incubation with a detection antibody (2 steps), or the surface can be incubated with the mixture of sample and detection antibody (1 step). In the latter case, the detection antibody needs to bind to a different epitope of the analyte than the capture antibody. In general, it is advantageous to select antibodies with low dissociation rates because it is a non-competitive test format. The read-out can be either label-free, for example, SPR (3a in Figure 6.7b) or fluorescence-based (3b in Figure 6.7b) for example, scanning or TIRF. The combination with fluorescence-based detection allows a very low limit of

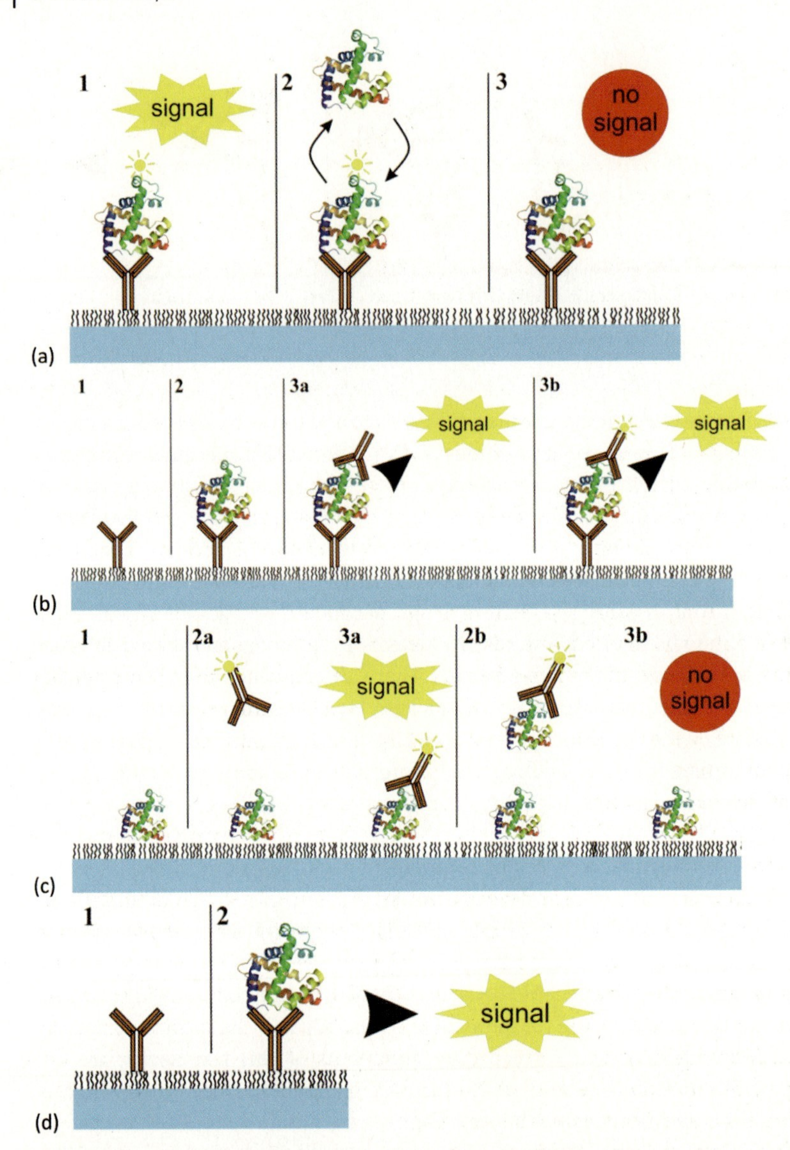

Figure 6.7 (a) Replacement assay, (b) sandwich assay (c) binding inhibition assay (d) direct test format.

detection. Especially for the sandwich assay, there is the additional advantage of very high selectivity because of the involved detection- and secondary antibody. Furthermore, this test format in the labeled mode is preferable when resistance against matrix compounds (e.g., dissolved organic carbon) of environmental samples is important. The only relevant limitation of this test format exists for small molecules which do not offer two epitopes that can be bound by two antibody molecules simultaneously, for example, because of steric hindrance. This is often the case for small chemical compounds, like pesticides or endocrine-disrupting chemicals. Furthermore, the regeneration of the sensor surface is problematic, as mentioned for the replacement assay, because of the low chemical stability of the capture antibody. This might be overcome in the future by using more stable recognition elements (e.g., scaffolds).

A test format which overcomes the limitations described for the sandwich assay in terms of small molecule detection, and to some extent the stability of the sensor surfaces, is the binding inhibition test (Figure 6.7c). A drawback is the complexity of the necessary automation. This is due to the step-wise assay procedure which starts with a pre-incubation step of the antibody with the sample. According to the law of mass action, the paratops can bind to their epitopes. In the subsequent incubation phase each antibody molecule with at least one free binding site can bind to an immobilized antigen derivative at the sensor surface. This test format is non-competitive, because the incubation step with the modified surface is mass transport controlled (due to flow conditions the equilibrium in the bulk solution is not changed) due to a very high surface loading with antigen derivative molecules. If this test format is performed with high affinity antibodies (low-off rates) together with a TIRF set-up for read out, extremely low limits of detection can be achieved. Label-free detection is also applicable and offers the advantage to use recognition elements, which cannot easily be labeled with a fluorescence dye. A further big advantage is the easy adaption towards multianalyte immunoassays for small molecules, when the surface modification is based on a spatially resolved chemistry protocol of antigen derivatives (e.g., derivatives of pesticides or pharmaceuticals). Many of these antigen derivatives are robust towards the chemical conditions of regeneration procedures, which allows several hundreds of measurement cycles with the same sensor surface.

Restricted only to label-free transduction principles, but superior in simplicity, is the direct test format (Figure 6.7d). This non-competitive testformat employs a sensor surface with immobilized recognition elements. In a one-step procedure the sample incubated with this surface and the binding can be directly recorded, for example, with SPR or RIFS. However, there are many critical conditions to be aware of. The antibody, or any other recognition element, should provide very low off-rates (high affinity constants) to the analyte (antigen), and during the immobilization process it is necessary to bind many recognition elements (high surface loading) in such a way that the binding sites are intact and accessible for the analyte. A limitation for any label-free transduction technology is the rather high limit of detection due to the mass sensitivity, which correlates with the molecular weight of the analyte. Furthermore, non-specific binding of matrix compounds to the sensor surface can disturb the measurement, which is the main source of artefacts. However, the direct test format is a very good choice for fast and reliable detection of particles, for example, pathogens and bigger molecules such as allergens even in complex matrices, if combined with sophisticated surface-shielding polymers.

6.5
Applications

6.5.1
Chemical Contaminants, EDCs, Pharmaceuticals and Toxins

Environmental monitoring of chemical contaminants covers a wide field of chemical species and sample matrices, such as air, water or soil. Furthermore, there is a requirement to analyze directly the immissions of industrial processes or effluents from wastewater treatment plants. The focus of this section will be on technologies for on-site analytics of chemical contaminants in ground water and surface water bodies. In general, there are different approaches for the quantification of individual compounds up to effect-directed analytics. Besides the introduced optical detection technologies and assays, the recognition elements play an essential role in the performance of the monitoring system.

Because drinking water, as one of our most important resources, depends on the quality of the above-mentioned water bodies, it is necessary to take into account the European Water Framework Directive (WFD) and the related Marine Strategy Framework Directive which give maximum values for allowed concentrations of a set of chemical contaminants. Furthermore, the list of priority substances (EU decision No. 2455/2001/EC) lists numerous very important chemical contaminants with thresholds. These values are important for water analytical methods, because the method applied has to meet, at least, these standards. For example, the threshold for a single pesticide is at 0.1 ppb (equivalent to $0.1\,\mu g\,l^{-1}$) and for the sum-concentration of pesticides in water samples the threshold is 0.5 ppb. A recently published study clearly indicates the need for long-term monitoring data as one of the future challenges of the WFD implementation [141].

The analytical requirements can become much more ambitious when taking into account some endocrine disrupting chemicals (EDC) – EDCs act like hormones in the endocrine system and disrupt the physiological function of endogenous hormones – the no-effect level in living animals or humans can be much lower. Different studies have linked endocrine disruptors to adverse biological effects in animals [142]. This has led to increasing concerns that low-level exposure might cause similar effects in human beings [143]. Furthermore, the quantification of single compounds does not give an answer to the much more complex situation when mixtures of compounds like EDCs act in a cumulative or amplified way. Aside from synthetic EDCs, naturally occurring toxins, such as the resorcyclic acid lactones (RALs) are also known to have endocrine-disrupting potential. The most prevalent compound of the RALs is the mycotoxin zearalenone (ZON), which is orders of magnitude more potent than notorious synthetic EDCs such as DDT, BPA or atrazine. In addition to the above-described compounds, natural toxins, like saxitoxin from bluegreen algae, are of great importance. Although not official, a threshold concentration for saxitoxin of $3\,\mu g\,l^{-1}$ is under discussion.

Another class of emerging chemical contaminants are pharmaceutical residues, which can be found in drinking water and represent an up-to-date topic in water quality [144]. The human body excretes applied medication in the original or metabolized form. Although wastewater is treated before it is discharged into reservoirs, rivers or lakes, these measures do not remove all drug residues. Even if the concentrations of these pharmaceuticals are of the order of parts per billion or trillion, the presence of many pharmaceuticals causes worries among scientists about long-term consequences to human health. As with EDCs, residues of pharmaceuticals have been proven to cause major ecological damage [145]. Beside antibiotics or ethinylestradiol from the contraceptive pill, there is a long list of pharmaceuticals which can be found in water bodies, including radio-opaque substances, analgesics (e.g., diclophenac) or anti-epileptic and anti-anxiety medications.

Fully automated immunoassays based on the described test formats and hyphenated to optical transduction provide an interesting analytical strategy for on-site quasi-continuous quantification of chemical contaminants [13, 67, 146–148]. These systems should be designed for quick analysis without the need for labor-intensive sample pretreatments or enrichment steps, as is necessary for classical analytical methods. Monitoring systems offering such capabilities are useful additions to classical analytical methods. The development of multi-analyte immunoassays for water monitoring has not only to meet the above-mentioned threshold but also specific requirements related to practicability. For example, a quasi-continuous measurement mode will only be possible with a regeneration of the sensitive surface after each single measurement, combined with resistance toward bio-fouling and degradation. Therefore, it is necessary to create a specific surface coating including the recognition element, which is chemically stable during repeated complete measurement cycles. Especially, the research efforts in developing new robust and chemically stable recognition elements with tunable affinity show very promising results [149].

The replacement assay can look back on a long success story for environmental applications. It allows the quantification of small organic molecules but is limited in its regeneration capabilities for quasi-continuous measurements because the capture antibodies are not chemically stable. Furthermore, the antibodies have to be designed with reasonable dissociation rates to allow the replacement step during the incubation with the analyte-containing sample. On the other hand a too high off-rate will lead to a high degree of dissociation of the labeled analyte derivative. If set up in a proper way, including a flow injection analysis system and fluorescence transduction, it has been demonstrated that the replacement assay works with adequate LODs for pesticides [150, 151]. Variations of the competitive test format with fluorescence can also be found in the literature [152].

It was shown that the PASA system can handle the different test formats. For the detection of pesticides an indirect competitive immunoassay was combined with chemiluminescence detection. This test format is related to the binding inhibition assay performed under competitive conditions and allows multi-analyte detection with regeneration capabilities for monitoring applications [153]. The same system was used to quantify antibiotics [154].

The binding inhibition test performed under non-competitive conditions was used by the systems RIANA [155] and AWACSS [156]. This test format meets the analytical and performance requirements described above. The sensor surface is modified with an analyte derivative or with the hapten-carrier protein-complex used during antibody production. The advantage is the chemical stability of these molecules compared to antibodies. To achieve chemically stable sensitive layers it is advantageous to covalently attach these molecules to the transducer surface in combination with a biopolymer and standard microarray printing procedures for multi-analyte capability. The incubation phase can be performed, even with a mixture of specific antibodies, in one step, as for the indirect competitive assay. Although cross reactivity of the antibodies has to be taken into account it has been shown, that it is possible to use this test format in a fully automated TIRF biosensor for the parallel quantification of six analytes from different chemical classes (pesticides, pharmaceuticals and EDCs) [157]. For the AWACSS biosensor fully automated and unattended monitoring has been realized for in-field testing [158]. Together with the implemented intelligent communication to a data base, the system is able to monitor thresholds recommended by the EU legislation.

The binding inhibition test can also be combined with label-free read-out. For example Ref. [159] showed a SPR-based biosensor for the detection of bisphenol A in water samples. Reference [160] reports the detection of atrazine with a similar approach. The advantage compared to fluorescence-based technologies is obviously the prevention of the antibody labeling step. However, the achievable limit of detection is above labeled systems. The simpler direct test format, in combination with label-free read-out, is not a preferable choice due to the necessary low LODs and the low molecular weights of many analytes such as pesticides. Furthermore, antibody surfaces are not stable during the regeneration procedures. The same applies for the sandwich assay, which fails for these applications because of the small target molecules, offering in many cases only one accessible epitope. A completely different situation applies for several toxins, such as microcystin LR. This relatively large molecule is rather unstable when immobilized to a sensor surface. Therefore, the direct test format has been successfully used in combination with SPR transduction to quantify this toxin [105, 161]. Other toxins, such as botulinum, ricin or cholera toxin are especially interesting for security reasons. These molecules can be quantified in a multi-analyte approach using the sandwich immunoassay in combination with a TIRF-based detection [162].

Employing nuclear receptors as the analytical tool enables a new analytical strategy called effect-based analytics. Compared to conventional immunoassay-based biosensor technology the advantage is that the total effect on the nuclear receptor can be measured, instead of concentrations of single compounds, and that even currently unknown ligands are also found. Usually, the measured concentrations are given as hormone-equivalents, for example, as estrogen-equivalents in the case of the estrogen receptor. TIRF read-out, combined with the binding inhibition assay [163] employing the ERα-LBD as the analytical tool, is a promising approach for a next generation biosensor technology to be used in environmental monitoring. The recognition structure is derived from a natural nuclear receptor, which mediates the effects of estrogenic EDCs *in vivo*. Therefore, the assay gives a measure of the exogeneous influences on the hormone system, expressed in estrogen equivalents.

For the future it is important to develop ERα-LBD based assays which are able to distinguish between estrogenic and anti-estrogenic effects. A first step in this direction was reported in [164]. The developed label-free assay can quantify these different effects. The application of this knowledge for the development of a new set of TIRF assays can lead to high-sensitivity effect-based environmental assays for EDCs with estrogenic effects for the first time.

This problem can be overcome by introducing cell-based assays for effect-based analytics. However, the practicality differs extremely with the used cell-lines/bacteria [165]. Mammalian cell lines can report effects with a very high sensitivity, as is known from toxicity studies, for example, in pharma screening. However, the stability of these cells in monitoring systems is low and the handling needs special and expensive cultivation environments. Bacteria with reporter systems, such as the green fluorescence protein, are more robust and easy to use. Although there are commercial products already on the market [166], the comparability to quantifying analytical methods is still an ongoing discussion.

6.5.2 Pathogens

The detection of pathogens in water samples is a very challenging analytical problem. Following the Drinking Water Directive it is necessary to detect one single colony-forming unit (cfu) in 100 ml (the directive threshold value e.g., for *E. coli* is 0 per 100 ml). The classical test for bacteria is based on filtration and cultivation, as has been state-of-the-art for more than 100 years. This time- and labor-intensive method has the advantage that it is very reliable in detecting only viable bacteria. For virus particles the situation is even more complicated, because the virus particles cannot replicate autonomously.

One of the most promising strategies is nucleic acid-based analytical methods, which make use of the polymerase chain reaction (PCR) to amplify genomic DNA or RNA molecules. This technology offers the advantage that quantification of very low copy numbers is possible in a multi-analyte approach. However, the usability is also limited when the extremely low thresholds of the drinking water directive have to be applied. Furthermore, the technology can lead to false positive results if nucleic acid of dead pathogens is amplified. In practice, the analytical procedure is complex and time consuming and starts with cell lysis followed by nucleic acid extraction and PCR amplification. The detection is usually based on fluorescence measurements and can be performed in, for example, a sandwich hybridization format on microarrays, as a real-time PCR experiment, flow cytometry-based or as a fiber sensor array [24, 167–170].

Another strategy makes use of mammalian cell lines, which are sensitive to the presence of pathogens. In [171] it has been shown that several bacterial pathogens can be detected at very low cfu concentrations. A big advantage of this method is that only living pathogens are detected. However, the analytical procedure is very complicated and time-consuming. The introduction of immunoassays to pathogen detection in water samples is linked to some fundamental challenges: the need for extremely low

LODs, and the very low diffusion rate of these pathogens/particles during the incubation phase of an assay. Therefore, it is mandatory to include sample pre-concentration steps prior to the analytical procedure. Recently it has been shown that special filtration procedures can help to overcome these problems by a quick pre-concentration of the sample by a factor of 200 [172]. A further problem is the availability of antibodies or other recognition elements which allow species-specific detection, because of the high cross-reactivity between the groups or families of pathogens. Typically the applied test format is the sandwich immunoassay. This test format is especially helpful in increasing the specificity by the detection antibody. In a multi-analyte approach it is possible to detect several pathogens in parallel in microarray- or bead-based approaches [53, 54, 173–178]. It has also been shown that with a stopped-flow immunoassay it is possible to detect viable *E. coli* more effectively, because of the reduction of the above-mentioned hydrodynamic restrictions [179], or to combine detection and quantification by a nanoparticle-based sandwich ELISA [19].

Recently many strategies have been developed to enhance the performance of immunoassays for pathogen detection in water samples. One example is the liposome-based colorimetric sensor using a novel immobilization procedure [180] which allows the multiplexed detection of six pathogens. An overview of nanomaterial-enabled biosensors for pathogen detection employing different recognition elements, such as antibodies, aptamer carbohydrates, or even antimicrobial peptides can be found in [181]. Instrumental techniques for direct and indirect identification of bacteria, such as IR, flow cytometry, chromatography, and others have also been used to address this analytical task [182].

Label-free transduction can also be successfully used for pathogen detection in water samples. Different test formats have been described so far, but the most interesting is of course the direct test format. Here the detection of the pathogens is performed by a one step assay via a capture antibody, which is immobilized on the transducer surface [183]. The relatively low LOD of label-free methods is partially compensated by the enormous size of the pathogens compared to small molecules like pesticides [184, 185]. However, the limited specificity of the direct test format allows a quick first screening rather than a precise identification. This limitation can be overcome by methods which combine an immunoassay with the advantages of SERS. This detetction principle gives added value to the pathogen detection due to the species-specific SERS spectra as a whole-organism fingerprint [186–189]. A further elegant method for the identification of bacteria is provided by a detection principle that makes use of the different scattering properties of the pathogens. This sensor technology also enables the label-free detection of multiple bacterial pathogens [190].

6.6 Perspectives and Visions

Normally, just a specific interaction is used to monitor pollutants. However, some years ago so-called Quorum sensing [192] was introduced to environmental analysis to direct conventional analytics away from measuring the concentration of specific compounds to an effect-based "quantification", as in a dose measurement known in radioactivity control [131, 132]. The influence of some pollutants on health does not depend on concentration but rather on, for example, endocrine activity. Thus, this new approach of effect-based "quantification" opens some perspectives for better analysis. Another approach is to use spores in genetically engineered bacteria-based sensing systems [191] using their biospecific recognition capability. Recent activities in photonics will also lead to new instrumentation [193].

References

1 Skoog, D.A., Holler, F.J., and Crouch, S.R. (2007) *Principles of Instrumental Analysis*, 6th edn, Brooks/Cole Publishing Company, Pacific Grove.
2 Radojević, M. and Bashkin, V.N. (2009) *Practical Environmental Analysis*, Royal Society of Chemistry, Cambridge.
3 Alloway, B.J. and Ayres, D.C. (1993) *Chemical Principles of Environmental Pollution*, Blackie Academic & Professional, London.
4 Fifield, F.W. and Haines, P.J. (2000) *Environmental Analytical Chemistry*, John Wiley & Sons.
5 Capdeville, M.J. and Budzinski:, H. (2011) Trace-level analysis of organic contaminants in drinking waters and groundwaters. *TrAc, Trends Anal. Chem.*, **30**, 586–606.
6 Urban Waste Water Treatment Directive (91/271/EEC).
7 Water Framework Directive (2000/60/EC).
8 Community Strategy for Endocrine Disrupters COM (99)706, COM (2001) 262.
9 Regulation (EC) No 178/2002 of the European Parliament and of the Council of 28 January 2002 laying down the general principles and requirements of food law, establishing the European Food Safety Authority and laying down procedures in matters of food safety, OJ L 31, 1.2.2002, p. 1–24.
10 Commission Regulation (EC) No 1881/2006 of 19 December 2006 setting maximum levels for certain contaminants in foodstuffs, OJ L 364, 20.12.2006, p. 5.
11 http://water.epa.gov/scitech/methods/cwa/bioindicators/biological_index.cfm, http://water.epa.gov/scitech/drinkingwater/labcert/analyticalmethods.cfm#approved.
12 http://www.dvgw.de/fileadmin/dvgw/wasser/recht/1104gerhardy.pdf.
13 Namour, P., Lepot, M., and Jaffrezic-Renault, N. (2010) Recent trends in monitoring of European Water Framework Directive priority substances using micro-sensors: A 2007–2009 review. *SENSORS*, **10**, 7947–7978.
14 Farre, M., Kantiani, L., Perez, S., and Barcelo, D. (2009) Sensors and biosensors in support of EU directives. *TrAC, Trends Anal. Chem.*, **28**, 170–185.
15 EPA Method 1682: Salmonella spp. in Biosolids by Enrichment, Selection, and Biochemical Characterization http://water.epa.gov/scitech/methods/cwa/bioindicators/upload/2008_11_25_methods_method_biological_1682-bio.pdf.
16 Ferguson, C., Kaucner, C., Krogh, M., Deere, D., and Warnecke, M. (2004) Comparison of methods for the concentration of Cryptosporidium oocysts and Giardia cysts from raw waters. *Can. J. Microbiol.*, **50**, 675–682.
17 Morales-Morales, H.A., Smith, G.B., Vidal-Martinez, G. *et al.* (2003) Optimization of a reusable hollow-fiber ultrafilter for simultaneous concentration of enteric bacteria, protozoa, and viruses from water. *Appl. Environ. Microbiol.*, **69**, 4098–4102.
18 Peskoller, C., Niessner, R., and Seidel, M. (2009) Cross-flow microfiltration

system for rapid enrichment of bacteria in water. *Anal. Bioanal. Chem.*, **393**, 399–404.

19 Pappert, G., Rieger, M., Niessner, R., and Seidel, M. (2010) Immunomagnetic nanoparticle-based sandwich chemiluminescence-ELISA for enrichment and quantification of *E. coli. Microchem. Acta*, **168**, 1–8.

20 Peskoller, C., Niessner, R., and Seidel, M. (2009) Development of an epoxy-based monolith used for the affinity capturing of *Eschericha coli* bacteria. *J. Chromatogr. A*, **1216**, 3794–3801.

21 Aprodu, I., Walcher, G., and Schelin, J. (2011) Advanced sample preparation for the molecular quantification of Staphylococcus aureus in artificially and naturally contaminated milk. *Int. J. Food Microbial.*, **145**, S61–S65.

22 Krämer, N., Löfström, Ch., Vigre, H., Hoorfar, J., Bunge, C., and Mlorny, B. (2011) A novel strategy to obtain quantitative data for modelling: Combined enrichment and real-time PCR for enumeration of salmonellae from pig carcasses. *Int. J. Food Microbiol.*, **145**, S86–S95.

23 Reidt, U., Geisberger, B., Heller, C., and Friedberger, A. (2011) Automated immunomagnetic processing and separation of *Legionella pneumophila* with manual detection by sandwich ELISA and PCR amplification of the ompS Gene. *J. Lab. Automation*, **16**, 157–164.

24 Donhauser, S., Niessner, R., and Seidel, M. (2011) Sensitive quantification of Escherichia coli O157:H7, salmonella enterica, and campylobacter jejuni by combining stopped polymerase chain reaction with chemiluminescence flow-through DNA microarray analysis. *Anal. Chem.*, **83**, 3153–3160.

25 Advances in label-Free Detection Technologies, Frost&Sullivan, June30, 2008 – Pub ID: MC2018124 (www.marketresearch.com/product/display.asp).

26 Tallury, P., Malhotra, A., Byrne, L.M., and Santra, S. (2010) Nanobioimaging and sensing of infectious diseases. *Adv. Drug Deliv. Rev.*, **62**, 424–437.

27 [http://www.eugris.info/FurtherDescription.asp?Ca=2&Cy=0&T=Emerging%20Pollutants&e=95].

28 Ligler, F. and Taitt, Ch.R. (eds) (2008) *Optical Biosensors: Today and Tomorrow*, Elsevier, Amsterdam.

29 Borisov, S.M. and Wolfbeis, O.S. (2008) Optical biosensors. *Chem. Rev.*, **108**, 423–461.

30 Nagl, S., Wolfbeis, O.S., and Otto, S. (2008) Classification of chemical sensors and biosensors based on fluorescence and phosphorescence. in Springer Series on Fluorescence, Vol. **5** *Standardization and Quality Assurance in Fluorescence Measurements I*, 325–346, Springer.

31 Stich, M.I.J., Fischer, L.H., and Wolfbeis, O.S. (2010) Multiple fluorescent chemical sensing and imaging. *Chem. Soc. Rev.*, **39**, 3102–3114.

32 Gauglitz, G. (2005) Direct optical sensors: principles and selected applications. *Anal. Bioanal. Chem.*, **381**, 141–155.

33 Fan, X.-D., White, I.M., Shopova, S.I., Zhu, H.-Y., Suter, J.D., and Sun, Y.-Z. (2009) Sensitive optical biosensors for unlabeled targets: A review. *Anal. Chim. Acta*, **620**, 8–26.

34 Gauglitz, G. (2010) Direct optical detection in bioanalysis: an update. *Anal. Bioanal. Chem.*, **398**, 2363–2372.

35 van de Hulst, H.C. (1957) *Light Scattering by Small Particles*, DoverPublications, New York.

36 Nobbmann, U. (2003) Characterization down to nanometers: Light scattering from proteins and micelles in Mesoscale phenomena in fluid systems. *ACS Symp. Ser.*, **861**, 44–59.

37 Bergquist, P.L., Hardiman, E.M., Ferrari, B.C., and Winsley, T. (2009) Applications of flow cytometry in environmental microbiology and biotechnology. *Extremophiles*, **13**, 389–401.

38 Comas-Riu, J. and Rius, N. (2009) Flow cytometry applications in the food industry. *J. Ind. Microbiol. Biotechnol.*, **36**, 999–1011.

39 Rajwa, B., Dundar, M.M., Akova, F., Bettasso, A., Patsekin, v., Hirleman, E.D., Bhunia, A.K., and Robinson, J.P. (2010) Discovering the unknown: detection of emergin pathogens using a label-free light-scattering system. *Cytometry, Part A*, 77, 1103–1112.

40 http://www.nanosight.com/technology/nanosights-technology.

41 Bhunia, A.K., Banada, P., Banerjee, P., Valadez, A., and Hirleman, E.D. (2007) Light scattering fiber optic- and cell-based sensors for sensitive detection of foodborne pathogens. *J. Rapid Methods Autom. Microbiol.*, **15**, 121–145.

42 Llorent-Martínez, E.J., Ortega-Barrales, P., Fernández-de Córdova, M.L., and Ruiz-Medina, A. (2011) Trends in flow-based analytical methods applied to pesticide detection: A review. *Anal. Chim. Acta*, **684**, 30–39.

43 Förster, Th. (1948) Zwischenmolekulare Energiewanderung und Fluoreszenz. *Ann. Physik*, **2**, 55–75.

44 Van der Meer, B.W., Coker, G. III, and Chen, S. (eds) (1994) *Resonance Energy Transfer – Theory and Data*, VCH Publishers Inc., Weinheim.

45 Roda, A., Guardigli, M., Michelini, E., and Mirasoli, M. (2007) Nanobioanalytical luminescence: Förster-type energy transfer methods. *Anal. Bioanal. Chem.*, **393**, 109–123.

46 Varghese, S., Zhu, Y., Davis, T.J., and Trowell, S.C. (2010) FRET for lab-on-a-chip devices — current trends and future prospects. *Lab Chip*, **10**, 1355–1364.

47 Algar, W.R. and Krull, U.J. (2010) New opportunities in multiplexed optical bioanalyses using quantum dots and donor-accceptor interactions. *Anal. Bioanal. Chem.*, **398**, 2439–2449.

48 Daunert, S. and Deo, S.K. (eds) (2006) *Photoproteins in Bioanalysis, Chapter 6: Bioluminescence Resonance Energy Transfer in Bioanalysis*, Wiley-VCH GmbH, Weinheim.

49 Park, J., Park, S., and Kim, Y.-K. (2010) Multiplex detection of pathogens using an immunochromatographic assay strip. *BioChip J.*, **4**, 305–312.

50 Chouhan, R.S., Vivek Babu, K., Kumar, M.A., Neeta, N.S., Thakur, M.S., Amitha, B.E., Pasha, A., Karanth, N.G.K., and Karanth, N.G. (2006) Detection of methyl parathion using immuno-chemiluminescence based image analysis using charge coupled device. *Biosens. Bioelectron.*, **21**, 1264–1271.

51 Roda, A., Guardigli, M., Michelini, E., Mirasoli, M., and Pasini, P. (2003) The possibility of detecting a few molecules using bioluminescence and chemiluminescence is exciting, especially in the context of miniaturized analytical devices. *Anal. Chem.*, **75**, 462A–470A.

52 Aslan, K. and Geddes, Ch.D. (2009) Metal-enhanced chemiluminescence: advanced chemiluminescence concepts for the 21st century. *Chem. Soc. Rev.*, **38**, 2556–2564.

53 Karsunke, X.Y.Z., Niessner, R., and Seidel, M. (2009) Development of a multichannel flow-through chemiluniescence microarray chip for parallel calibration and detection of pathogenic bacteria. *Anal. Bioanal. Chem.*, **395**, 1623–1630.

54 Wolter, A., Niessner, R., and Seidel, M. (2008) Detection of *Escherichia coli* O157: H7, *Salmonella typhimurium*, and *Legionella pneumophila* in water using a flow-through chemiluminescence microarray read-outsystem. *Anal. Chem.*, **80**, 5854–5863.

55 Hecht, E. (2001) *Optics*, Pearson Addison Wesley, München.

56 Gauglitz, G. and Nahm, W. (1991) Observation of spectral interferences for the determination of volume and surface effects of thin polymer films. *Fresenius' J. Anal. Chem.*, **341**, 279–283.

57 Bürk, J., Conzen, J.P., and Ache, H.J. (1992) A fiber optic evanescent field absorption sensor for monitoring contaminants in water. *Fresenius' J. Anal. Chem.*, **342**, 421–430.

58 Bürk, J., Zimmermann, B., Mayer, J., and Ache, H.J. (1996) Integrated optical NIR-evanescent wave absorbance sensor for chemical analysis. *Fresenius' J. Anal. Chem.*, **354**, 284–329.

59 Gauglitz, G. (2010) Direct optical detection in bioanalysis: an update. *Anal. Bioanal. Chem.*, **398**, 2363–2372.

60 Hunt, H.K. and Armani, A.M. (2010) Label-free biological and chemical sensors. *Nanoscale*, **2**, 1544–1559.

61 Mehne, J., Markovic, G., Pröll, F., Schweizer, N., Zorn, S., Schreiber, F., and Gauglitz, G. (2008) Characterisation of morphology of self-assembled PEG monolayers: a comparison of mixed and pure coatings for biosensor

applications. *Anal. Bioanal. Chem.*, **391**, 1783–1791.

62 Albrecht, C., Fechner, P., Honcharenko, D., Baltzer, L., and Gauglitz, G. (2010) A new assay design for clinical diagnostics based on alternative recognition element. *Biosens. Bioelectron.*, **25**, 2302–2308.

63 Chudy, M., Grabowska, I., Ciosek, P., Filipowicz-Szymanska, A., Stadnik, D., Wyzkiewicz, I., Jedrych, E., Juchniewicz, M., Skolimowski, M., Ziolkowska, K., and Kwapiszewski, R. (2009) Miniaturized tools and devices for bioanalytical applications: an overview. *Anal. Bioanal. Chem.*, **395**, 647–668.

64 Bally, M., Halter, M., Vörös, J., and Grandin, H.M. (2006) Optical microarray biosensing techniques. *Surf. Interface Anal.*, **38**, 1442–1458.

65 Yu, X.-B., Xu, D.-K., and Cheng, Q. (2006) Label-free detection methods for protein microarrays. *Proteomics*, **6**, 5493–5503.

66 Lopez, C.A., Daaboul, G.G., Vedula, R.S., Özkumur, E., Bergstein, D.A., Geisbert, T.W., Fawcett, H.E., Goldberg, B.B., Connor, J.H., and Ünlü, M.S. (2011) Label-free multiplexed virus detection using spectral reflectance imaging. *Biosens. Bioelectron.*, **26**, 3432–3437.

67 Seidel, M. and Niessner, R. (2008) Automated analytical microarrays: a critical review. *Anal. Bioanal. Chem.*, **391**, 1521–1544.

68 Schäferling, M. and Nagl, S. (2006) Optical technologies for the read-outand quality control of DNA and protein mircoarrays. *Anal. Bioanal. Chem.*, **385**, 500–517.

69 Rasooly, A. (2008) Food microbial pathogen detection and analysis using DANN microarray technologies. *Foodborne Pathog. Dis.*, **5**, 531–550.

70 Chandra, H., Reddy, P.J., and Srivastava, S. (2011) Protein microarrays and novel detection platforms. *Expert Rev. Proteomics*, **8**, 61–79.

71 Ivnitski, D., Abdel-Hamid, I., Atanasov, P., and Wilkins, E. (1999) Biosensors for detection of pathogenic bacteria. *Biosens. Bioelectron.*, **14**, 599–624.

72 Medintz, I.L., Uyeda, H.T., Goldman, E.R., and Mattoussi, H. (2005) Quantum dot bioconjugates for imaging, labelling and sensing. *Nature Materials*, **4**, 435–446.

73 Sanvicens, N., Pastells, C., Pascual, N., and Marco, M.-P. (2009) Nanoparticle-based biosensors for detection of pathogenic bacteria. *TrAC, Trends Anal. Chem.*, **28**, 1243–1252.

74 Wang, L.-B., Ma, W., Xu, L.-G., Chen, W., and Zhu, Y.-Y. (2010) Nanoparticle-based environmental sensors. *Mater. Sci. Eng., R*, **70**, 265–274.

75 Asefa, T., Duncan, C.T., and Sharma, K.K. (2009) Recent advances in nanostructured chemosensors and biosensors. *Analyst*, **134**, 1980–1990.

76 Nichkova, M., Dosev, D., Gee, S.J., Hammock, B.D., and Kennedy, M. (2005) Microarray immunoassay for phenoxybenzoic acid using polymer encapsulated Eu: Gd2O3 nanoparticles as fluorescent labels. *Anal. Chem.*, **77**, 6864–6873.

77 http://www.biocompare.com/ProductCategories/192/Microarrays.html.

78 Sapsford, K.E. and Ligler, F.S. (2004) TIRF array biosensors for environmental monitoring. in Springer Series on Chemical Sensors and Biosensors, Vol. **1** *Optical Sensors* 359–439, Springer.

79 Nagl, S., Schaeferling, M., and Wolfbeis, O.S. (2005) Fluorescence analysis in microarray technology. *Microchim. Acta*, **151**, 1–21.

80 Yu, F., Yao, D.F., and Knoll, W. (2003) Surface plasmon field-enhanced fluorescence spectroscopy studies of the interaction between an antibody and its surface-coupled antigen. *Anal. Chem.*, **75**, 2610–2617.

81 Liebermann, T. and Knoll, W. (2003) Parallel multispot detection of target hybridization to surface-bound probe oligonucleotides of different base mismatch by surface-plasmon field-enhanced fluorescence microscopy. *Langmuir*, **19**, 1567–1572.

82 Tallury, P., Malhotra, A., Byrne, L.M., and Santra, S. (2010) Nanobioimaging and sensing of infectious diseases. *Adv. Drug Deliv. Rev.*, **62**, 424–437.

83 Hecht, E. (1997) *Optics*, 3rd edn, Addison Wesley, London.

84 Vo-Dinh, T. and Allain, L. (2003) Biosensors for medical applications, in *Biomedical Photonics* (ed. V.-D. Tuan), CRC Press, Boca Raton, 20/1-20/40.

85 Cooper, M.A. (2002) Optical biosensors in drug discovery. *Nat. Rev. Dru. Disc.*, **1**, 515–528.

86 Clerc, D. and Lukosz, W. (1994) Integrated optical output grating couper as biochemical sensor. *Sens. Actuator, B*, **19**, 581.

87 Dubendorfer, J. and Kunz, R.E. (1998) Compact integrated optical immunosensor using replicated chirped grating coupler sensor chips. *Appl. Opt.*, **37**, 1890.

88 Grego, S., McDaniel, J.R., and Stoner, B.R. (2008) Wavelength interrogation of grating-based optical biosensors in the input coupler configuration. *Sensor Actuator B*, **131**, 347–355.

89 Cush, R., Cronin, J.M., Stewart, W.J. *et al.* (1993) The resonant mirror – a novel optical biosensor for sensing of biomolecular interaction. 1. Principle of operation and associated instrumentation. *Biosens. Bioelectron.*, **8**, 347–353.

90 Goddard, N.J., Pollard-Knight, D., and Maule, C.H. (1994) Real-time biomolecular interaction analysis using the resonant mirror sensor. *Analyst*, **119**, 583–588.

91 de Tommasi, E., De Stefano, L., Rea, I., Di Sarno, V., Rotiroti, L., Arcari, P., Lamberti, A., Sanges, C., and Rendina, I. (2008) Porous silicon based resonant mirrors for biochemical sensing. *Sensors*, **8**, 6549–6556.

92 Kretschmann, E. and Raether, H. (1968) Radiative decay of non radiative surface plasmons excited by light. *Z. Naturforsch. A*, **23**, 2135–2136.

93 Liedberg, B., Nylander, C., and Lundström, I. (1983) Surface plasmon resonance for gas detection and biosensing. *Sens. Actuators*, **4**, 299–304.

94 Liedberg, B., Nylander, C., and Lundstrom, I. (1995) Biosensing with surface plasmon resonance–how it all started. *Biosens. Bioelectron.*, **10**, i–ix.

95 Frank, R. (2005) *Reflektometrische und integriert optische Sensoren für die Bioanalytik*, Dissertation Eberhard-Karls-Universität Tübingen, http://tobias-lib.uni-tuebingen.de/volltexte/2005/2095/pdf/Diss_Ruediger_Frank.pdf.

96 Laib, Th. (2008) Anwendung moderner Auswerteverfahren in der Chemometrie und Geostatistik, Definition von Kenngrößen von Bioanalytsystemen und Charakterisierung eines Mach-Zehnder-Interferometers für die Bioanalytik, Dissertation, Eberhard-Karls-Universität Tübingen, http://tobias-lib.uni-tuebingen.de/volltexte/2008/3471/pdf/Anwendung_ moderner_Verfahren_in_der_Chemometrie_und_ Geostati.pdf.

97 Washburn, A.L. and Bailey, R.C. (2011) Photonics-on-a-chip: recent advances in integrated waveguides as enabling detection elements for real-world, lab-on-a-chip biosensing applications. *Analyst*, **136**, 227–236.

98 http://www.biacore.com/lifesciences/company/index.html.

99 Homola, J., Yee, S.S., and Gauglitz, G. (1999) Surface plasmon resonance sensors. *Sens. Actuators, B*, **54**, 3–15.

100 Petz, M. (2009) Recent Applications of surface plasmon resonance biosensors for analyzing residues and contaminants in food. *Monatsh. Chem.*, **140**, 953–964.

101 Dostalek, J., Pribyl, J., Homola, J., and Skladal, P. (2007) Multichannel SPR biosensor for detection of endocrine-disrupting compounds. *Anal. Bioanal. Chem.*, **389**, 1841–1847.

102 Situ, C., Buijs, J., Mooney, M.H., and Elliott, C.T. (2010) Advances in Surface Plasmon resonance biosensor technology towards high-throughput, food-safety analysis. *TrAC, Trends Anal. Chem.*, **29**, 1305–1315.

103 Fernandez, F., Hegnerova, K., Piliarik, M., Sanchez-Baeza, F., Homola, J., and Marco, M. (2011) A label-free and portable multichannel surface plasmon resonance immunosensor for on site analysis of antibiotics in milk samples. *Biosens. Bioelectron.*, **26**, 1231–1238.

104 Zordan, M.D., Grafton, M.M.G., and Leary, J.F. (2011) An integrated microfluidic biosensor for the rapid screening of foodborne pathogens by

surface plasmon resonance imaging. *Proc. SPIE*, **7888** (Frontiers of Biological Detection) 1–10.

105 Herranz, S., Bockova, M., Marazuela, M.D., Homola, J., and Moreno-Bondi, M.C. (2010) An SPR biosensor for the detection of microcystins in drinking water. *Anal. Bioanal. Chem.*, **398**, 2625–2634.

106 Huang, Ch.-J., Dostalek, J., Sessitsch, A., and Knoll, W. (2011) Long-range surface plasmon-enhanced fluorescence spectroscopy biosensor for ultrasensitive detection of E. coli O157:H7. *Anal. Chem.*, **83**, 674–677.

107 Liebermann, T. and Knoll, W. (2003) Parallel multispot detection of target hybridization to surface-bound probe oligonucleotides of different base mismatch by surface-plasmon field-enhanced fluorescence microscopy. *Langmuir*, **19**, 1567–1572.

108 Azzam, R.M.A. and Bashara, N.M. (1989) *Ellipsometry and Polarised Light*, North Holland, Amsterdam.

109 Jin, G., Tengvall, P., Lundstrom, I., and Arwin, H. (1995) A biosensor concept based on imaging ellipsometry for visualization of biomolecular interactions. *Anal. Biochem.*, **232**, 69–72.

110 Gauglitz, G. and Nahm, W. (1991) Observation of spectral interferences for the determination of volume and surface effects of thin polymer films. *Fresenius' J. Anal. Chem.*, **341**, 279–283.

111 Brecht, A., Gauglitz, G., and Nahm, W. (1992) Interferometric measurements used in chemical and biochemical sensors. *Analusis*, **20**, 135–140.

112 Frank, R., Möhrle, B., Fröhlich, D., and Gauglitz, G. (2005) A transducer-independent optical sensor system for the detection of biochemical binding reactions. *Proc. SPIE*, **5993**, 49–59.

113 Gauglitz, G., Brecht, A., Reichl, D., and Seemann, J. (1999) Apparatus and methods for detection using physical or chemical effects. Ger. Offen., DE 19830727 A1 19990114.

114 Pröll, F., Markovic, G., Schweizer, N., and Gauglitz, G. (2008) Imaging reflectometric interference spectroscopy (iRIfS): a versatile tool for high throughput biomolecular interaction analysis. Proceedings of The TenthWorld Congress on Biosensors, Shanghai, China.

115 Brecht, A. and Gauglitz, G. (1997) Label free optical immunoprobes for pesticide detection. *Anal. Chim. Acta*, **347**, 219–233.

116 Fechner, P., Pröll, F., Albrecht, C., and Gauglitz, G. (2011) Kinetic analysis of the estrogen receptor alpha using RIfS. *Anal. Bioanal. Chem.*, **400**, 729–735.

117 Schluecker, S. (ed) (2011) *Surface Enhanced Raman Spectroscopy: Analytical, Bioanalytical and Life Science Applications*, Wiley VCH Verlag, Weinheim, p. 331.

118 Vo-Dinh, T. (1995) Surface-enhanced raman scattering, in *Photonic Probes of Surfaces* (ed. P. Halevi), Elesevier Scince.

119 http://www.enwaveopt.com/EZRaman-MSeries.htm?gclid=CKPA4t_6nqkCFUqIzAodDTyZuA.

120 http://www.ramansystems.com/?gclid=CPvNwbX7nqkCFUe_zAodmWactw.

121 Popp, J. and Mayerhöfer, Th. (2010) Surface-enhanced Raman spectroscopy. (*Anal. Bioanal. Chem.* **394**, 1717–1718.

122 Álvarez-Puebla, R.A. and Liz-Marzán, L.M. (2010) Environmental applications of plasmon assisted Raman scattering. *Energy Environ. Sci.*, **3**, 1011–1017.

123 Bossecker, A., Popp, J., Meisel, S., Roeschl, P., and Stoeckel, St. (2010) Tracing of pathogenic bacteria by light. Raman spectroscopic detection of bacteria in complex matrices. *Bioforum*, **33**, 22–24.

124 Angela, W., Maerz, A., Schumacher, W., Roesch, P., and Popp, J. (2011) Towards a fast, high specific and reliable discrimination of bacteria on strain level by means of SERS in a microfluidic device. *Lab on a Chip*, **11**, 1013–1021.

125 Zengerle, R. (1987) Light propagation in singly and doubly periodic planar waveguides. *J. Mod. Opt.*, **34**, 1589–1617.

126 Sakoda, K. (2005) *Optical Properties of Photonic Crystals, Springer Series in Optical Sciences*, 2nd edn, vol. **80**, Springer-Verlag GmbH, Heidelberg, XIV, p. 253.

127 Kashyap, R. (1999) *Fiber Bragg Gratings*, Academic Press, San Diego.

128 Cerqueira S. Jr., A. (2010) Recent progress and novel applications of photonic crystal fibers. *Rep. Prog. Phys.*, **73**, 1–21. doi: 10.1088/0034-4885/73/2/024401

129 Nair, R.V. and Vijaya, R. (2010) Photonic crystal sensors: an overview. *Prog. Quantum Electron.*, **34**, 89–134.

130 Ricciardi, C., Canavese, G., Castagna, R., Digregorio, G., Ferrante, I., Marasso, S.L., Ricci, A., Alessandria, V., Rantsiou, K., and Cocolin, L.S. (2010) Online portable microcantilever biosensors for *Salmonella enteric* serotype *Enteritidis* detection. *Food Bioprocess. Technol.* doi: 10.1007/s11947-010-0362-0

131 Fechner, P., Gauglitz, G., and Gustafsson, J.A. (2010) Nuclear receptors in analytics - a fruitful joint venture or a wasteful futility? *TrAC*, **29**, 297–305.

132 Fechner, P., Pröll, F., and Carlquist, M. (2009) An advanced biosensor for the prediction of estrogenic effects of endocrine-disrupting chemicals on the estrogen receptor alpha. *Anal. Bioanal. Chem.*, **393**, 1579–1585.

133 Henares, T.G., Mizutani, F., and Hisamoto, H. (2008) Current development in microfluidic immunosensing chip. *Anal. Chim. Acta*, **611**, 17–30.

134 Borrebaeck, C.A.K. (2000) Antibodies in diagnostics – from immunoassays to protein chips. *Immunol. Today*, **21**, 379–382.

135 Ruzicka, J. and Hansen, E.H. (1975) Flow injection analysis. Part I. A new concept of fast continuous flow analysis. *Anal. Chim. Acta*, **78**, 145–152.

136 Price, C.P. and Newman, D.J. (1997) *Principles and Practice of Immunoassay*, Stockton Press, New York.

137 Fintschenko, Y. and Wilson, G.S. (1998) Flow injection immunoassay: a review. *Mikrochim. Acta*, **129**, 7–18.

138 Hansen, E.H. (1996) Principles and applications of flow injection analysis in biosensors. *J. Mol. Recognit.*, **9**, 316–325.

139 Cerda, V., Estela, J.M., Forteza, R., Cladera, A., Becerra, E., Altimira, P., and Sitjar, P. (1999) Flow techniques in water analysis. *Talanta*, **50**, 695–705.

140 Krämer, P. and Schmid, R. (1991) Flow Injection Immunoanalysis (FIIA) - a new immunoassay format for the determination of pesticides in water. *Biosens. Bioelectron.*, **6**, 239–243.

141 Hering, D., Borja, A., Carstensen, J., Carvalho, L., Elliott, M., Feld, C.K., Heiskanen, A.S., Johnson, R.K., Moe, J., Pont, D., Solheim, A.L., and de Bund, W. (2010) The european water framework directive at the age of 10: a critical review of the achievements with recommendations for the future. *Sci. Total Environ.*, **409**, 4007–4019.

142 Thorpe, K.L., Maack, G., Benstead, R., and Tyler, C.R. (2009) Estrogenic wastewater treatment works effluents reduce egg production in fish. *Environ. Sci. Technol.*, **43**, 2976–2982.

143 Schell, L.M. and Gallo, M.V. (2010) Relationships of putative endocrine disruptors to human sexual maturation and thyroid activity in youth. *Physiol. Behav.*, **9**, 246–253.

144 Oetken, M., Nentwig, G., Löffler, D., Ternes, T., and Oehlmann, J. (2005) Effects of Pharmaceuticals on Aquatic Invertebrates. *Arch. Environ. Contam. Toxicol.*, **49**, 353–361.

145 Cunningham, V.L., Binks, S.P., and Olson, M.J. (2009) Human health risk assessment from the presence of human pharmaceuticals in the aquatic environment. *Regul. Toxicol. Pharmacol.*, **53**, 39–45.

146 Rodriguez-Mozaz, S., Lopez de Alda, M.J., and Barceló, D. (2006) Biosensors as useful tools for environmental analysis and monitoring. *Anal. Bioanal. Chem.*, **386**, 1025–1041.

147 Farréa, M., Péreza, S., Gon¸8calves, C., Alpenduradab, M.F., and Barceló, D. (2010) Green analytical chemistry in the determination of organic pollutants in the aquatic environment. *TrAC, Trends Anal. Chem.*, **29**, 1347–1362.

148 Farréa, M., Kantiania, L., Péreza, S., and Barceló, D. (2009) Sensors and biosensors in support of EU Directives. *TrAC, Trends Anal. Chem.*, **28**, 170–185.

149 Van Dorst, B., Mehta, J., Bekaert, K., Rouah-Martin, E., De Coen, W., Dubruel, P., Blust, R., and Robbens, J. (2010) Recent advances in recognition elements of food and environmental biosensors:

a review. *Biosens. Bioelectron.*, **26**, 1178–1194.

150 Krämer, P.M. and Schmid, R.D. (1991) Automated quasi-continuous immunoanalysis of pesticides with a flow injection system. *Pestic. Sci.*, **32**, 451–462.

151 Kusterbeck, A.W. and Blake, D.A. (2008) Flow Immunosensors, in *Optical Biosensors: Today and Tomorrow*, 2nd edn (eds F.S. Ligler and C.R. Taitt), Elsevier, Washington.

152 González-Martíneza, M.A., Puchadesa, R., Maquieiraa, A., Ferrerc, I., Marcob, M.P., and Barceló, D. (1999) Reversible immunosensor for the automatic determination of atrazine. Selection and performance of three polyclonal antisera. *Anal. Chim. Acta*, **286**, 201–210.

153 Weller, M.G., Schuetz, A.J., Winklmair, M., and Niessner, R. (1999) Highly parallel affinity sensor for the detection of environmental contaminants in water. *Anal. Chim. Acta*, **393**, 29–41.

154 Knecht, B.G., Strasser, A., Dietrich, R., Märtlbauer, E., Niessner, R., and Weller, M.G. (2004) Automated microarray system for the simultaneous detection of antibiotics in milk. *Anal. Chem.*, **76**, 646–654.

155 Tschmelak, J., Proll, G., and Gauglitz, G. (2004) Verification of performance with the automated direct optical TIRF immunosensor (River Analyser) in single and multi-analyte assays with real water samples. *Biosens. Bioelectron.*, **20**, 743–752.

156 Tschmelak, J., Proll, G., Riedt, J., Kaiser, J., Kraemmer, P., Bárzaga, L., Wilkinson, J.S., Hua, P., Hole, J.P., Nudd, R., Jackson, M., Abuknesha, R., Barceló, D., Rodriguez-Mozaz, S., López de Alda, M.J., Sacher, F., Stien, J., Slobodník, J., Oswald, P., Kozmenko, H., Korenková, E., Tóthová, L., Krascsenits, Z., and Gauglitz, G. (2005) Automated water analyser computer supported system (AWACSS) - Part I: project objectives, basic technology, immunoassay development, software design & networking. *Biosens. Bioelectron.*, **20**, 1499–1508.

157 Tschmelak, J., Proll, G., Riedt, J., Kaiser, J., Kraemmer, P., Bárzaga, L., Wilkinson, J.S., Hua, P., Hole, J.P., Nudd, R., Jackson, M., Abuknesha, R., Barceló, D., Rodriguez-Mozaz, S., López de Alda, M.J., Sacher, F., Stien, J., Slobodník, J., Oswald, P., Kozmenko, H., Korenková, E., Tóthová, L., Krascsenits, Z., and Gauglitz, G. (2005) Automated water analyser computer supported system (AWACSS) - Part II: intelligent, remote-controlled, cost-effective, on-line, water-monitoring measure-ment system. *Biosens. Bioelectron.*, **20**, 1509–1519.

158 Proll, G., Tschmelak, J., Kaiser, J., Kraemmer, P., Sacher, F., Stien, J., and Gauglitz, G. (2006) Advanced environmental biochemical sensor for water monitoring: automated water analyser computer supported system (AWACSS), in *Soil and Water Pollution Monitoring, Protection and Remediation, IV, Earth and Environmental Sciences NATO Science Series*, Vol. **69** (eds I. Twardowska, H.E. Allen, and M.M. Häggblom), Springer, pp. 131–145.

159 Hegnerováa, K. and Homola, J. (2010) Surface plasmon resonance sensor for detection of bisphenol A in drinking water. *Sens. Actuator B-Chem.*, **151** (1), 177–179.

160 Farré, M., Martínez, E., Ramón, J., Navarro, A., Radjenovic, J., Mauriz, E., Lechuga, L., Marco, M.P., and Barceló, D. (2010) Part per trillion determination of atrazine in natural water samples by a surface plasmon resonance immunosensor. *Anal. Bioanal. Chem.*, **388** (1), 207–214.

161 Vinogradova, T., Danaher, M., Baxter, A., Moloney, M., Victory, D., and Haughey, S.A. (2011) Rapid surface plasmon resonance immunobiosensor assay for microcystin toxins in blue-green algae food supplements. *Talanta*, **84** (3), 638–643.

162 Taitt, C.R., Shriver-Lake, L.C., Ngundi, M.M., and Ligler, F.S. (2008) Array biosensor for toxin detection: continued advances. *Sensors*, **8**, 8361–8377.

163 Le Blanc, A., Albrecht, C., Bonn, T., Fechner, P., Proll, G., Pröll, F., Carlquist, M., and Gauglitz, G. (2009) A novel analytical tool for quantification of estrogenicity in river water based on fluorescence labeled estrogen receptor α. *Anal. Bioanal. Chem.*, **395** (6), 1769–1776.

164 Fechner, P., Pröll, F., and Carlquist, M. (2009) An advanced biosensor for the prediction of estrogenic effects of endocrine-disrupting chemicals on the estrogen receptor alpha. *Anal. Bioanal. Chem.*, **393** (6–7), 1579–1585.

165 Eltzov, E. and Marks, R.S. (2011) Whole-cell aquatic biosensors. *Anal. Bioanal. Chem.*, **400** (4), 895–913.

166 van der Linden, S.C., Heringa, M., Puijker, L., Van der Burg, B., and Brouwer, A. (2007) Application of a steroid receptor-based battery of CALUX bioassays for water quality control analysis. *Organohalogen Compounds*, **69**, 1908–1911.

167 Donhauser, S.C., Niessner, R., and Seidel, M. (2009) Quantification of *E. coli* DNA on a flow-through chemiluminescence microarray read-out system after PCR amplification. *Anal. Sci.*, **25**, 669–674.

168 Corrie, R., Lawrie, G.A., Battersby, B.J., Ford, K., Rühmann, A., Koehler, K., Sabath, D.E., and Trau, M. (2008) Quantitative data analysis methods for bead-based DNA hybridization assays using generic flow cytometry platforms. *Cytometry A*, **73A** (5), 467–476.

169 Song, L., Ahn, S., and Walt, D.R. (2006) Fiber-optic microsphere-based arrays for multiplexed biological warfare agent detection. *Anal. Chem.*, **78** (4), 1023–1033.

170 Elsholz, B., Wörl, R., Blohm, L., Albers, J., Feucht, H., Grunwald, T., Jürgen, B., Schweder, T., and Hintsche, R. (2006) Automated detection and quantitation of bacterial RNA by using electrical microarrays. *Anal. Chem.*, **78** (14), 4794–4802.

171 Banerjee, P. and Bhunia, A.K. (2010) Cell-based biosensor for rapid screening of pathogens and toxins. *Biosens. Bioelectron.*, **26** (1), 99–106.

172 Peskoller, C., Niessner, R., and Seidel, M. (2009) Cross-flow microfiltration system for rapid enrichment of bacteria in water. *Anal. Bioanal. Chem.*, **393**, 399–404.

173 Pappert, G., Rieger, M., Niessner, R., and Seidel, M. (2010) Immunomagnetic nanoparticle-based sandwich chemiluminescence-ELISA for the enrichment and quantification of *E. coli*. *Microchim. Acta*, **168**, 1–8.

174 Sapsford, K.E., Rasooly, A., Taitt, C.R., and Ligler, F.S. (2004) Detection of campylobacter and Shigella species in food samples using an array biosensor. *Anal. Chem.*, **76** (2), 433–440.

175 Taitt, C.R., Shubin, Y.S., Angel, R., and Ligler, F.S. (2004) Detection of Salmonella enterica serovar typhimurium by using a rapid, array-based immunosensor. *Appl. Environ. Microbiol.*, **70** (1), 152–158.

176 Dunbar, S.A., Vander Zee, C.A., Oliver, K.G., Karem, K.L., and Jacobson, J.W.J. (2003) Quantitative, multiplexed detection of bacterial pathogens: DNA and protein applications of the Luminex LabMAP system. *J. Microbiol. Methods.*, **53** (2), 245–252.

177 Jing, W., Ya, C., Yuanyuan, X., and Li, G. (2009) Colorimetric multiplexed immunoassay for sequential detection of tumor markers. *Biosens. Bioelectron.*, **25** (2), 532–536.

178 Huelseweh, B., Ehricht, R., and Marschall, H.-J. (2006) A simple and rapid protein array based method for the simultaneous detection of biowarfare agents. *Proteomics*, **6** (10), 2972–2981.

179 Langer, V., Niessner, R., and Seidel, M. (2011) Stopped-flow microarray immunoassay for the detection of viable *E. coli* by use of chemiluminescence flow-through microarrays. *Anal. Bioanal. Chem.*, **399**, 1041–1050.

180 Park, C.H., Kim, J.P., Lee, S.W., Jeon, N.L., Yoo, P.J., and Sim, S.J. (2009) A direct, multiplex biosensor platform for pathogen detection based on cross-linked polydiacetylene (PDA) supramolecules. *Adv. Funct. Mat.*, **19** (23), 3703–3710.

181 Vikesland, P.J. and Wigginton, K.R. (2010) Nanomaterial enabled biosensors for pathogen monitoring - a review. *Environ. Sci. Technol.*, **44** (10), 3656–3669.

182 Ivnitski, D., Abdel-Hamid, I., Atanasov, P., and Wilkins, E. (1999) Biosensors for detection of pathogenic bacteria. *Biosens. Bioelectron.*, **14** (7), 599–624.

183 Bergwerff, A.A. and van Knapen, F. (2006) Surface plasmon resonance biosensors for detection of pathogenic microorganisms: strategies to secure food and environmental safety. *J. AOAC Int.*, **89** (3), 826–831.

184 Dudak, F.C. and Boyaci, I.H. (2009) Rapid and label-free bacteria detection by surface plasmon resonance (SPR) biosensors. *Biotechnol. J.*, 4 (7), 1003–1011.

185 Leonarda, P., Heartya, S., Brennana, J., Dunnea, L., Quinna, J., Chakrabortyc, T., and O'Kennedy, R. (2003) Advances in biosensors for detection of pathogens in food and water. *Enz. Microb. Tech.*, **32** (1), 3–13.

186 Knauer, M., Ivleva, N., Liu, X., Niessner, R., and Haisch, C. (2010) Surface-enhanced Raman scattering-based label-free microarray read-outfor the detection of micro-organism. *Anal. Chem.*, **82**, 2766–2772.

187 Ivleva, N., Wagner, M., Horn, H., Niessner, R., and Haisch, C. (2010) Raman microscopy and SERS for in situ analysis of biofilms. *J. Biophotonics*, **3**, 548–556.

188 Sengupta, A., Mujacic, M., and Davis, E.J. (2006) Detection of bacteria by surface-enhanced Raman spectroscopy. *Anal. Bioanal. Chem.*, **386** (5), 1379–1386.

189 Jarvis, R.M. and Goodacre, R. (2008) Characterisation and identification of bacteria using SERS. *Chem. Soc. Rev.*, **37**, 931–936.

190 Banada, P.P., Huff, K., Rajwa, B., Aroonnual, A., Bayraktar, B., Adil, A., Robinson, J.P., Hirleman, E.D., and Bhunia, A.K. (2010) Label-free detection of multiple bacterial pathogens using light-scattering sensor. *Biosens. Bioelectron.*, **24**, 1685–1692.

191 Knecht, L.D., Pasini, P., and Daunert, S. (2011) Bacterial spores as platforms for bioanalytical and biomedical applications. *Anal. Bioanal. Chem.*, **400**, 977–989.

192 Kumari, A., Pasini, P., and Daunert, S. (2008) Detection of bacterial quorum sensing N-acyl homoserine lactones in clinical samples. *Anal. Bioanal. Chem.*, **391**, 1619–1627.

193 http://www.optischetechnologien.de/forschung.

Part Three
Security Applications

Handbook of Biophotonics. Vol.3: Photonics in Pharmaceutics, Bioanalysis and Environmental Research, First Edition.
Edited by Jürgen Popp, Valery V. Tuchin, Arthur Chiou, and Stefan Heinemann

7
Body Scanner

Torsten May and Hans-Georg Meyer

7.1
Introduction

Many recent terrorist assaults on public places – such as airports or train stations – would have been prevented if a bomb hidden by the attacker underneath his clothing could have been detected. It is obvious that current security technologies, like walk-through metal detectors, are not adequate for such scenarios; as a matter of principle they can only detect electric conductors, for example, firearms or metallic knifes. Therefore, the idea of "body scanning", meaning the electromagnetic visualization of hidden threats against the human body, became the concept of choice for future security screening aimed at prevention of terrorist attacks.

Reviewing suggested or realized body scanner technologies shows that all of them have a common approach: the object to be detected has electromagnetic properties (absorption, reflection, transmission) which differ from those of the human body. In an electromagnetic image, they become visible if the clothing above them is at least partly transparent in the used spectrum. Both preconditions limit the usable bands for imaging. Basically, appropriate technologies can be divided into two major categories: high photon energy bands above the visible part of the electromagnetic spectrum, and at the opposite extreme, low energy bands below the visible spectrum. The first category is addressed by X-ray technology, which relies on longstanding developments for medical application. The latter – low photon energy methods – include a variety of different methods, ranging from RADAR techniques to thermal far-infrared imaging. A categorization is difficult, since many of the suggested approaches overlap in terms of operation bands and components used. Therefore, the classification chosen in this review is somewhat arbitrary. The frequency band of operation covers the section between approx. 30 GHz ($\lambda = 1$ cm) and 1 THz ($\lambda = 300\,\mu$m). Hence, it ranges from bands used for microwave techniques to optical bands in the far-infrared region. Although a synergy between these two traditional technology areas is observable, for competitiveness a fuzzy line can be drawn at a wavelength of about 1 mm. Longer waves (millimeter waves) are mostly used by body scanners based on electronic technologies. At shorter wavelength (sub-millimeter

Handbook of Biophotonics. Vol.3: Photonics in Pharmaceutics, Bioanalysis and Environmental Research, First Edition.
Edited by Jürgen Popp, Valery V. Tuchin, Arthur Chiou, and Stefan Heinemann

waves) electronic technologies become less competitive in technical terms, and at the same time they become costly. Consequently, body scanners for sub-millimeter waves make use of emerging photonic technologies.

7.2 X-Ray Techniques

7.2.1 Overview

The concept of creating an image of a human body (or at least of parts of it) using X-rays is as old as the discovery of these rays by Wilhelm Conrad Roentgen in 1895, who soon thereafter published the first X-ray image of his wife's left hand. Ever since, this technique has evolved into a unique tool for medical imaging, with a remarkable impact on medical sciences [1]. It was obvious to implement these developed technologies also in other fields of social life, like security technologies. However, although X-ray imaging is in use for security screening of inanimate objects (e.g., baggage at an airport check-in), for a long time the screening of humans was excluded because of the radiation exposure and the implied health risk.

Within the last decades, increasing attention has been paid to the minimization of radiation exposure during medical examination. Partly because of emerging semiconductor X-ray detector technologies – which nowadays have almost completely replaced the traditional photosensitive film – it became possible to create X-ray images using only a fraction of the previous radiation dose.

Moreover, sophisticated detectors have been proven to be sensitive enough to record not only transmitted radiation but also the weak backscattered response from an illuminated human body. The backscatter technique became the concept of choice for security body scanning, because the required radiation dose is very low, allowing several hundred exposures per year within the legal limits of radiation safety. With it, on the other hand, one loses the unique feature of X-ray imaging of being the only technique able to reveal objects *inside* a human body. Although the transmission technique is in use in some security-sensitive places, the considerable radiation exposure justifies its employment only under extreme circumstances.

7.2.2 Physical and Technical Background

X-rays have been defined as electromagnetic waves having wavelength from 10^{-8} m (photon energy 100 eV) up to 10^{-12} m (1 MeV). The spectrum is strongly overlapping with gamma rays, which are differentiated from X-rays not by band definition but by their origin (electron vs. nuclear processes). In imaging, several interaction processes between X-rays and matter (soft tissue of the human body, the material of the object to be detected) are exploited. Relevant for security imaging are the photoelectric effect and the *Compton* scattering process [2].

The photoelectric effect is the absorption of an X-ray photon by electrons from the inner shells of an atom. In consequence, this electron is knocked out, having energy of the original photon energy minus its original binding energy. The cross section of this process is proportional to Z^5 (where Z is the atomic number). In contrast, in *Compton* scattering the X-ray photon elastically impacts free or weakly bound electrons of the illuminated matter. In this process, the X-ray photon loses energy and is reflected into all solid angles. The amount of backscattered radiation is a function of the photon energy and the matter composition, mostly the electron density; it is linearly dependent Z.

For security, as for medical imaging, typical operating bands are between 50 and 150 keV. Here, *Compton* scattering is dominant for the soft tissue of the human body, resulting in a high backscatter response. In contrast, for example, metal objects (high Z) absorb strongly because of the dominating photoelectric effect, consequently showing only a weak response, and hence a high contrast to the human body. For explosives, which are often based on organic compounds chemically similar to biological matter, the contrast is rather low, which makes their detection difficult.

For illumination, as in almost all medical devices, X-ray sources based on a metallic cathode in a vacuum tube are used [3], with varying materials and configurations depending on the intended energy range, output power and beam size. Recently published measurement techniques (see below) use a scanning configuration, with the source shuttered into a line or needle beam.

The detection of X-rays can be accomplished with a variety of different methods. As state of the art, solid state detectors have replaced established photosensitive film technologies, which are still competitive in terms of resolution. However, for technical reasons, especially the speed of image construction, electronic technologies have been favored. In general, these detectors can be divided into two major groups: direct and indirect [4]. An indirect detector uses a scintillating material to convert the incoming X-ray photon into visible light, which is detected by a charge coupled device (CCD) camera or by an array of photodiodes whose electrical charge response is read by thin film transistors (TFT) [5]. In contrast, an X-ray photoconductor (e.g., amorphous selenium) can directly convert the radiation into electrical charges, again read by a TFT.

The achievable spatial resolution is defined not by the diffraction limit of X-rays but by the digital resolution of the scanning technique. A reasonable goal, and hence the design rule for X-ray security imaging devices, is a spatial resolution of about 1 mm. An even better spatial resolution of below 0.1 mm would require an extremely high sampling of the image (20 000 × 10 000 pixels for a full body scan), rendering the system unnecessarily complex, and thus costly.

In the following, two system approaches are briefly introduced, both relying on the X-ray technologies described above.

7.2.3 Backscatter Imaging

A recent implementation of the X-ray backscatter technique is the *Smartcheck*™ by the US American company American Science and Engineering Incorporation [6].

It is based on scientific work from the 1990s, patented by ASEs predecessor, the IRT Corporation. The current realization is already in use by the Transportation Security Agency (TSA, Department of Homeland Security), which has installed at least 94 machines at various airports in the United States. The worldwide dissemination is slow, because of European objections to the use of X-rays for non-medical applications [7]. However, machines can be found at least in the United Kingdom (London and Manchester airports), since currently they are still the most effective security solution.

The manufacturer specifies his machine as operating at photon energies of about 50 keV, creating a backscatter image of one side of a human body in approx. 3 s. The scan is performed using a mechanically steered needle beam in the center of the set-up, whilst two large-volume scintillator detectors at each side of the device integrate the backscattered response over a notable solid angle. For security purposes, a subsequent second image is required from the opposite side, although some installations use a master-slave set-up with a second device for parallel recording. The radiation exposure during the scan is specified by the manufacturer as 0.1 μSv per single-side scan, compared to the typical 5 μSv for dental radiography [8].

The machine is shown in Figure 7.1, alongside a typical X-ray backscatter image. The example demonstrates the suitability for security applications: the clothing of the subject is almost invisible, and hidden objects become apparent with high resolution, allowing object identification by shape. This answers one of the crucial problems of body scanner technologies: because it is almost impossible to determine the actual material of a detected object, only a telling shape would allow its identification. The displayed X-ray image has sufficient spatial resolution to detect for example, the belt buckle as a non-dangerous object, where the longwave technologies described later, due to their restricted spatial resolution, would probably create a false alarm with the same object.

(a)

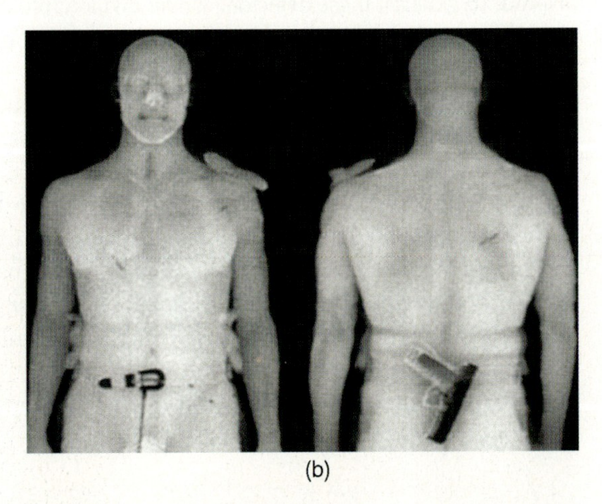

(b)

Figure 7.1 American Science and Engineering Inc, System *Smartcheck*™ at a US airport (a), X-ray backscatter image recorded by the device (b). William J. Baukus. X-ray Imaging for On-The-Body Contraband Detection Presented to 16th Annual Security Technology Symposium & Exhibition Session V: Technology Forum Focus Group II Transportation Security Technologies and Tools June 28, 2000.

7.2.4
Transmission Imaging

The transmission technique goes one step further in making use of X-rays. Recent implementations are for example, the *CONPASS* by Braun and Company Limited, the *Soter RS* by OD security, the *BS16HR* by Smith Detection and the *Scannex* by DebTech (De Beers).

As an example, the *CONPASS* device uses a narrow monochromatic X-ray beam and utilizes a linear array of semiconductor scintillation detectors for image scanning. Choosing between two different modes (LD and HR, corresponding to 0.25 μSv and 3 μSv per scan, respectively) images can be recorded [9]. Thereby, radiation passing through the inspected person is recorded, thus revealing – as the only body scanning technique – objects hidden inside the body or even behind the back, which would make a second scan from the opposite direction unnecessary (see Figure 7.2). During a scan, which takes about 8 s, the person is moved by a platform whilst the X-ray beam scans vertically. The reconstructed image with 2688 × 1000 pixels allows a spatial resolution of better than 2 mm (better than 0.2 mm for high contrast objects such as metals, according to the manufacturer's information).

Despite the undeniable security value of detecting objects inside the human body, the use of this technology did not become common in public places, mostly due to the obvious ethical issues involved. However, the manufacturer's literature reveals that in locations with restricted personal rights (e.g., prisons, diamond mines) its use is valid. Moreover, under the precondition of reasonable suspicion for example, of smuggling, even European border control officials are using this technology.

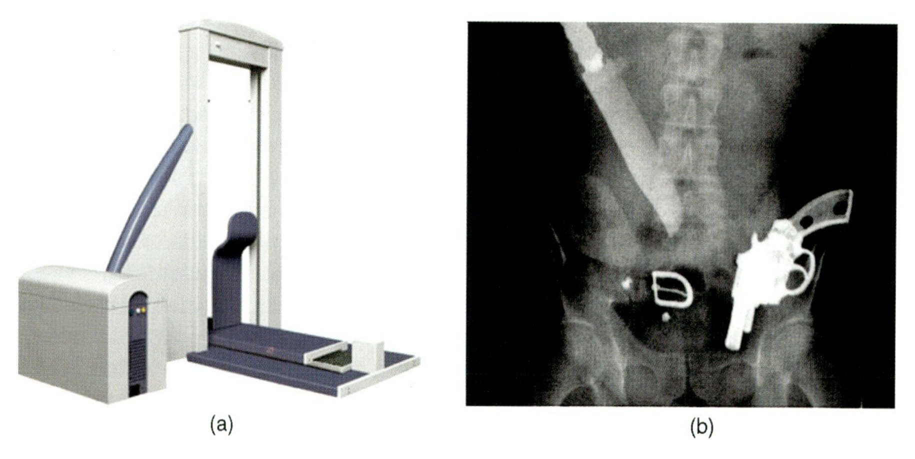

Figure 7.2 Braun and Company Limited, System "*CONPASS*" (a), X-ray transmission image revealing a Teflon knife behind the person (b) [10].

7.3
Millimeter Wave Electronic Techniques

7.3.1
Overview

In general, the basic idea of millimeter wave body scanning is the same as for X-ray imaging: the recorded electromagnetic image reveals objects because they differ in reflectivity and/or emissivity, whilst the covering cloth is at least partly transparent. The most crucial difference is the by 6 orders of magnitude larger wavelength compared to X-rays, making almost every system approach diffraction limited. Moreover, the maximum useful wavelength is limited. Although one could easily conceive a centimeter wave body scanner (the physics would be almost identical), in practice the achievable spatial resolution would be too low for security applications.

The term millimeter wave refers to electromagnetic waves in the range between 1 mm and 1 cm, corresponding to the definition for "extremely high frequency, EHF" (30 GHz to 300 GHz) for radio waves. Although it is strongly overlapping with the so-called sub-millimeter or terahertz waves in terms of physical properties, it is described separately because of the different technologies used for building body scanners in the respective wavelength band. For millimeter wave bands, technical components evolving from microwave technologies are in use.

7.3.2
Physical and Technical Background

The human body thermally generates millimeter wave photons according to Planck's equation; however, in the long wave Rayleigh limit its spectral emission is very low. Consequently, thermal imaging, as established in infrared bands, is technically ambitious. Hence, for visualization, in most cases a secondary photon source is required. Fittingly, here the human skin acts as an adequate reflector [11], whilst for the shorter sub-millimeter waves the reflectivity drops substantially [12] (see Figure 7.3). In conclusion, at millimeter wavelength it is reasonable to create a radiometric image using reflected secondary photons as contrast amplification.

At the same time, waves longer than 1 mm can easily penetrate clothes [13], so an object with different reflectivity becomes visible in contrast to the skin. Likewise, the atmospheric attenuation of millimeter waves is also acceptable for security screening [14], which is typically performed from distances not longer than a few meters. However, simply because many of the technical components used (emitters and receivers) have been developed for parallel use in other applications such as telecommunications, all recent body scanners still favor operation in atmospheric windows, such as the K_a band (27–40 GHz) or the W band (75–110 GHz) [15].

One of the first approaches to realizing millimeter wave imaging was to use the help of a cold background in order to increase the natural radiometric contrast between the warm human body and its surroundings. That can be accomplished because the atmosphere features an equivalent noise temperature well below

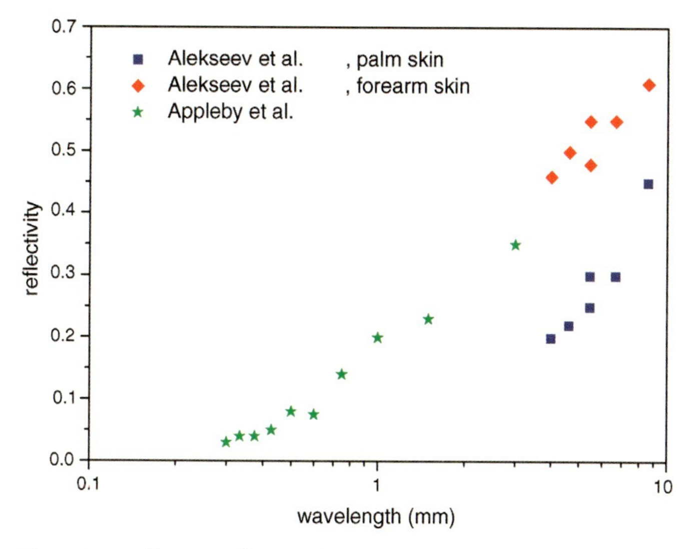

Figure 7.3 Reflectivity of human skin, according to Alekseev *et al.*: and Appleby *et al.*

100 K [16], increasing the indoor contrast of typically 15 K (35 °C skin temperature of human body vs. 20 °C room temperature) by about one order of magnitude. Consequently, many of the systems described later can operate without a technical source in outdoor operation, and are therefore described – somehow misleadingly – as passive, although they are not sensitive enough to effectively resolve the weak thermal millimeter wave radiation.

Applications which require indoor operation (e.g., security scanning at airports) are dependent on technical sources. As a result of the long heritage of microwave technology, there exists a variety of different emitter concepts, among them for example, varactor/varistor-mode frequency multipliers [17], backward-wave oscillators [18], Gunn oscillators [19], and more. Microwave sources in general share some physical similarities, allowing general statements on millimeter wave sources: typically they feature a relatively narrow output band, not larger than ±10% full-width-half-maximum (FWHM) of the center wavelength. Some concepts allow tuning the center emission to various bands. State of the art for the achievable output power is of the order of several watts at 10 mm wavelength; exponentially decaying to hundreds of milliwatts at 1 mm. Recently, the concepts of millimeter wave sources have been extended to sub-millimeter wavelength, and in the extreme to beyond 1 THz frequency (< 0.3 mm wavelength). However, the output power becomes increasingly small [20], making their use reasonable only for scientific experiments.

Similar performance trends are observable for millimeter wave detectors. They can be categorized into coherent and incoherent, the latter sensing only the power (or amplitude) of the incoming signal instead of the full wave information. Basically, millimeter wave detectors utilize an electronic device with a highly nonlinear characteristic, allowing for rectification (i.e., conversion of a high frequency signal into a dc voltage response) or mixing with a known source (i.e., heterodyne principle) [21]. One of the most widespread devices is the *Schottky* diode [22], which has been in use for radio frequency applications for almost half a century. A major leap was the development of thin-film semiconducting *Schottky* diodes, which made them suitable for direct detectors as well as for mixers. Even today there is still notable research activity [23] on the *Schottky* effect. One of the reasons for its success is the compatibility with the so-called monolithic microwave integrated circuit (MMIC) technology [24], which allows cost-effective manufacturing of microwave devices in an integrated semiconductor fabrication technology. Please note, that the technical term MMIC is misleadingly used without suffix in different contexts, since there exist MMIC microwave sources as well as MMIC direct (incoherent) detectors and MMIC (coherent) mixers.

7.3.3 Active Imaging

The most mature system is the *ProVision* portal (see Figure 7.4), patented by the US American company Safeview (now L3 Communications) [25]. The device uses two columns with linear arrays of millimeter wave transmitters and receivers operating in the K_a band (with an actual bandwidth of 27–33 GHz) to scan a person by moving the columns around the subject. The procedure takes about 2 s. A frequency modulated continuous wave technique (FMCW) [26] is used to measure delays, in order to reconstruct a three-dimensional image of a human body from the reflected millimeter waves.

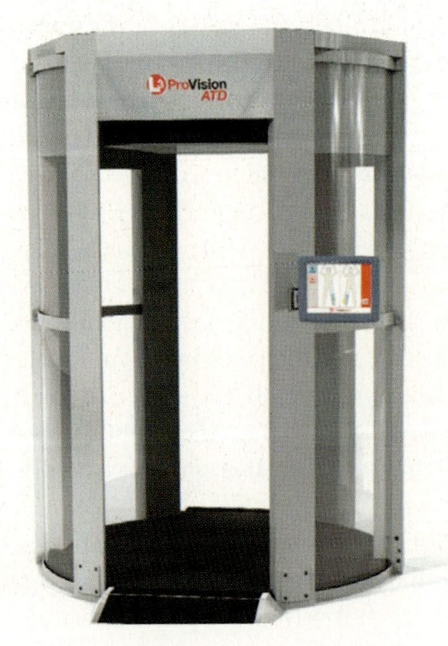

Figure 7.4 *L3 Communication Provision ATD* portal [27].

L3 devices were among the first body scanners deployed in real application. TSA has been using them since 2007 at various US airports. Since 2009, European airports have followed, favoring the millimeter wave technology over the much disputed X-ray scanners. They are now in test use in many European countries. A still existing obstacle is caused not by technical problems but by the political constraint to use the devices in a fully automated mode without raw data display. Thus the challenge to implement an algorithm for automated threat detection arises. Taking into account typical features of millimeter wave images (artifacts from clothing due to partial absorption, limited spatial resolution), a separation between potential threats and non-dangerous objects is ambitious. To date, some countries (e.g., Italy and Germany) are still working on this software problem, delaying the routine use of the device.

A newer system, the *eqo* by Smith Detection, favors a scan procedure with a stationary receiver. In the set-up shown in Figure 7.5, the emitters and receivers (ISM band 24–24.25 GHz) are installed in the column of the walk-through portal, whereas the panel behind contains an array of steerable patch antennae [28] which are used to scan a 3D volume in between. *eqo* delivers an output video stream with 10 to 25 Hz frame rate, with an image quality comparable to the *Provision ATD*. The specified spatial resolution is of the order of 1 cm and below for high contrast objects.

The current state of the art in active millimeter wave imaging is represented by the ongoing German research project *QPASS*, using a multistatic sparse array of transmitters and receivers for 77 GHz [30]. The modular system allows a full 3D image reconstruction with about 2 mm spatial resolution, gaining from the three

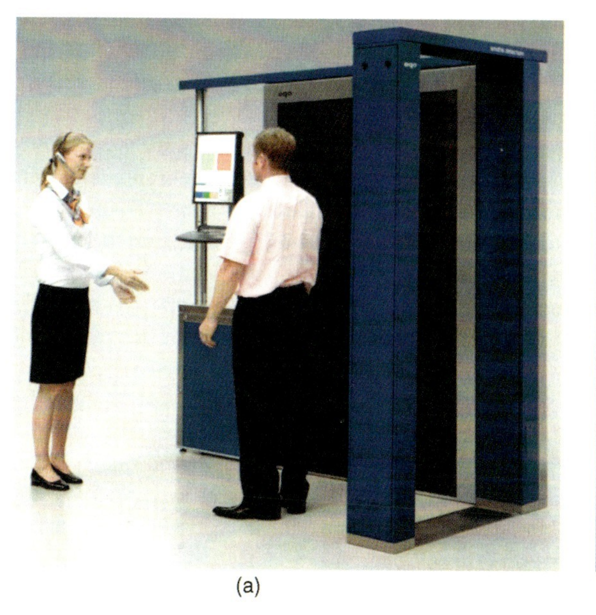

(a)

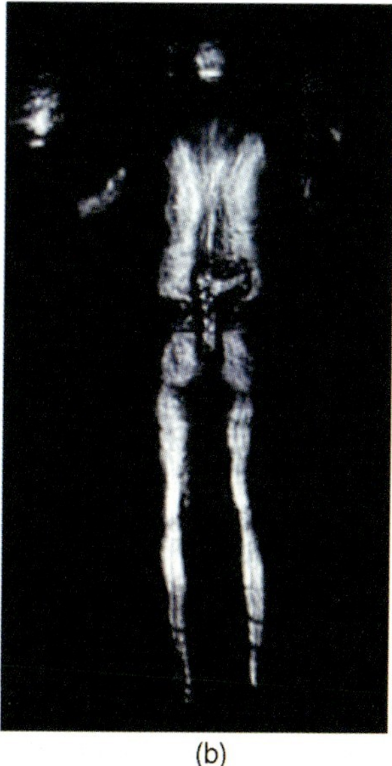

(b)

Figure 7.5 Smith Detection *eqo* [29] (a), and recorded millimeter wave image taken from a video stream with 10 Hz frame rate (b) courtesy of Smith Detection.

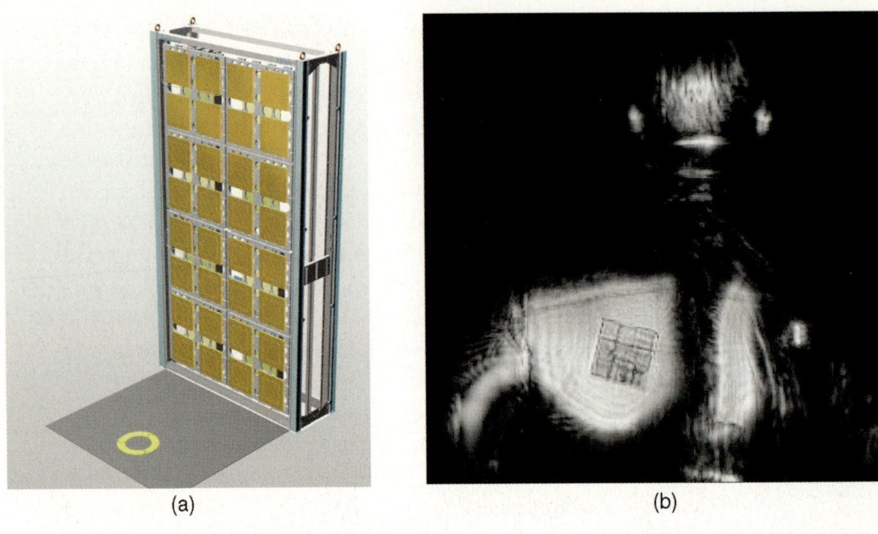

Figure 7.6 *Andromeda* flat-panel detector under construction (courtesy by Rohde & Schwarz Germany) (a), and millimeter wave image [31] recorded in a scanning test set-up(b).

times smaller wavelength compared to the L3 system, Figure 7.6. In this way the problem of automated threat detection could be answered, since the achievable spatial resolution approaches the reference of X-ray imagery, allowing classification of detected objects by their shape. Moreover, the relatively high bandwidth of the system allows direct determination of the dielectric constant of the detected object, potentially enabling a material identification.

7.3.4
Passive Imaging

As mentioned above, the available energy of thermal millimeter wave radiation from a human body is weak. However, some systems address a passive operation, especially in the case of intended stand-off operation. That is mainly because it becomes increasingly difficult to effectively illuminate subjects from a distance. A directed radiation from a point source will be reflected back in a relatively small solid angle; whereas a distributed illumination is elaborate and would result in a small reflected energy per area.

Obviously, an intended passive operation requires highly sensitive detectors. Flat panel architectures with a large number of state-of-the-art receivers are still too expensive; consequently, all systems described below are based on reasonably small arrays (10 to 100 pixels) in combination with an opto-mechanical scanner. However, even with the most sophisticated detectors one still has to trade-off a passive operation for less frame rate or less resolution.

Recent implementations have demonstrated different approaches to solve this compromise. The most elaborate system has been developed by the UK's QinetiQ, who have been building millimeter wave imagers for military application for the past

(a)

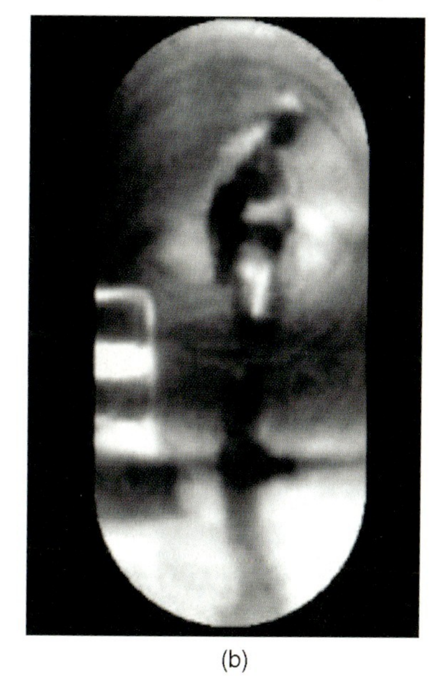

(b)

Figure 7.7 QinetiQ *SPO 30* stand-off imager for outdoor screening (a), passive millimeter wave video (b) [33].

20 years. The highlight is a 3 mm (94 GHz) helicopter-borne imager [32] with 150 MMIC receivers, achieving a 25 Hz video frame rate.

Modified implementations of this system's technology have been used to demonstrate the usability for body scanning; however, for effectual indoor operation the sensitivity of the receivers is too low. Therefore, devices such as the *SPO 20* (see Figure 7.7) are in use for stand-off explosive detection in outdoor operation, where they are still subjected to extensive trials by the US and British armies.

In order to achieve a passive operation indoors, the US Company Millivision has followed an unconventional approach. In their *Portal 350* [34] (see Figure 7.8), they

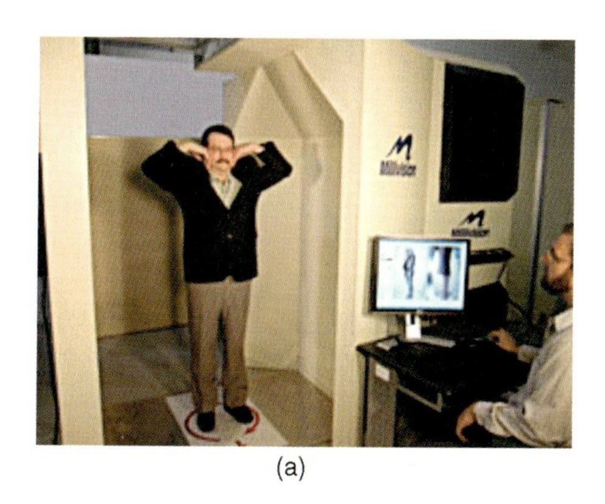

(a)

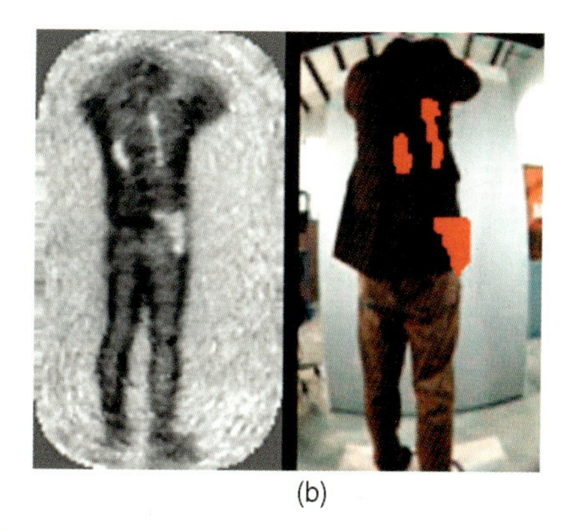

(b)

Figure 7.8 Millivision *Portal 350* (a), and recorded still image from a 10 Hz video stream (b).

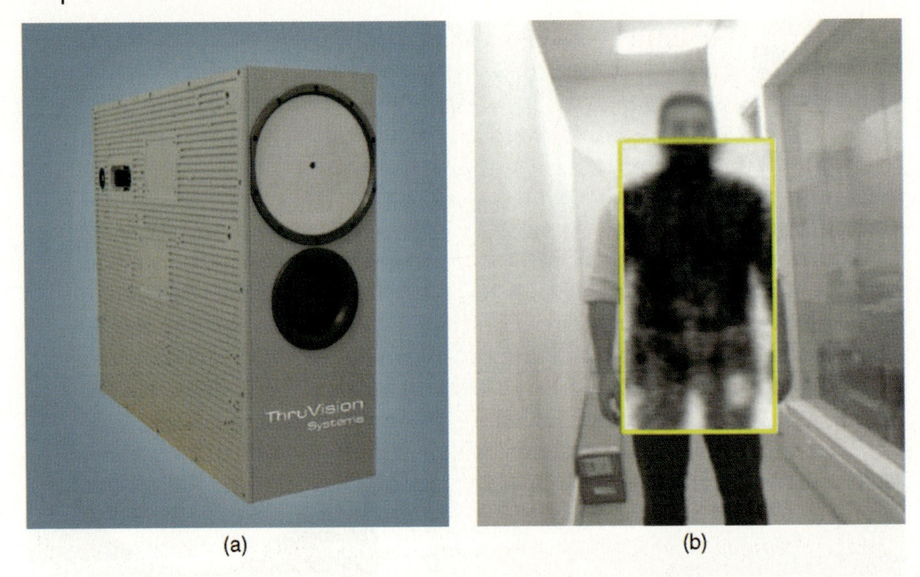

Figure 7.9 Thruvision *TS4* (a), passive millimeter wave image recorded from 5 m distance (b) [35].

achieve a reasonable spatial resolution and a high frame rate (10 Hz) by increasing the radiometric contrast with a cooled wall inside the inspection cabin. In that way, there is no obstacle of artificial illumination by a technical source, whilst the cooled wall allows the system to operate in the same environmental conditions as in outdoor operation.

Another approach is followed by the British company Thruvision: their systems are based on technology developed in the ESA Startiger program. Stretching the millimeter wave technology to the limit, namely to a wavelength of about 1.2 mm (220 GHz), the useable thermal energy is already by a factor of 5 higher compared to the QinetiQ system at 3 mm (94 GHz). In combination with a sophisticated heterodyne detection it allows for reasonable spatial resolution at frame rates of about 6 Hz in indoor operation (see Figure 7.9).

7.4 Sub-Millimeter Wave (Terahertz) Photonic Techniques

7.4.1 Overview

Security scanning at sub-millimeter bands extends the general idea of using low-energy photons for imaging. The physical reason to enhance the millimeter wave technique to shorter wavelengths is the diffraction effect, which limits the achievable spatial resolution of security relevant images. However, despite the strong connection between millimeter and sub-millimeter bands, crucial parameters are different, consequently changing the design constraints.

In general, sub-millimeter wave systems are investigated because of the intended capability to perform a body scan from a certain distance of a few meters up to 100 m. This is motivated by the wish to free the person under test from inspection cabins. At the same time, it would allow a security check to be carried out from a safe distance, preventing, for example, a suicide bomber from approaching the site to be protected.

In such application scenarios, millimeter wave systems could achieve a reasonable spatial resolution only by using oversized optics. On the other hand, simply scaling the millimeter wave technology to shorter wavelength would result in a disproportional increase in system complexity and costs because of the technical constraints of the electronic devices used. Therefore, the sub-millimeter band is dominated by emerging photonic technologies which are more closely related to infrared technologies.

7.4.2
Physical and Technical Background

Compared to millimeter wave bands, the issue of atmospheric and clothing attenuation becomes much more pronounced. In general, every sub-millimeter wave system is a compromise between spatial resolution and transparency, which narrows the band of suitable wavelengths to a range between 1 mm and a minimum of 0.3 mm. Even in this restricted band, the trend for worsening transparencies of atmosphere and clothing (see Figure 7.10) strongly influences the performance, so prototypes have been demonstrated only in atmospheric windows at 0.9 mm (350 GHz), 0.45 mm (660 GHz) and 0.35 mm (850 GHz). Although even shorter wavelengths would be very attractive, especially for explosive detection (see Chapter 8), the inadequate clothing transparency makes a "real" terahertz system ($f > 1$ THz, $\lambda < 0.3$ mm) almost useless for security applications.

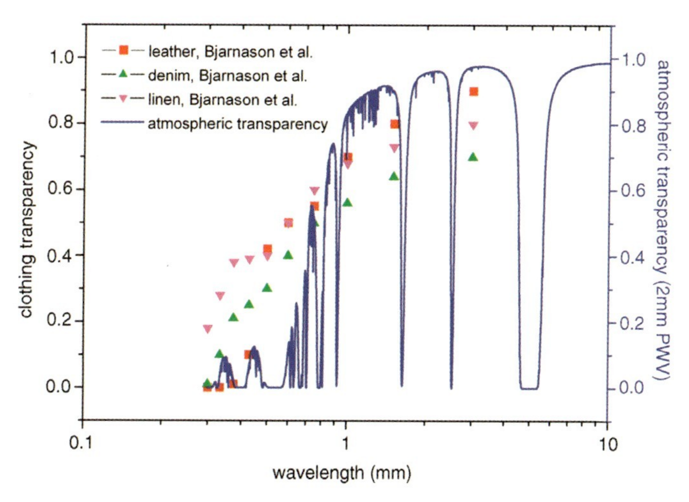

Figure 7.10 Comparison of clothing transparencies (Bjarnason *et al.* [13]) with a calculated atmospheric spectrum using the APEX transmission calculator [36].

The question whether to build an active or passive system is also answered differently compared to millimeter wave bands. The thermal radiation of a human body is notably stronger, so a passive operation is feasible with an acceptable technical effort. At the same time, the reflectivity of a human body is small (compare Figure 7.3). This brings up a problem for an active system, since for example, direct reflections from metallic objects would result in a – by orders of magnitude – larger signal compared to the weak response from the human body. Consequently, any detector for an active sub-millimeter body scanner is required to achieve a high dynamic range ($>10^3$), which constitutes a technical challenge.

Eligible technical sources for sub-millimeter radiation can be categorized into derivations of the already described microwave technologies, and photonic sources, for example, lasers. As mentioned above, electronic sources suffer from an inevitable loss in performance with decreasing wavelength, although substantial progress has been achieved by recent research [37]. From the opposite side of the sub-millimeter spectrum, sources based on photonic infrared concepts are extended to longer waves, facing the inverse problem of decreasing efficiency with increasing wavelength. Among these are quantum cascade lasers (QCL) [38], terahertz pulse lasers pumped by femtosecond infrared lasers [39], photomixers [40], and more. Up to now, the use of sub-millimeter sources for active imaging has been demonstrated only in research projects; there is no commercial system based on this technology. The development is driven mainly by the intended use for terahertz spectroscopy, which, in a security context, could be used to identify explosives.

The field of detector technologies for sub-millimeter waves is quite heterogenic. First, concepts for millimeter wave detectors have been enhanced by modern research to stay competitive. A completely different branch is occupied by broadband incoherent (power or energy) detectors based on bolometric principles, typically cooled to low temperatures for high sensitivity [41]. They have been shown to achieve unprecedented performance, making them the first choice for passive sub-millimeter wave imaging, but at the same time raising the constraint to implement cooling technologies in the system. In the case of active methods, detectors of choice are coherent, in order to gain full wave information for spectroscopy or 3D image reconstruction. Here, two major categories are available: heterodyne mixer technologies, for quantum-limited performance based on superconducting elements [42], and electro-optical concepts which are directly adapted to time domain spectroscopy with femtosecond lasers [43].

7.4.3 Active Imaging

An example of the further enhancement of millimeter wave technologies is a US project (lead by the Pacific Northwest National Laboratory for the Department of Energy) aiming for an active 350 GHz system [44]. This system uses planar GaAs *Schottky* diodes in a MMIC multiplier set-up. To date it has been demonstrated that a 3D image can be reconstructed at distances up to 10 m, at a frame rate of 10 s.

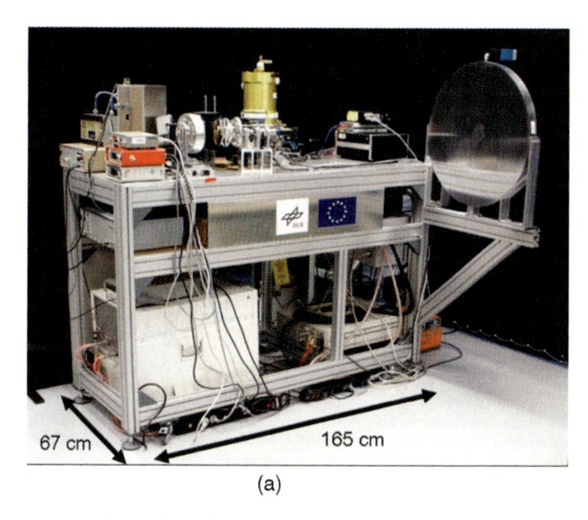

(a)

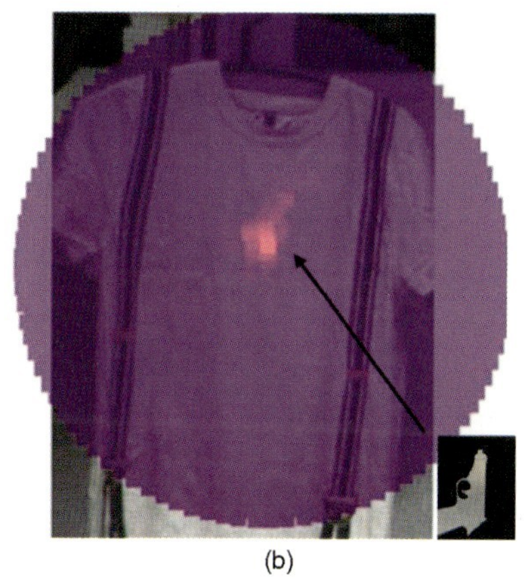

(b)

Figure 7.11 *TERASEC* system demonstrator (a) and detected handgun mock-up from 18 m distance, recorded within 0.5 ms (b).

In Europe, the project *TERASEC* has been accomplished in frame of the Preparatory Action for Security Research (PASR) program, started in 2004. Here, at the German Aerospace Centre (DLR) a real-time scanner (see Figure 7.11) has been demonstrated, which is able to detect objects from distances up to 20 m [45]. An optically pumped THz gas laser was used as transmitter, combined with a single superconducting hot-electron bolometric mixer as detector. To date, this is still the largest stand-off distance demonstrated.

These two projects are only two of a lively research field. They demonstrate the trend to use continuous wave sources combined with heterodyne detection for 3D imaging. The acclaimed time-domain spectroscopy method has not made it into security imaging, mainly because of the issue of diverging pulses over a stand-off distance. It is still the method of choice for nondestructive testing or spectroscopy, for a review see [46].

7.4.4 Passive Imaging

As mentioned above, building a passive system for sub-millimeter wave bands is much more promising compared to the millimeter wave spectrum: thermal radiation is notably stronger, so it is possible to conceive a passive imager whose designed spatial and thermal resolution is not limited by technical constraints but by physical limits.

As for infrared thermal imagers, the figure of merit is the background limit of inevitable signal noise (Poisson-distributed fluctuations in received photon number) [47]. Recent research has demonstrated that, using cryogenic incoherent detectors, it is possible to achieve this ideal mode of operation (background limited photometry, BLIP). Two groups have followed this approach; a European

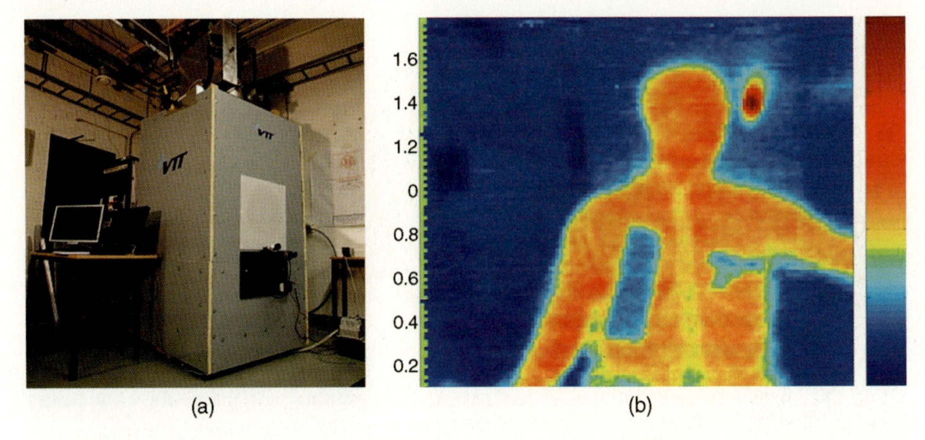

Figure 7.12 Current realization of a passive terahertz imager at VTT Finland (a), and typical image recorded at 8 m distance (b).

and US consortium led by the Technical Research Centre of Finland (VTT) in Helsinki, and a German consortium lead by the Institute of Photonic Technologies (IPHT) in Jena.

The Helsinki team has chosen antenna-coupled micro-bolometers, cooled to about 4 K by a commercial pulse tube cooler (PTC). The PTC technology is based on a *Stirling* process utilizing pressure impulses. Cryogenic temperatures well below 4 K are achievable by emerging two-stage devices, rather than single-stage implementations which have been in routine use for many years, for example, for cooling infrared focal plane arrays [48].

The current VTT system (see Figure 7.12) uses a linear array of 64 bolometers, in combination with a conical scanner as part of a folded Schmidt optics [49]. In contrast to almost all body scanners described in this chapter, the system features a notably large optical bandwidth between 1 and 0.3 mm (300 GHz up to 1 THz). Consequently, it integrates over the corresponding atmospheric windows, which in some sense constitutes an inherent fusion of bands with high clothing transparency and bands with higher spatial resolution. At the same time it eases the demands on the detectors, because the available energy from different bands is notably larger than for single-band operation.

At the same time, the Jena group has independently followed a parallel path [50] by using a smaller number of detectors which feature a higher sensitivity beause they are cooled to even lower temperatures of 0.3 K. This allows the system to be operated in a single atmospheric window, in this case 0.9 mm (350 GHz).

Again, the operation temperature is achieved by a two-stage PTC, featuring an additional closed-cycle ^{3}He evaporation cooler for thermal stability. The built system (see Figure 7.13) uses a receiver with 20 transition-edge bolometers, which to date is the limit for an economically priced scanning system. The scanning is accomplished on a spiral trace by simultaneously tilting and rotating the secondary mirror of the Cassegrain optics, allowing for video frame rates up to 10 Hz.

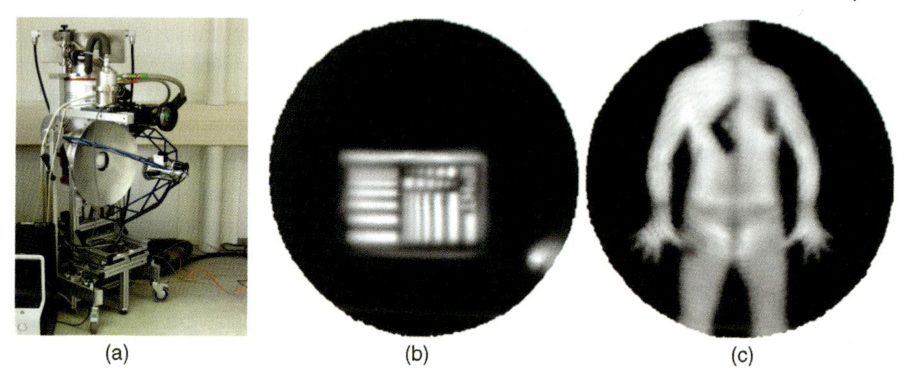

Figure 7.13 System demonstrator built at IPHT Jena (a) and images recorded from 8 m distance (resolution test (b) person with handgun and explosive mock-ups (c)).

7.5 Conclusion and Outlook

Body scanning for security applications has evolved into a highly dynamic market with enormous growth potential, driven by the permanent threat of terrorist attacks. The first systems made use of X-ray technologies, relying on the heritage of medical imaging. However, because of ethical issues and health effects, X-ray imaging did not become widely accepted, at least in the presence of an existing alternative.

The gap is being filled by emerging millimeter wave technologies. The enhancement of traditional microwave techniques into the millimeter wave band has reached a high state of maturity, enabling the production of the first commercial systems which can compete with X-ray imaging at an almost negligible exposure level to harmless millimeter wave radiation. Moreover, by using techniques related to the RADAR approach, the unique feature of 3D image reconstruction of the person under test allows an improved detection capability, overcoming the obstacle of fully automated object recognition, as required by security officials.

Thinking beyond established security concepts, the further enhancement of security technologies could, by using camera-like sub-millimeter wave systems able to perform security checks *en passant*, remove the need for the people under test to be examined in inspection cabins. This is the subject of recent research, which is additionally driven by the vision of material identification, for example, explosives, by means of terahertz spectroscopy.

References

1 Bushburg, J., Seibert, A., Leidholdt, E., and Boone, J. (2002) *The Essential Physics of Medical Imaging*, Lippincott, Williams & Wilkins, USA.

2 Towe, B.C. and Jacobs, A.M. (1981) X-Ray Compton Scatter Imaging Using a High Speed Flying Spot X-Ray Tube. *IEEE Trans. Bio.-Med. Eng.*, **28** (10), 717–721.

3 Huda, W. and Slone, R.M. (2003) *Review of Radiological Physics*, Lippincott, Williams & Wilkins, USA.

4 Chotas, H.G., Dobbins, J.T., and Ravin, C.E. (1999) Principles of digital radiography with large-area, electronically readable detectors: a review of the basics. *Radiology*, **210**, 595–599.

5 Kotter, E. and Langer, M. (2002) Digital radiography with large-area flat-panel detectors. *Eur. Radiol.*, **12**, 2562–2570.

6 http://www.as-e.com/products_solutions/smart_check.asp.

7 council directive 96/29/EURATOM of the European Atomic Energy Community (1996).

8 Brenner, D.J. and Hall, E.J. (2007) Computed tomography—an increasing source of radiation exposure. *N. Engl. J. Med.*, **357** (22), 2277–2284.

9 http://www.brauninternational.com/conpass-body-scanner-2-37-217.php.

10 Company Presentation of the Braun Full-body scanner Part No BRAU1950, company website, http://www.brauninternational.com.

11 Alekseev, S.I. and Ziskin, M.C. (2007) Human Skin Permittivity Determined by Millimeter Wave Reflection Measurements. *Bioelectromagnetics*, **28**, 331–339.

12 Appleby, R. and Wallace, H.B. (2007) Standoff Detection of Weapons and Contraband in the 100GHz to 1 THz Region. *IEEE Trans. Antenn. Propag.*, **55** (11), 2944–2956.

13 Bjarnason, J.E., Chan, T.L.J., Lee, A.W.M., Celis, M.A., and Brown, E.R. (2004) Millimeter-wave, terahertz, and mid-infrared transmission through common clothing. *Appl. Phys. Lett.*, **85** (4), 519–521.

14 Liebe, H.J. and Layton, D.H. (1987) *Millimeter-Wave Properties of the Atmosphere Laboratory Studies and Propagation Modelling*, U.S. Dept. of Commerce, National Telecommunications and Information Administration, Washington.

15 Whitaker, J.C. (2005) *Standard Handbook of Broadcast Engineering*, McGraw-Hill Professional, USA.

16 Shambayati, S. (2008) Atmosphere attenuation and noise temperature at microwave frequencies, in *Low-Noise Systems in the Deep Space Network* (ed. M.S. Reid), Wiley.

17 Faber, M.T., Chramiec, J., and Adamski, M.E. (1995) *Microwave and Millimeter-Wave Diode Frequency Multipliers*, Artech House Microwave Library.

18 Kantorowicz, G. and Palluel, P. (1979) Backward wave oscillators, in *Infrared and Millimeter Waves* (ed. K. Button), Academic Press.

19 Chen, J.C., Pao, C.K., and Wong, D.W. (1987) Millimeter-wave monolithic gunn oscillators. *Microwave and Millimeter-Wave Monolithic Circuits*, **87** (1), 11–13.

20 http://vadiodes.com/index.php.

21 Hagen, J.B. (1996) *Radio-Frequency Electronics: Circuits and Applications*, Cambridge University Press, Technology & Engineering.

22 Pozar, D.M. (1993) *Microwave Engineering*, Addison-Wesley Publishing Company.

23 Pardo, D., Grajal, J., Pérez, S., Mencía, B., Mateos, J., and González, T. (2011) Analysis of noise spectra in GaAs and GaN Schottky barrier diodes. *Semicond. Sci. Tech.*, **26** (5), 55023–55033.

24 Siweris, H.J., Werthof, A., Tischer, H., Schaper, U., Schafer, A., Verweyen, L., Grave, T., Bock, G., Schlechtweg, M., and Kellner, W. (1998) Low-cost GaAs pHEMT MMIC's for millimeter-wave sensor applications. *IEEE Trans. Microw. Theory*, **46** (12), 2560–2567.

25 Grudkowski, T.W., Rowe, R.L., Blasing, R.R., Trosper, S.T., Trawick, T.W., Meyer, K.A., Genske, D.J., Van Eyck, G.J., and Brinkerhoff, M.D. (2006) Millimeter-wave active imaging system, US patent application 6992616.

26 Federici, J.F., Schulkin, B., Huang, F., Gary, D., Barat, R., Oliveira, F., and Zimdars, D. (2005) THz imaging and sensing for security applications and explosives, weapons and drugs. *Semicond. Sci. Tech.*, **20** (7), 266–280.

27 http://www.sds.l-3com.com/advancedimaging/provision-at.htm.

28 Sugiura, S. (2009) A Review of Recent Patents on Reactance-Loaded Reconfigurable Antennas. *Recent Patents Electr. Eng.*, **2**, 200–206.

29 http://www.smithsdetection.com/eqo.php.

30 Ahmed, S.S., Schiessl, A., and Schmidt, L.P. (2009) Near field mm-wave imaging

with multistatic sparse 2D-arrays. Proceedings of the 6th European Radar Conference 2009, Rome.

31 Ahmed, S.S., Ostwald, O., and Schmidt, L.P. (2009) Automatic detection of concealed dielectric objects for personnel imaging. Proceedings of the IEEE MTT-S International Microwave Workshop on Wireless Sensing, Local Positioning, and RFID.

32 Appleby, R., Anderton, R.N., Thomson, N.H., and Jack, J.W. (2004) The design of a real-time 94-GHz passive millimetre-wave imager for helicopter operations. *Proc. SPIE*, **5619**, 38–46.

33 Appleby, R. (2006) Passive Millimetre Wave Imaging what it can and cannot do, presentation at International Forum on Terahertz Spectroscopy and Imaging, March 2 2006.

34 http://www.millivision.com/.

35 http://www.thruvision.com.

36 http://www.apex-telescope.org/sites/chajnantor/atmosphere/transpwv/.

37 Siegel, P.H. (2002) Terahertz technology. *IEEE Trans. Microw. Theory*, **50** (3), 910–928.

38 Williams, B.S. (2007) Terahertz quantum-cascade lasers. *Nat. Photon.*, **1**, 517–525.

39 Hu, B.B., Zhang, X.-C., and Auston, D.H. (1991) Terahertz radiation induced by subband-gap femtosecond optical excitation of GaAs. *Phys. Rev. Lett.*, **67**, 2709–2712.

40 Vergheses, S., McIntosch, K.A., Calawa, S., Dinatale, W.F., Duerr, E.K., and Molvar, K.A. (1998) Generation and detection of coherent terahertz waves using two photomixers. *Appl. Phys. Lett.*, **73** (26), 3824–3826.

41 Walcott, T.M. (ed.) (2011) *Bolometers: Theory, Types and Applications*, Nova Science.

42 Zmuidzinas, J. and Richards, P.L. (2004) Superconducting Detectors and Mixers for Millimeter and Submillimeter Astrophysics. *Proc. IEEE*, **92** (10), 1597–1616.

43 Wu, Q. and Zhang, X.C. (1996) Ultrafast electro-optic field sensors. *Appl. Phys. Lett.*, **68** (12), 1604–1606.

44 Sheen, D.M., Hall, T.E., Severtsen, R.H., McMakin, D.L., Hatchell, B.K., and Valdez, P.L.J. (2009) Active wideband 350 GHz imaging system for concealed-weapon detection. *Proc. SPIE*, **7309**, 73090I-1.

45 Hübers, H.W., Semenov, A.D., Richter, H., and Böttger, U. (2007) Terahertz imaging system for stand-off detection of threats. *Proc. SPIE*, **6549**, 65490A.

46 Chan, W.L., Deibel, J., and Mittleman, D.M. (2007) Imaging with terahertz radiation. *Rep. Prog. Phys.*, **70** (8), 1325–1379.

47 Benford, D.J., Hunter, T.R., and Phillips, T.G. (1998) Noise Equivalent Power of Background Limited Thermal Detectors at Submillimeter Wavelengths. *Int. J. Infrared Millim.*, **19** (7), 931–938.

48 Radebaugh, R. (2000) Pulse Tube Cryocoolers for Cooling Infrared Sensors. *Proc. SPIE*, **4130**, 363–379.

49 Grossman, E., Dietlein, C., Ala-Laurinaho, J., Leivo, M., Gronberg, L., Gronholm, M., Lappalainen, P., Rautiainen, A., Tamminen, A., and Luukanen, A. (2010) Passive terahertz camera for standoff security screening. *Appl. Opt.*, **49** (19), 245–259.

50 Heinz, E., May, T., Zieger, G., Born, D., Anders, S., Thorwirth, G., Zakosarenko, V., Schubert, M., Krause, T., Starkloff, M., Krueger, A., Schulz, M., Bauer, F., and Meyer, H.G. (2010) Passive Submillimeter-wave Stand-off Video Camera for Security Applications. *J. Infrared Millim. Terahertz W.*, **31** (11), 1355–1369.

8
Detection of Explosives

Wolfgang Schade, Rozalia Orghici, Mario Mordmüller, and Ulrike Willer

8.1
Introduction

Nowadays, much attention is given to the detection of explosives in order to prevent terrorist activities and assure security at airports, aviation, public transportation, convention halls, concerts, and so on. The location of undetonated and buried landmines is also a matter of great importance that could help to minimize fatalities and injuries among civilians caused by mines detonating. Another problem related to the use of explosives is the fact that nitrogen-based explosives, such as trinitrotoluene (TNT), are toxic and able to rapidly penetrate the skin, leading to significant health problems [1, 2]. For this reason, the sensing of hazardous substances may be of great interest at many facilities, where explosive materials were/are still manufactured and deposited, in order to avoid contamination of the soil or groundwater with the toxic compounds that can result from improper waste disposal, and to ensure the safety of both workers and nearby residents.

As can be seen, the detection of explosives is a significant task in the field of security and environmental analysis. Over the last several years major efforts have been focused on developing innovative sensing devices, capable of detecting hazardous species. In contrast to commonly used methods, such as ion mobility spectrometry (IMS), gas chromatography, or mass spectrometry, that require sampling for performing analysis and are relatively cost-intensive, optical methods are ideally suited for a fast online and *in situ* detection of hazardous substances due to their contact-free and nondestructive operation.

However, at the present time, there is no sensing system available which can identify all explosive compounds, promising high sensitivity, selectivity, and low false-alarm rate, simultaneously. Depending on the nature of the applications, sensing devices based on different laser spectroscopic methods have been developed, each of them offering high efficiency for a specific type of explosive material. A reliable detection of explosive traces in the vapor phase or in the form of particles, essential for security applications, can only be achieved by combining the sensing methods and their respective features. In the following, an overview of several laser-

Handbook of Biophotonics. Vol.3: Photonics in Pharmaceutics, Bioanalysis and Environmental Research, First Edition.
Edited by Jürgen Popp, Valery V. Tuchin, Arthur Chiou, and Stefan Heinemann

based detection techniques will be given, including examples of the first photonic devices and their practical application for the detection of explosives.

8.2
Optical Methods for the Detection of Explosives – Overview

Common optical stand-off, remote detection methods of trace amounts of samples have been shown to provide detection of explosives and their stimulants under specific conditions. These include laser-induced breakdown spectroscopy (LIBS), fragmentation via laser photolysis (LP) combined with fragment detection via laser-induced fluorescence (LIF), laser ionization in combination with mass spectroscopy (MS), mid-infrared (MIR) spectroscopy and spontaneous Raman scattering [3–7]. These methods suffer, at least to some extent, from one (or both) of the two following major drawbacks: false alarms and low sensitivity, since they do not efficiently discriminate between the "real" and the background signals, and since they rely on noncoherent emissions. At the present time, several approaches are being discussed and investigated but a suitably reliable, highly sensitive optical stand-off, remote detection method is not available. For point detection the situation is different, here several techniques do exist. However, a multi-species analysis is still problematic. In the following, existing laser-based methods will be discussed and evaluated with respect to the detection of explosives in near- and far-field applications.

Vibrational spectroscopy is well suited for identification of molecular species, providing spectral signatures and recognition of specific compounds. However, vibrational spectroscopy based on near- infrared (NIR) and MIR absorption is often barely applicable, due to weak overtone absorption bands and/or the possible interference of strong NIR/MIR absorbers, for example, water on a surface, which can obscure significant portions of the spectrum.

Spontaneous Raman spectroscopy has been more successful, despite the relatively low Raman scattering cross sections that lead to weak signals. Indeed, portable chemical sensors, combining nanosecond (ns) pulses in the ultraviolet (UV) or visible laser excitation for light detection and ranging (LIDAR) have been developed and evaluated [7–10]. These sensors were applied for remote sensing to identify the vibrational "fingerprints" of substances on surfaces. In principle, these systems have the potential for stand-off detection of contaminant films several micrometers thick at distances of up to ten meters, and bulk quantities of substances at distances of hundreds of meters. Nevertheless, detection sensitivity is still of major concern.

One way to enhance the sensitivity is by replacing spontaneous Raman with a two- or four- wave coherent, parametric process, namely, stimulated Raman scattering (SRS) [11, 12] and coherent anti-Stokes Raman spectroscopy (CARS) [11–13]. Both SRS and CARS are powerful nonlinear scattering processes, providing unique spectroscopic tools promising high sensitivity, molecular specificity, and detailed structural information based on Raman signatures of materials. In spontaneous Raman scattering, a laser "pump" beam at a frequency ω_p illuminates the sample, and, due to inelastic scattering, a signal is generated at the Stokes and anti-Stokes frequencies, ω_S and ω_{AS}, respectively. In conventional SRS and CARS, however, two laser beams at ω_p and ω_S coincide on the sample. For SRS, when the difference frequency, $\Delta\omega = \omega_p - \omega_S$, matches a particular molecular vibrational frequency, efficient vibrational excitation is induced. In the CARS process, a particular Raman transition is coherently driven by the ω_p and ω_S incident laser beams, and subsequently the vibrational coherence is probed by a third laser beam (usually ω_p), giving rise to the (anti-Stokes) CARS signal, designated as ω_{CARS}. Due to their nonlinear nature, SRS and CARS share the benefits of multiphoton processes, including significant signal enhancement over spontaneous Raman, as well as enhanced depth penetration. One very significant advantage of SRS over other excitation processes is its extremely high efficiency: as an example, 18–28% of the population of the vibrational ground state was transferred to the $v'' = 1$ stretch of C–D in $CDCl_3$ [14] and up to 60% to the C–H stretch in the ring of 2-phenylethylamine. For CARS, the ratio between the intensity of its signal and that of spontaneous Raman under comparable conditions is typically several orders of magnitude, depending on the molecules at hand [12, 13]. Moreover, the Raman signal is red-shifted relative to the exciting frequency, whereas CARS is blue-shifted, therefore, when applying CARS, scattering and fluorescence produced by the exciting light can easily be filtered and false alarms minimized. A scheme of SRS and CARS is presented in Figure 8.1.

Indeed, due to these benefits there has been extensive use of SRS [15–18] and CARS [19–21] in spectroscopic studies and in a variety of applications, including the use of the latter for bacterial spores [22] and for stand-off detection of various species [23]. Very recently, coherent control techniques, applying single beam, shaped femtosecond (fs) pulses [24], as well as narrowband nanosecond excitation [25], were employed for obtaining backscattered-CARS (B-CARS) (resulting from diffuse reflections of the forward-generated CARS, occurring in the sample itself) to detect solid particles of explosives and explosive-related compounds. It has been shown that the positions of the major spectral features of spontaneous Raman scattering of each compound remain essentially the same in B-CARS, but the latter allows much higher detection sensitivity [25].

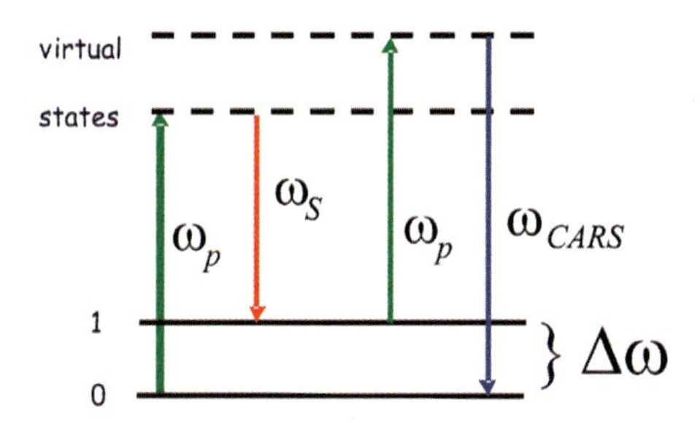

Figure 8.1 SRS and CARS schemes, where two lasers of frequencies ω_p and ω_S are used; $\Delta\omega$ is the frequency by which the vibrational state is excited by the SRS process induced by the ω_p and ω_S beams represented by the two arrows on the left. ω_p also induces the CARS transition shown on the right.

In principle, SRS and CARS spectroscopy can be implemented by applying pulses ranging from femtoseconds to nanoseconds. Whereas for fs excitation a broadband single beam was applied [24], for narrowband ns excitation [25] the spectra was produced by the time-consuming process of tuning ω_S over the excitation frequencies. However, it is anticipated that applying an ultrabroadband ns laser source for the ω_S beam will enable simultaneous excitation in a wide vibrational range, and hence capturing of multiplex SRS and CARS spectra in a short time, using a relatively simple and robust device.

Another recent development relevant to the sensitive and selective detection of explosives in the gas phase is the quartz-enhanced photoacoustic spectroscopy (QEPAS), a modification of conventional photoacoustic spectroscopy (PAS) [26]. PAS is well known as a very selective and sensitive detection technique for measuring gas concentrations down to the ppt (parts per trillion) level for specific species. However, in the past this technique required a bypass through a cell where the spectroscopic analysis has to be performed. In standard linear PAS, the gas in the cell is selectively excited by laser radiation tuned in resonance to a molecular vibrational or rotational absorption line. This is followed by vibrational–translational or rotational–translational energy transfer and generation of an acoustic wave that is detected by a microphone mounted on the cell walls. One significant disadvantage of conventional PAS is the background acoustic signals generated by the environment. Using a piezoelectric quartz tuning fork as detector, the exciting laser beam is focused between the two prongs of the tuning fork. Tuning the laser radiation on resonance with a vibrational line of a species to be investigated induces an acoustic wave which drives the tuning fork, and in the case of piezoelectric materials induces a piezo voltage in the tuning fork.

QEPAS has very recently been applied to explosive detection [27], allowing the development of an extremely compact and ultra-sensitive laser sensor. This technique is not limited to piezoelectric materials for the tuning fork since optical readout of a QEPAS device was demonstrated very recently. This allows, for the first time, fabrication of a QEPAS sensor device on a silicon chip applying well known techniques from integrated optics [28].

For gas phase detection, the combination of SRS and QEPAS, via single beam fs laser or broadband ns excitation, offers a novel approach and, up to now, nearly unexplored possibilities for engineering a new generation of multi-compound gas sensor devices that have high potential to find interesting applications in civil security. An additional advantage of these devices is the potential to combine them to build a system capable of detecting explosives in both the condensed and gas phases. Since the lasers for detecting explosives in both types of phases can be the same, it is feasible that just one system can be devised for their detection. This will have important consequences for the following reasons: it is well known that, on the one hand, some explosives have very low vapor pressure and environmental conditions (e.g., wind) make their detection very difficult, but, on the other hand, the vapor may form a thin film or particle layer on the surface of its container and, moreover, particles often adhere to the container or to the clothes of the person preparing, delivering or carrying the explosives [4]. In addition, the approach of laser hole drilling in sealed samples followed by spectroscopic analysis will also offer new possibilities for *in situ* explosive detection. Highly sensitive, multi-compound stand-off detection systems in both the condensed and gas phases will be an important step forward in the means for defense against terrorism.

8.2.1
State of the Art Spectroscopic Methods

Laser spectroscopic methods all rely on the interaction of light and matter. One of the great advantages, in contrast to, for example, chemical analysis, is the fact that no sample preparation is necessary. Therefore, the performance of measurements online and *in situ*, and often in a stand-off configuration is possible. The methods are based on the different light–matter interactions and are highlighted in the following, together with some examples with respect to the application for the detection of explosives.

8.2.1.1 Absorption Spectroscopy

Conventional laser absorption spectroscopy is a method that allows the *in situ*, contact-free and nondestructive detection of gaseous substances. Especially in the MIR spectral region, selective and sensitive detection is possible, because nearly all molecules possess distinctive absorption lines there, that can be addressed with narrowband laser sources.

The major difficulty for the detection of explosives, however, is their low vapor pressure. For nitro-aromatic explosives, it ranges between 10^{-6} and 10^{-9} mbar, leading to a concentration at ambient conditions in the ppb to ppt range. In contrast, peroxide-based explosives, for example, triacetone triperoxide (TATP), which is an explosive favored by terrorists because of its high explosive force and relative simplicity in production, have much higher vapor pressures ($p = 7$ Pa for TATP [29] at ambient conditions), leading to concentrations in the range of tens of ppm (parts per million). The equilibrium concentrations for different explosives extracted from [30, 31] are given in Figure 8.2. It can clearly be seen, that the equilibrium concentrations for different explosives vary dramatically. So, not only the different optical properties, but also the different vapor pressures call for distinctive methods for single classes of explosives.

In the case of TATP direct absorption spectroscopy is possible in the MIR spectral region. Calculation of vibrational spectra has been performed by Oxley *et al.* [32] and Brauer *et al.* [33]. First results for a differential absorption LIDAR system for the stand-off detection of TATP by Pal *et al.* are very promising, and yield a sensing limit of 52 ppm for an absorption path of 30 cm [34].

However, for nitro-aromatics, methods have to be used to enable detection even with the much lower equilibrium concentrations at ambient conditions. This can be done using a pre-concentration set-up: the gas containing the explosive is led over a trap that consists of a cooled surface or a filter on which the molecules are adsorbed. Subsequently, the trap is heated; the molecules desorb and are present in a much higher concentration within the gas that is used to flush the trap. With this kind of

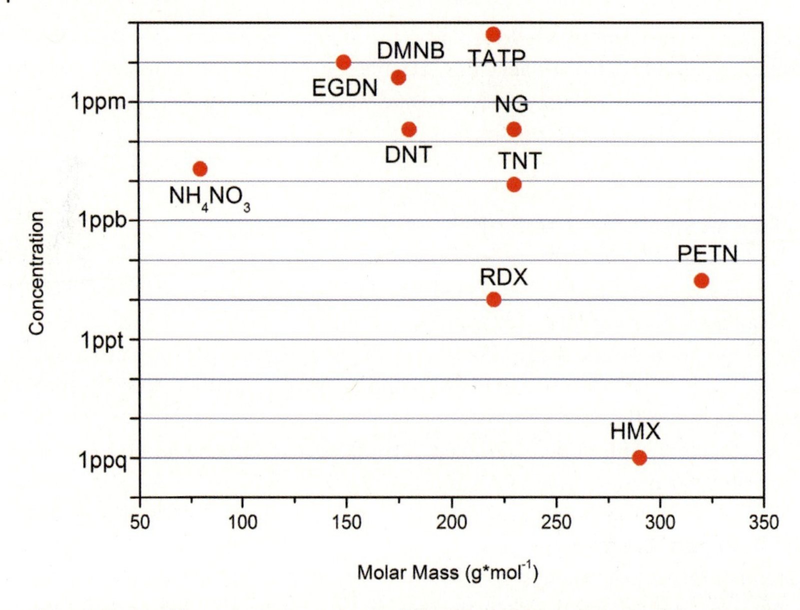

Figure 8.2 Equilibrium concentrations for different explosives as extracted from [29–31].

pre-concentration used for cavity ring down spectroscopy, detection of TNT in the ppt concentration range is expected [35].

8.2.1.2 Detection of Decomposition Products and Fragments

A different approach is to detect decomposition products rather than the explosive itself. The vapor phase concentration of the natural decay products can be several orders of magnitude higher than that of their parent explosive and the special composition of the vapors of the explosive and its decomposition products might be the key for the amazing sensitivity of canines for the detection of traces of explosives [36]. The concentrations of the decay products are dependent on the ambient temperature and pressure. Nadezhdinski *et al.* evaluated the possibility of detecting explosives by their natural decomposition products with tunable diode laser spectroscopy (TDLS) [37]. The decay products are smaller molecules which are easier to address with narrow line width tunable diode lasers than the explosives themselves, which are mostly broad-band absorbers.

Another possibility is the active production of fragments of the explosive, for example, by the use of photofragmentation. This technique is used in different spectral regions and the decomposition products can be detected with different techniques: laser induced fluorescence spectroscopy (LIF) [38], resonance enhanced multiphoton ionization (REMPI) [39–41], chemiluminescence [42] or absorption spectroscopy of the photofragments [6] (see Figure 8.3). Investigations have shown that a different ratio of NO/NO_2 production is present for explosives and non-energetic material, for example, plastics. Therefore, the possibility of the discrimination of

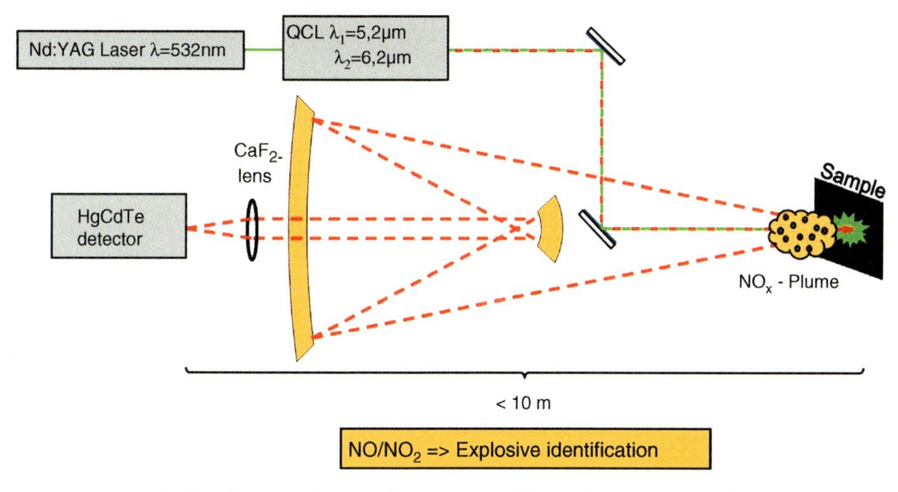

Figure 8.3 Stand-off configuration by photofragmentation followed by mid-infrared QCL detection of NOx fragments in the plume.

explosives and harmless materials is investigated by simultaneous absorption spectroscopy of the NO and NO_2 concentration after photofragmentation (Figure 8.4).

8.2.1.3 **Laser Induced Breakdown Spectroscopy (LIBS)**

Laser induced breakdown spectroscopy is an established method for surface investigation of bulk materials, for example, different kinds of metals and alloys [43]. With

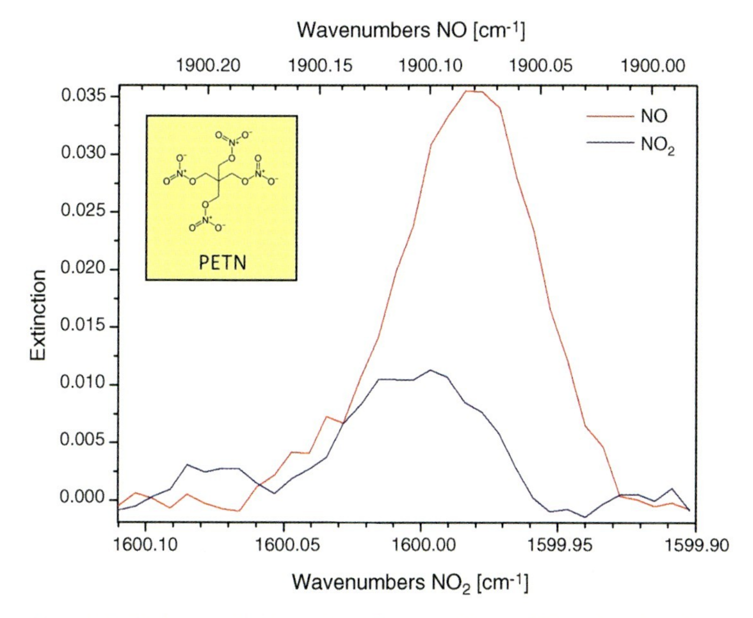

Figure 8.4 Mid-infrared QCL stand-off detection of NO and NO_2 after photofragmentation of the explosive PETN.

an intense laser pulse, a plasma is generated on the surface of a sample. This leads to different subsequent or simultaneous processes: plasma expansion, build-up of a shock wave, continuum emission (Bremsstrahlung), de-excitation of the ions, atoms and molecules through optical emission. These emission lines are detected for LIBS. Collisions represent a competing de-excitation process, inverse Bremsstrahlung and reabsorption alter the photon field. The spectral position of the emission lines, their intensity ratio and, for time-resolved LIBS, their decay rates, are parameters that are used for the identification of the substance.

LIBS is also discussed for the identification of landmines, explosives and biological warfare agents. Especially, the army research lab (ARL, USA) investigated the applicability of LIBS for the detection of bulk explosives and trace amounts on different surfaces [44–46]. The detection of trace amounts of explosives with LIBS is difficult because of matrix effects. These manifest not only in the simultaneous measurement of emission lines of the background material; as a consequence of different absorption behavior of the explosive and the matrix, different trace amounts of explosives lead to different plasma temperatures and, therefore, the intensities of lines associated with the explosive and their ratios are altered.

The surrounding atmosphere (gas composition and the pressure) and the efficiency of the collecting optics also have great influence on the signal. Therefore, data analysis is of great importance. Principal component analysis (PCA) [47] as well as partial least squares discriminant analysis (PLS-DA) [48] has proven to be powerful for the identification of explosives, even in the presence of matrix effects.

Effects of the surrounding gas atmosphere can be reduced by use of double-pulse excitation [49, 50]. With a first laser pulse, a plasma is generated which expands and locally reduces the air pressure. The plasma generated by a second, timely synchronized laser pulse then expands in a more rarefied gas atmosphere, thus the influence of the atmosphere on this plasma is reduced. Based on this technique, portable systems have been developed, for example, "TeleLis" by the Fraunhofer Institut für Lasertechnik (Germany) and "MP-LIBS" and "ST-LIBS" in the collaboration of the ARL and Ocean Optics. LIBS was also used to upgrade the function of a conventional mine prodder for the humanitarian search for anti-personnel mines. For this purpose fiber optics have been integrated into the prodder and the combination of LIBS with a miniaturized microchip laser and neural network for data analysis enables reliable identification of different mine casings [51, 52]. The concept for the fiber-based LIBS prodder system and experimental results from a field campaign for the identification of different casing materials in soil are shown in Figure 8.5.

8.2.1.4 **Raman Spectroscopy**

Raman spectroscopy relies on the inelastic scattering of photons. During this instantaneous process some energy is lost to the target molecule. Therefore, the scattered light possesses a different wavelength with respect to the incident. The change in wavelength corresponds to the differences in the vibrational energy levels of the target molecule, thus providing a means to identify it by its characteristic vibrational modes. A drawback of the method is the weak signal intensity since spontaneous Raman scattering has a very low cross section.

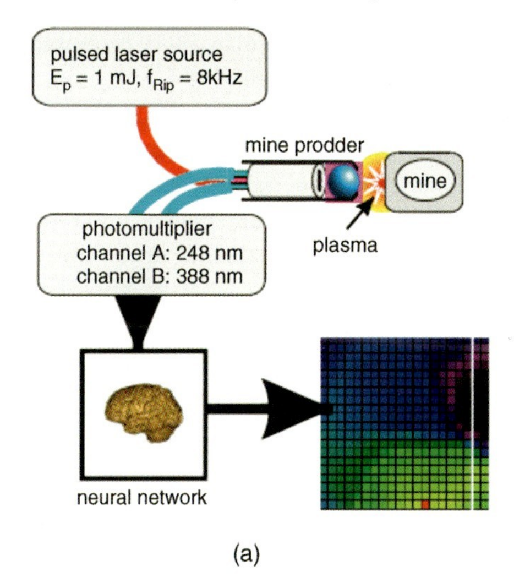

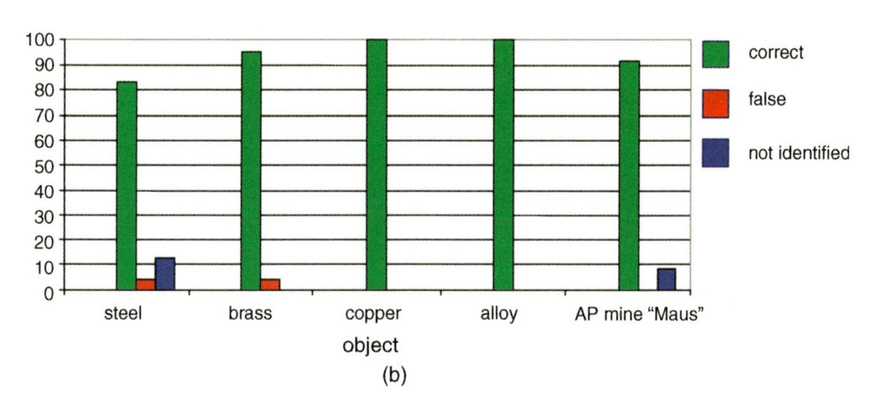

Figure 8.5 (a) Concept of a fiber coupled LIBS set-up for mine detection. (b) Results from field measurements for identification of different materials in soil by the LIBS prodder [52].

Lewis *et al.* investigated the Raman spectra of 32 different nitro-containing explosives and were able to group them into three classes with respect to their spectra: nitrates esters, nitro-aromatics and nitramines [53].

An advantage of the method is that it can be applied to small sample amounts and, in principle, through glass vials, transparent plastic bags and thin translucent materials. A limiting factor is the luminescence of background materials that might mask the weak Raman signals. Eliasson *et al.* demonstrated the noninvasive detection of hydrogen peroxide within plastic and glass containers [54]. Nagli *et al.* [55, 56] investigated the feasibility of different excitation wavelengths with respect to the strength of the Raman signal and suppression of luminescence. They found that with an excitation wavelength of $\lambda = 248$ nm it is possible to identify even Semtex, which was impossible with excitation at $\lambda = 532$ nm due to fluorescence [56].

Extensive work for the application of Raman spectroscopy to the detection of explosives has been done by FOI of Sweden [57]. Carter *et al.* demonstrated the detection of RDX, TNT, PETN and nitrate- and chlorate-containing materials in a stand-off configuration using CARS [58]. KNO_3, urea-nitrate and RDX were detected by Portnov *et al.* with backward coherent anti-Stokes Raman spectroscopy [25]. A mobile Raman stand-off system has been demonstrated by Wu *et al.* for the detection of chemicals [59]. Miniaturization enabled the set up of even portable stand-off systems, as described by Carter *et al.* [60], Ray *et al.* [8], and Sharma *et al.* [9].

There are some commercially available portable and hand-held devices relying on Raman spectroscopy. They were recently reviewed by Moore and Scharff with respect to their applicability for the detection of explosives [7]. They conclude that good Raman spectra are obtainable for white powder explosives. For colored explosives there are two issues to be noted: (i) the signal acquisition is more difficult because of fluorescence (thus demanding at least longer acquisition times) and (ii) materials that are highly absorbent for the excitation laser might pose a safety risk [7, 61].

Surface enhanced Raman spectroscopy (SERS) overcomes the low intensity levels of the signal by substantial enhancement of the effective Raman cross section of spatially confined molecules within high electromagnetic fields. Such fields are present at nanostructured noble metal surfaces. Practical application of this method is hampered by the fact that the analyte must be close to the metal surface. However, Stuart *et al.* have demonstrated the detection of a vapor phase chemical warfare agent simulant which might lead the way towards real-time detection [62]. Zhang *et al.* showed the detection of an Anthrax biomarker with SERS within the same time frame as with immunoassays [63].

8.2.1.5 Photoacoustic Spectroscopy of Explosives

Photoacoustic spectroscopy (PAS) is a very sensitive method for the measurement of gaseous species. Due to the modulated irradiation of the sample molecules with laser radiation, periodic absorption occurs, leading to the build-up of a sound wave which can be detected with a microphone. The signal depends mainly on the absorption coefficient of the molecule under investigation and on the incident laser power.

Patel recently summarized different gases for which photoacoustic sensing at ppb/sub-ppb level was achieved [64]. Investigation of the photoacoustic spectra of TNT and RDX have been performed by Prasad *et al.* [65]. The practical applicability of the method might be impaired by interferences with surrounding gases. Webber *et al.* analyze the problem of identifying explosives and warfare agents under real-world conditions [66] and Patel *et al.* report detection of TNT at a level of 0.1 ppb and TATP at $\sim$1 ppb by use of multiple wavelengths and reduced pressure [64, 67].

A very important step for the set-up of a miniaturized and rugged sensor was the invention of quartz-enhanced photoacoustic spectroscopy (QEPAS) [26]. For QEPAS, the conventional microphone is replaced by a commercially available quartz tuning fork and the excitation laser is focused between its prongs while modulated at the resonance frequency of the tuning fork (Figure 8.6). Upon absorption, a sound wave is formed which can be sensitively measured by the piezo current that is produced by the tuning fork.

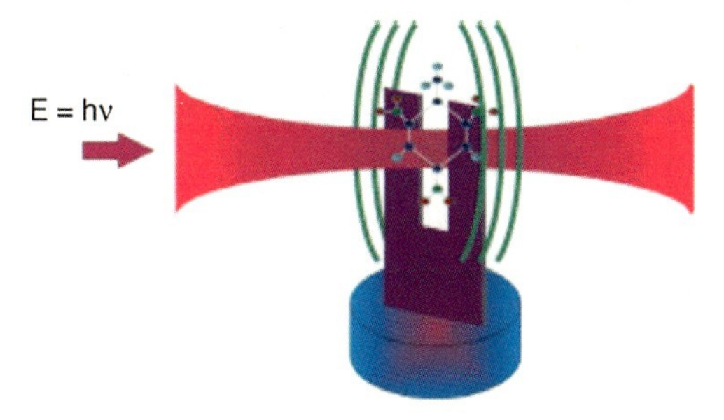

Figure 8.6 Concept of QEPAS.

As an example, the QEPAS method is demonstrated for detection of the explosive triacetone triperoxide (TATP). Figure 8.7a shows a FTIR spectrum of atmospheric water and for TATP under room temperature conditions. For the QEPAS measurements a pulsed external cavity quantum cascade laser operating between 1155 and 1220 cm^{-1} was used. The tuning position of the laser is marked within Figure 8.7b, here "water-free" detection of TATP is possible. The laser was amplitude modulated at a repetition rate of $f = 32.756$ kHz and a duty cycle of 5%, which results in an output power of $P = 5$ mW. The TATP was synthesized as described by Wolfenstein [68]. A small amount of TATP (<500 μg) was placed inside the cell and after a few seconds it could be detected via QEPAS. Figure 8.8a shows the response time for acetone detection while Figure 8.8b shows the recorded TATP spectrum. The two C–O stretch bands near $\tilde{\nu} = 1200\ cm^{-1}$ can clearly be seen. Assuming the equilibrium concentration of ~65 ppm for TATP in ambient air, a theoretical detection limit of 1 ppm TATP can be estimated with this set-up at ambient pressure and temperature conditions [69].

Based on this technique, a portable sensor with hand-held sensor head was designed, as shown in Figure 8.9. The QCL, together with all the electronics and coupling optics, is enclosed in a 19 inch box. The laser radiation is coupled into a hollow core fiber which is connected to the sensor head. There the light is focused

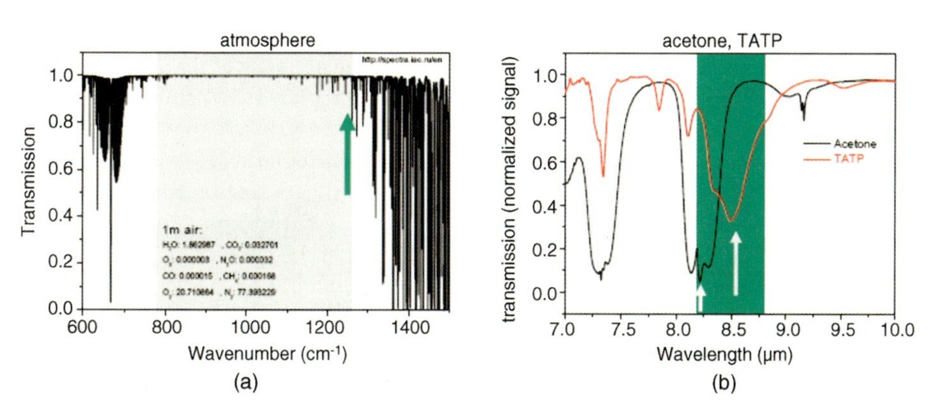

Figure 8.7 Calculated MIR spectra of the atmosphere including water vapor from Hitran database (a) and FTIR spectra for the explosive TATP and its precursor acetone (b). The arrow indicates the spectral position for "water free" detection of TATP.

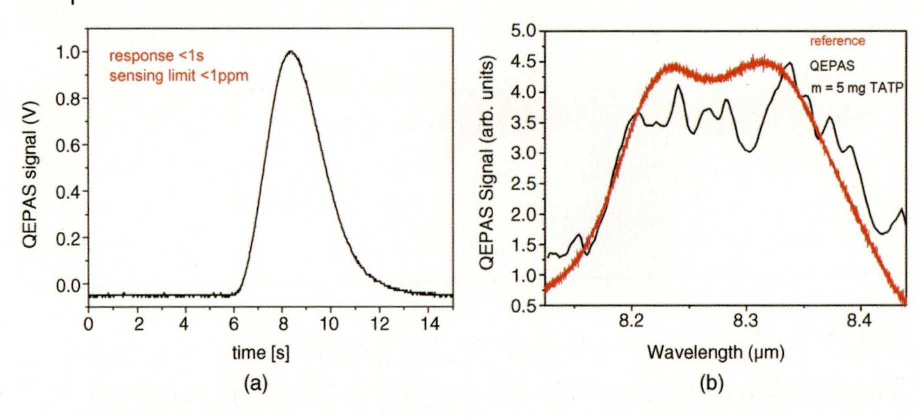

Figure 8.8 (a) Response time for acetone detection [69]. (b) Mid-infrared QEPAS recording of TATP [100].

between the prongs of the tuning fork. With a pump it is possible to generate a gas flow through the sensor head. First measurements with acetone (whose absorption can also be addressed with the QCL) are promising. The timely behavior of the QEPAS signal is shown in Figure 8.8a where the sensor head was positioned over a cloth moist with acetone and air was sucked through the sensor head.

Very recently, the optical readout of QEPAS was successfully demonstrated [28]. Thus, the design of the tuning forks is not restricted only to piezo-electric materials. In particular, this allows chip sensors for gas detection to be manufactured from silicon, making them extremely compact and ultra-sensitive. A design example is shown in Figure 8.10. Furthermore, this silicon technology allows mass production of a QEPAS chip.

8.2.1.6 Cavity Ring Down Spectroscopy

Cavity ring down spectroscopy (CRDS) is a very sensitive method for the determination of gas concentrations. For this method, the temporal behavior rather than the intensity of light that couples out of an optical cavity is measured. Therefore, fluctuations in laser intensity do not influence the accuracy of the measurement. For explosives, sub-ppb detection limits have been reported [70] using CRDS in the

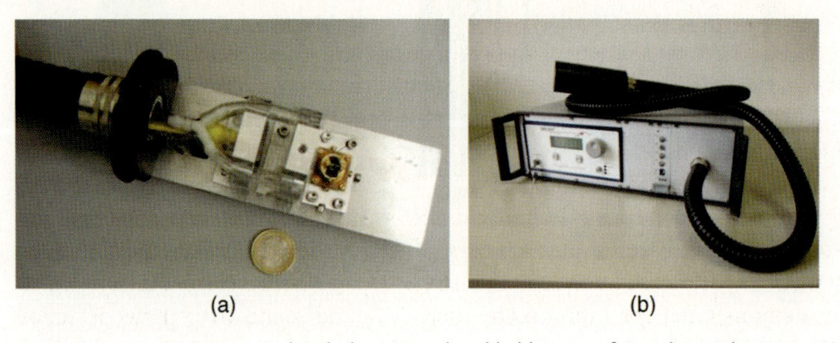

Figure 8.9 (a) QEPAS sensor head. (b) QEPAS hand-held system for explosive detection [100].

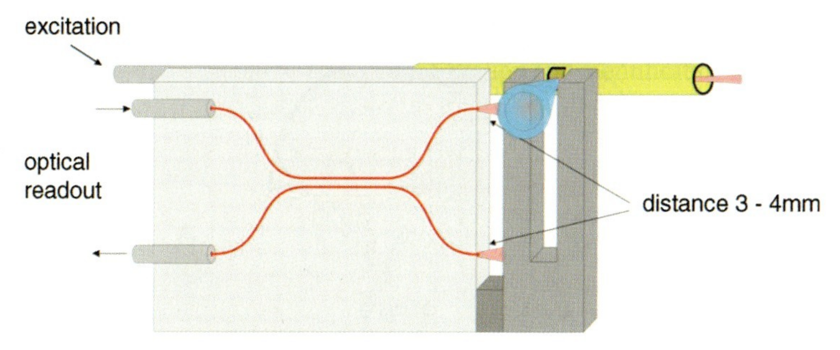

Figure 8.10 Chip-design of an optically integrated QEPAS sensor in silicon.

UV spectral range. Despite the high sensitivity they discuss the disadvantage of comparing poor selectivity in the UV range and possible interference with ozone, and propose the combination of their sensor with a second, less sensitive but more selective one [70]. Todd *et al.* use CRDS in the MIR spectral region with an optical parametric oscillator as laser source and report ppb level detection for TNT, TATP, RDX, PETN, and Tetryl, and anticipate the possibility of detecting 75 ppt TNT with pre-concentration [35]. Since the high reflectivity of the cavity mirrors is of the utmost importance, the adsorption of explosives on them has to be prevented. Therefore, heating of the optical cavity is needed.

8.2.2 Novel Approaches

Current optical stand-off and remote detection methods of trace amounts of samples suffer from several drawbacks. Especially, novel methods and recent developments from some research facilities in Israel and Europe have tried to overcome these drawbacks. Major novel development concentrates on miniaturization, simple laser excitation sources and multi-species analysis.

8.2.2.1 Femtosecond Coherent Control

The Weizmann Institute in Israel has pioneered the field of femtosecond coherent control of Raman scattering processes, in particular with connection to CARS [24, 71–76]. By using pulses shorter than the molecular vibration period, it is possible to coherently excite molecular vibrations with a single pulse. The main advantage of this approach is the simplicity of the source (a single, fixed frequency ultrashort laser oscillator), and the inherent spatial overlap of the different excitation wavelengths of the single beam, as necessary in conventional CARS schemes. Single-pulse excitation with broadband pulses by itself is unsuitable for spectroscopy, as it suffers from poor spectral resolution. Nonetheless, by controlling the phase and polarization of the single broadband pulse, it is possible both to restore the spectral resolution and to selectively excite a specific vibrational level of interest. Vibrational spectroscopy by a shaped femtosecond pulse was experimentally demonstrated for both microscopic [71] and stand-off remote detection application [24, 76], and, in particular, selective excitation of a given Raman level out

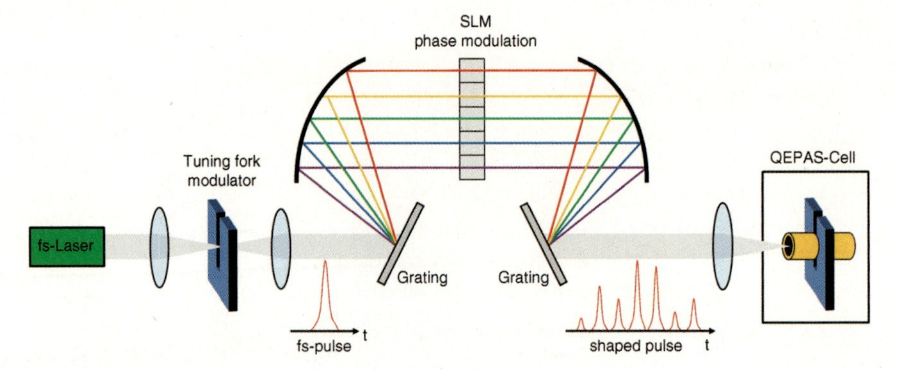

Figure 8.11 Femtosecond pulse shaping and QEPAS signal detection.

of the several vibrational levels that lie within the excitation pulse bandwidth [71]. In the work reported in Refs. [24, 76] a single-beam, stand-off (>10 m) detection and identification of various materials including minute amounts of explosives under ambient light conditions was demonstrated. It was obtained by multiplex CARS using a single femtosecond phase-shaped laser pulse. The strong nonresonant background was exploited for amplification of the backscattered resonant CARS signals by employing a homodyne detection scheme. It is thus demonstrated that this simple and highly sensitive spectroscopic technique has a potential for detection applications of hazardous materials, especially when QEPAS is used for signal detection [77]. The basic concept for such an experiment is shown in Figure 8.11. Here, the first tuning fork is used for the intensity modulation of the laser beam while the QEPAS signal detection is performed by the second tuning fork in the cell.

In addition, the Ben-Gurion University in Israel has pioneered the application of several methods for remote detection of various compounds. These methods include blue shifted LP/LIF [78–81], LIBS [82, 83] and the method of detection of particles of explosives via B-CARS, applying relatively narrowband nanosecond laser pulses [25], that has been extended to remote detection applying femtosecond pulses [24, 76]. It has been demonstrated that visible CARS in the nanosecond time domain allows favorable detection compared to spontaneous Raman spectroscopy, which has been applied for stand-off detection but with limited success due to the inherently weak Raman scattering. These first measurements were conducted at short distances, but the observed enhancement of several orders of magnitude of CARS versus spontaneous Raman indicate that the distance can be scaled to stand-off detection, in particular by applying higher power lasers. It was estimated that even by using energies of 10 mJ of the narrowband nanosecond pump and Stokes beams, detection of samples from distances of meters to hundreds of meters will be possible, depending on the specific compound.

8.2.2.2 **THz-Spectroscopy**

The terahertz spectral region is of great interest for the detection of drugs, hazardous materials and weapons since the radiation can penetrate easily through many nonmetallic, nonpolar materials. Explosives, pharmaceuticals and drugs show characteristic spectra and can selectively be detected with THz radiation. Shen *et al.* show the detection of explosives via THz imaging [84] and Federici *et al.* give an excellent overview of THz imaging and sensing for security applications [85]. With increasing availability of powerful sources and detectors for the THz spectral region, scanning velocities reach practical values. The energy level of the radiation is harmless for people; however, broad application for passenger screening is hindered by ethical concerns. Despite this limitation, wide fields of application, for example, postal screening, are open for this promising technique.

8.2.2.3 Photonic Ring Resonator Sensors

The detection of various species in the gaseous or fluid phases by using silicon photonic microring resonators is an emerging sensing technology increasingly gaining recognition due to the high sensitivity, mechanical stability, robustness and cost efficiency of these sensors. Their miniaturized design and compatibility with fiber networks or other photonic devices makes them suitable for field operation, which is an important feature for real-world applicability.

Microring resonator based sensors are commonly used for different biological and chemical applications [86–91]. Combining this sensing technology with specifically synthesized receptors, the molecular identification of hazardous substances like nitroaromatic explosives becomes possible. Trinitrotoluene (TNT), for example, belongs to this class of explosives and is the most commonly used compound for military and industrial purposes. It possesses a good thermal and chemical stability, enabling safe handling, and, due to its widespread availability, has been used in several terrorist incidents. Due to its very low vapor pressure in the gas phase, ($\sim$5 ppb [29, 92] at ambient conditions), the detection of TNT still remains one of the challenging tasks.

A new sensor concept consisting of a photonic silicon microring resonator sensitized with receptor molecules based on triphenylene-ketals, which enables the detection of TNT in the low parts per billion range, has been recently developed by the Fraunhofer Heinrich Hertz Institute, Goslar, in collaboration with the Johannes Gutenberg-University Mainz, both in Germany [93]. Figure 8.12 shows a scanning electron microscope (SEM) image of the silicon nitride microring resonator. The ring resonator designed in the form of a race track is positioned close to a straight waveguide that couples light in and out of the resonator via evanescent tunneling, since both waveguides are only 1 μm apart.

The resonance condition of the ring resonator is given by $m\lambda = n_{\mathrm{eff}} 2\pi r$ where m is an integer, n_{eff} is the effective refractive index, and r is the microring radius. As can be seen, the resonant wavelength of the ring resonator is directly proportional to the effective index of refraction and to the circumference of the ring. The light evanescing into the ring returns to the straight waveguide and exhibits a phase shift (which is exactly π at resonance) relative to the light propagating down the waveguide. Due to the destructive interference caused by the phase shift, the resonance is measured as a dip in the transmitted light intensity. Highly sensitive detection of molecular species using this photonic device is achieved by monitoring the resonance shift of the ring resonator, induced by changes in the local refractive index at the microring surface

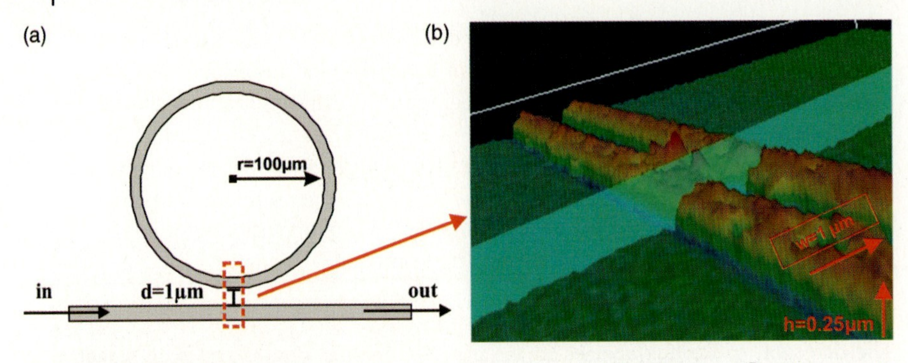

Figure 8.12 (a) Concept of a photonic microring resonator. (b) SEM image of the silicon nitride microring resonator and the coupling area between the ring and the straight waveguide [93].

that lead to changes in the effective refractive index of the guided mode inside the waveguides [94]. An increase in effective index causes a shift of the resonance towards a longer wavelength, whereas a decrease in effective index causes a shift towards a shorter wavelength.

However, this detection modality is not selective for a special species. Selectivity is gained only by functionalizing the sensor surface with specially developed coatings or receptor molecules. Therefore, the surface of the microring, that is, the sensor chip, was coated with a receptor film based on triphenylene-ketals that enables selective binding of TNT molecules in a reversible manner and via a key–lock-principle. The general synthesis of these molecules is described in detail by Schopohl *et al.* [95]. When TNT molecules are adsorbed onto the receptor layer an intensively colored intercalation complex is formed, exhibiting a charge-transfer band in the spectral region between 420 and 650 nm. In addition to the specific color change, the refractive index within the layer is changed, influencing the propagation of the light within the microring and leading to resonance shifts that can be measured by scanning the wavelength.

First investigations regarding the optical detection of the intercalation between receptor and TNT were performed via evanescent field spectroscopy by using a bare silica multimode fiber coated with the synthesized receptor molecules [96]. The sensing of species using this spectroscopic method is based on the interaction between the light guided within an optical waveguide, for example, fiber, and the surrounding medium, which occurs upon total internal reflection at the interface between two media with different index of refraction. For this purpose, a frequency doubled passively Q-switched $Cr^{4+}:Nd^{3+}$: YAG microchip laser (emission wavelength $\lambda_1 = 1064$ nm and frequency doubled to $\lambda_2 = 532$ nm) was used as light source, enabling the simultaneous measurement at both wavelengths and showing the wavelength sensitivity and selectivity of this sensor concept, since changes in the signal intensity were measured only by the interaction of the green light with TNT, whereas the intensity of the infrared light remained constant. The registered changes in the signal can be attributed to the formation of the colored charge-transfer complex between the receptor and TNT molecules that has its absorption

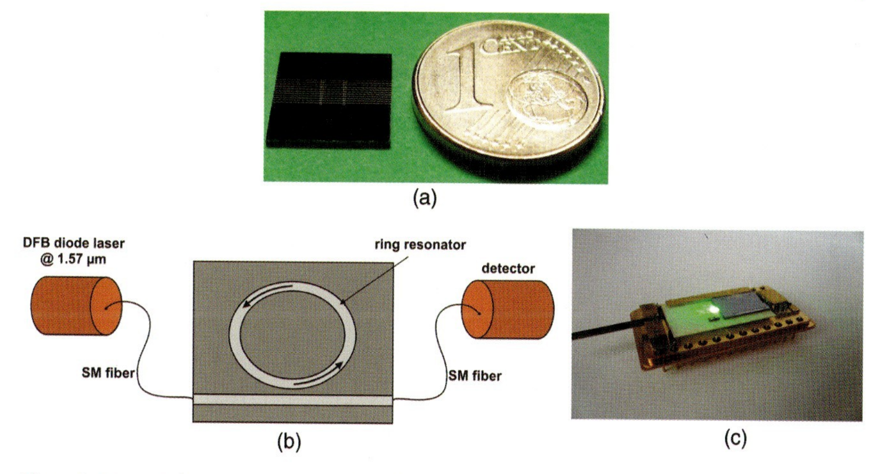

Figure 8.13 (a) Silicon ring resonator chip. (b) Schematic of the photonic ring resonator sensor. (c) Optically-integrated sensor chip.

peak around 532 nm [96]. However, even though this sensor concept allows the identification of TNT, the implementation of optical fibers as detection devices is not very convenient for field applications, because their sensitivity depends on the penetration depth of the evanescent field, which is connected to the acceptance and additionally the coupling angle of the fiber. Furthermore, due to the unknown modal distribution of the light within the fiber, changes in the transmitted intensity can occur when the fiber is bent or moved from its initial position, so that the determination of the concentration of absorbed molecules requires subsequent calibration measurements.

The situation is different for ring resonators where the whole assembly is fiber coupled and the concentration of species is determined by measuring the shift in the resonance dip. Figure 8.13b shows the experimental set-up which includes a standard telecommunication fiber coupled DFB laser diode with central wavelength at $\lambda = 1.57\,\mu m$, the microring resonator sensor and a detector for measuring the transmitted intensity through the sensor. The linear waveguide is coupled to the laser source and detector, respectively, via two single mode fibers.

The sensor surface is coated with the specially designed receptor molecules using the electrospray technique [97, 98]. By the incorporation of TNT into the receptor molecules, changes in the effective refractive index occur, inducing a shift of the microring resonant wavelength. The resultant shift can be accurately measured by scanning the wavelength around 1.57 µm, allowing in this way the estimation of the TNT concentration in the gaseous medium surrounding the sensor. Figure 8.14 shows typical resonance spectra of the microring resonator sensor measured at different concentrations of TNT under ambient conditions. As can be seen, the curves are consistently shifted with larger TNT concentrations. The limit of detection for TNT is in the range of 0.5 ppb and can be further enhanced by increasing the Q factor of the ring resonator.

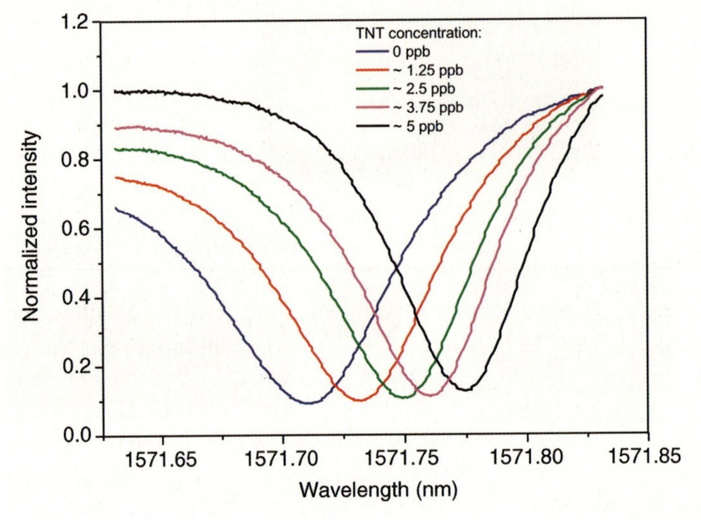

Figure 8.14 Shift of the resonance wavelength for different concentrations of the explosive TNT.

The adsorption of the TNT molecules onto the receptor layer is a reversible process, so that real time measurements can be performed. After contamination with the analyte, the sensor has to be flushed with air for desorption and can be reused within a few seconds, since the dip returns to its original position [93].

Since this sensor device is fiber coupled, it can easily be used as a hand-held device or adapted to a security gate for TNT detection in the gas phase, see Figure 8.15. The graph in Figure 8.15 shows the shift of the resonant wavelength when a person is passing the air lock security gate with and without TNT [101]. In addition, the multiplexing of several ring resonators coated with different kinds of receptors allows multi-species detection to be simplified [99]. Mass production of the sensor chips is feasible by applying standard silicon processing technology for manufacturing the sensor chips.

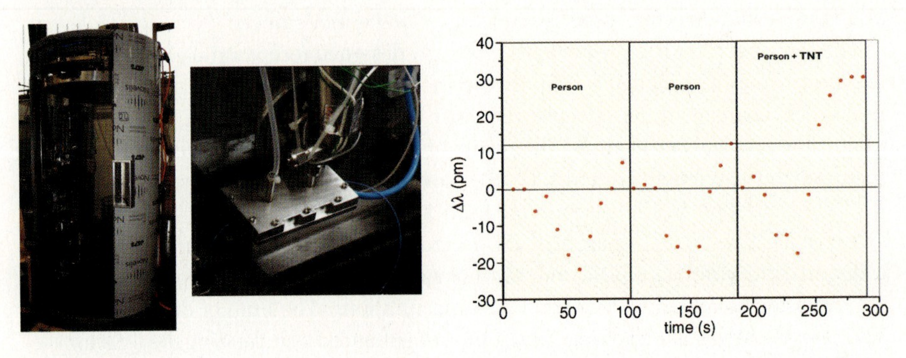

Figure 8.15 Detection of the explosive TNT in a security gate by the ring resonator sensor device [101].

8.3 Summary

Different spectroscopic methods that can be applied for the detection of explosives have been described, taking into account their specific advantages for the identification of special classes of hazardous materials. Some first generation sensing devices that are currently applied to sense explosives have been highlighted. The discussion shows that one method will not cover the identification of several explosives at the same time. Sensor fusion will be the next step to develop a spectroscopic system that allows multi-component analysis. The most important further steps besides this will be the development of novel laser sources that allow selective excitation over a wide range of molecules. In this context, shaped femtosecond laser excitation will provide a new and very interesting tool because broadband excitation in combination with pulse phase modulation will allow selective molecular excitation. On the other hand miniaturization will drive spectroscopic techniques towards applications. In particular, silicon photonics will offer new and very interesting possibilities in this field. In this context photoacoustics with microtuning forks and microring resonators will have a high potential for optically-integrated and sensitive photonic sensor devices of the next generation.

References

1 Rahal, A.Gh. and Moussa, L.A. (2011) Degradation of 2,4,6-trinitrotoluene (TNT) by soil bacteria isolated from TNT contaminated soil. *Aust. J. Basic Appl. Sci*, **5** (2), 8–17.

2 Neumann, H.-G., van Dorp, C., and Zwirner-Baier, I. (1995) The implications for risk assessment of measuring the relative contribution to exposure from occupation, environment and lifestyle: hemoglobin adducts from amino- and nitro-arenes. *Toxicol. Lett.*, **82/83**, 771–778.

3 Moore, D.S. (2004) Instrumentation for trace detection of high explosives. *Rev. Sci. Instrum.*, **75**, 2499–2512.

4 Napadensky, H.S. (ed.) (2004) *Existing and Potential Standoff Explosives Detection Techniques*, The National Academies Press, Washington, D.C.

5 Petryk, M.W.P. (2007) Promising spectroscopic techniques for the portable detection of condensed-phase contaminants on surfaces. *Appl. Spectrosc. Rev.*, **42**, 287–353.

6 Bauer, C., Sharma, A.K., Willer, U., Burgmeier, J., Braunschweig, B., Schade, W., Blaser, S., Hvozdara, L., Müller, A., and Holl, G. (2008) Potentials and limits of mid-infrared laser spectroscopy for the detection of explosives. *Appl. Phys. B*, **92**, 327–333.

7 Moore, D.S. and Scharff, R.J. (2009) Portable Raman explosives detection. *Anal. Bioanal. Chem.*, **393**, 1571–1578.

8 Ray, M.D., Sedlacek, A.J., and Wu, M. (2000) Ultraviolet mini-Raman lidar for stand-off, in situ identification of chemical surface contaminants. *Rev. Sci. Instrum.*, **71**, 3485–3489.

9 Sharma, S.K., Misra, A.K., Lucey, P.G., Angel, S.M., and Mckay, C.P. (2006) Pulsed Raman spectroscopy of inorganic and organic materials to a radial distance of 100 meters. *Appl. Spectrosc.*, **60**, 871–876.

10 Gaft, M. and Nagli, L. (2007) Standoff laser-based spectroscopy for explosives detection. *Proc. SPIE*, **6739**, 673903.

11 Bloembergen, N. (1967) The stimulated Raman effect. *Am. J. Phys.*, **35**, 989.

12 Valentini, J.J. (1987) *Laser Spectroscopy and its Applications*, vol. 11 (eds L.J. Radziemski, R.W. Solarz, and J.A.

Paisner), Optical Science and Engineering Series, Marcel Dekker, New York,
pp. 507–564.

13 Nibler, J.W. and Knighten, G.V. (1979) *Raman Spectroscopy of Gases and Liquids* (ed. A. Weber), Springer, Berlin, pp. 253.

14 Lanzisera, D.V. and Valentini, J.J. (1997) State-to-state dynamics of the $H + CDCl_3(v1''=1) \rightarrow HD(v',j') + CCl_3$ reaction. *J. Phys. Chem. A*, **101**, 6496–9503.

15 Owyoung, A. (1977) Sensitivity limitations for CW stimulated Raman spectroscopy. *Opt. Commun.*, **22**, 323–328.

16 Levine, B.F., Shank, C.V., and Heritage, J.P. (1979) Surface vibrational spectroscopy using stimulated Raman scattering. *IEEE J. Quantum. Electron.*, **15**, 1418–1432.

17 Kukura, P., McCamant, D.W., and Mathies, R.A. (2007) Femtosecond stimulated raman spectroscopy. *Annu. Rev. Phys. Chem.*, **58**, 461–488.

18 Freudiger, C.W., Min, W., Saar, B.G., Lu, S., Holtom, G.R., He, C., Tsai, J.C., Kang, J.X., and Xie, X.S. (2008) Label-free biomedical imaging with high sensitivity by stimulated raman scattering microscopy. *Science*, **322**, 1857–1861.

19 Grisch, F., Pealat, M., Bouchardy, P., Taran, J.P., Bar, I., Heflinger, D., and Rosenwaks, S. (1991) Real time diagnostics of detonation products from lead azide using coherent anti-stokes Raman scattering. *Appl. Phys. Lett.*, **59**, 3516–3518.

20 Zumbusch, A., Holton, G.R., and Xie, X.S. (1999) Three-dimensional vibrational imaging by coherent anti-stokes raman scattering. *Phys. Rev. Lett.*, **82**, 4142–4145.

21 Hashimoto, M. and Araki, T. (2000) Molecular vibration imaging in the fingerprint region by use of coherent anti-Stokes Raman scattering microscopy with a collinear configuration. *Opt. Lett.*, **25**, 1768–1770.

22 Petrov, G.I., Yakovlev, V.V., Sokolov, A.V., and Scully, M.O. (2005) Detection of Bacillus subtilis spores in water by means of broadband coherent anti-stokes Raman spectroscopy. *Opt. Expr.*, **13**, 9537–9542.

23 Li, H., Harris, D.A., Xu, B., Wrzesinski, P.J., Lozovoy, V.V., and Dantus, M. (2008) Coherent mode-selective Raman excitation towards standoff detection. *Opt. Expr.*, **16**, 5499–5504.

24 Katz, O., Natan, A., Silberberg, Y., and Rosenwaks, S. (2008) Standoff detection of trace amounts of solids by nonlinear Raman spectroscopy using shaped femtosecond pulses. *Appl. Phys. Lett.*, **92**, 171116.

25 Portnov, A., Rosenwaks, S., and Bar, I. (2008) Detection of particles of explosives via backward coherent anti-Stokes Raman spectroscopy. *Appl. Phys. Lett.*, **93**, 041115.

26 Kosterev, A.A., Bakhirkin, Y.A., Curl, R.F., and Tittel, F.K. (2002) Quartz-enhanced photoacoustic spectroscopy. *Opt. Lett.*, **27**, 1902–1904.

27 Bauer, C., Willer, U., Lewicki, R., Pohlkötter, A., Kosterev, A., Kosynkin, D., Tittel, F.K., and Schade, W. (2009) A Mid-infrared QEPAS sensor device for TATP detection. *J. Phys. Conf. Ser.*, **157**, 012002.

28 Köhring, M., Pohlkötter, A., Willer, U., Angelmahr, M., and Schade, W. (2010) Tuning fork enhanced interferometric photoacoustic spectroscopy: a new method for trace gas analysis. *Appl. Phys. B-Lasers O.*, **102**, 133–139.

29 Oxley, J.C., Smith, J.L., Shinde, K., and Moran, J. (2005) Determination of the vapor density of triacetone triperoxide (TATP) using a gas chromatography headspace technique. *Propell. Explos. Pyrot.*, **30**, 127–130.

30 Moore, D. (2007) Recent advances in trace explosives detection instrumentation. *Sens Imaging*, **8**, 9–38.

31 Thiesan, L., Hannum, D., Murray, D.W., and Parmeter, J.E. (2005) Survey of Commercially Available Explosives Detection Technologies and Equipment 2004, http://www.ncjrs.gov/App/Publications/abstract.aspx?ID=208861.

32 Oxley, J., Smith, J., Brady, J., Dubnikova, F., Kosloff, R., Zeiri, L., and Zeiri, Y. (2008) Raman and infrared fingerprint spectroscopy of peroxide-based explosives. *Appl. Spectrosc.*, **62**, 906–915.

33 Brauer, B., Dubnikova, F., Zeiri, Y., Kosloff, R., and Gerber, R.B. (2008) Vibrational spectroscopy of triacetone triperoxide (TATP): Anharmonic fundamentals, overtones and combination bands. *Spectrochim. Acta A*, **71**, 1438–1445.

34 Pal, A., Clark, C.D., Sigman, M., and Killinger, D.K. (2009) Differential absorption lidar CO2 laser system for remote sensing of TATP related gases. *Appl. Opt.*, **48**, B145–B150

35 Todd, M.W., Provencal, R.A., Owano, T.G., Paldus, B.A., Kachanov, A., Vodopyanov, K.L., Hunter, M., Coy, S.L., Steinfeld, J.I., and Arnold, J.T. (2002) Application of mid-infrared cavity-ringdown spectroscopy to trace explosives vapor detection using a broadly tunable (6–8 μm) optical parametric oscillator. *Appl. Phys. B-Lasers O.*, **75**, 367–376.

36 Furton, K.G. and Myers, L.J. (2001) The scientific foundation and efficacy of the use of canines as chemical detectors for explosives. *Talanta*, **54**, 487–500.

37 Nadezhdinskii, A.I., Ponurovskii, Y.Y., and Stavrovskii, D.B. (2008) Non-contact detection of explosives by means of a tunable diode laser spectroscopy. *Appl. Phys. B-Lasers O.*, **90**, 361–364.

38 Wu, D., Singh, J.P., Yueh, F.Y., and Monts, D.L. (1996) 2,4,6-Trinitrotoluene detection by laser-photofragmentation-laser-induced fluorescence. *Appl. Opt.*, **35**, 3998–4003.

39 Cabalo, J. and Sausa, R. (2005) Trace detection of explosives with low vapor emissions by laser surface photofragmentation-fragment detection spectroscopy with an improved ionization probe. *Appl. Opt.*, **44**, 1084–1091.

40 Heflinger, D., Arusi-Parpar, T., Ron, Y., and Lavi, R. (2002) Application of a unique scheme for remote detection of explosives. *Opt. Commun.*, **204**, 327–331.

41 Swayambunathan, V., Singh, G., and Sausa, R.C. (1999) Laser photofragmentation-fragment detection and pyrolysis-laser-induced fluorescence studies on energetic materials. *Appl. Opt.*, **38**, 6447–6454.

42 Monterola, M., Smith, B., Omenetto, N., and Winefordner, J. (2008) Photofragmentation of nitro-based explosives with chemiluminescence detection. *Anal. Bioanal. Chem.*, **391**, 2617–2626.

43 Vadillo, J.M. and Laserna, J.J. (2004) Laser-induced plasma spectrometry: truly a surface analytical tool. *Spectrochim. Acta B*, **59**, 147–161.

44 Gottfried, J.L., De Lucia, F.C. Jr., Munson, C.A., and Miziolek, A.W. (2008) Strategies for residue explosives detection using laser-induced breakdown spectroscopy. *J. Anal. Atom. Spectrom.*, **23**, 205–216.

45 Lopez-Moreno, C., Palanco, S., Laserna, J.J., De Lucia, F.C. Jr., Miziolek, A.W., Rose, J., Walters, R.A., and Whitehouse, A.I. (2006) Test of a stand-off laser-induced breakdown spectroscopy sensor for the detection of explosive residues on solid surfaces. *J. Anal. Atom. Spectrom.*, **21**,
55–60.

46 De Lucia, F.C. Jr., Samuels, A.C., Harmon, R.S., Walters, R.A., McNesby, K.L., LaPointe, A., Winkel, R.J. Jr., and Miziolek, A.W. (2005) Laser-induced breakdown spectroscopy (LIBS): a promising versatile chemical sensor technology for hazardous material detection. *Sensor J. IEEE*, **5**, 681–689.

47 De Lucia, F.C. Jr., Gottfried, J.L., Munson, C.A., and Miziolek, A.W. (2007) Double pulse laser-induced breakdown spectroscopy of explosives: Initial study towards improved discrimination. *Spectrochim. Acta B*, **62**, 1399–1404.

48 De Lucia, F.C. Jr., Gottfried, J.L., Munson, C.A., and Miziolek, A.W. (2008) Multivariate analysis of standoff laser-induced breakdown spectroscopy spectra for classification of explosive-containing residues. *Appl. Opt.*, **47**, G112–G121

49 Gottfried, J.L., De Lucia, F.C. Jr., Munson, C.A., and Miziolek, A.W. (2007) Double-pulse standoff laser-induced breakdown spectroscopy for versatile hazardous materials detection. *Spectrochim. Acta B*, **62**, 1405–1411.

50 Colao, F., Lazic, V., Fantoni, R., and Pershin, S. (2002) A comparison of single and double pulse laser-induced

breakdown spectroscopy of aluminum samples. *Spectrochim. Acta B*, **57**, 1167–1179.

51 Schade, W., Bauer, C., Orghici, R., Waldvogel, S., and Börner, S. (2008) Miniaturized photonic sensor devices for real time explosive detection, in *Detection of Liquid Explosives and Flammable Agents in Connection with Terrorism* (eds A. Kuznetsov, O.I. Osetrov, and H. Östmark), Springer Science + Business Media B.V., Dordrecht, pp. 215–227.

52 Bohling, C., Scheel, D., Hohmann, K., Schade, W., Reuter, M., and Holl, G. (2006) Fiber-optic laser sensor for mine detection and verification. *Appl. Opt.*, **45**, 3817–3825.

53 Lewis, I.R., Daniel, N.W., and Griffiths, P.R. (1997) Interpretation of Raman spectra of nitro-containing explosive materials. Part I: group frequency and structural class membership. *Appl. Spectrosc.*, **51**, 1854–1867.

54 Eliasson, C., Macleod, N.A., and Matousec, P. (2007) Noninvasive detection of concealed liquid explosives using Raman spectroscopy. *Anal. Chem.*, **79**, 8185–8189.

55 Nagli, L., Gaft, M., Fleger, Y., and Rosenbluh, M. (2008) Absolute Raman cross-sections of some explosives: Trend to UV. *Opt. Mater.*, **30**, 1747–1754.

56 Gaft, M. and Nagli, L. (2008) UV gated Raman spectroscopy for standoff detection of explosives. *Opt. Mater.*, **30**, 1739–1746.

57 Johansson, I., Norrefeldt, M., Pettersson, A., Wallin, S., and Östmark, H. (2008) Close-range and standoff detection and identification of liquid explosives by means of raman spectroscopy, in *Detection of Liquid Explosives and Flammable Agents in Connection with Terrorism* (eds A. Kuznetsov, O.I. Osetrov, and H. Östmark), Springer Science + Business Media B.V., Dordrecht, pp. 143–153.

58 Carter, J.C., Angel, S.M., Lawrence-Snyder, M., Scaffidi, J., Whipple, R.E., and Reynolds, J.G. (2005) Standoff detection of high explosive materials at 50 meters in ambient light conditions using a small raman instrument. *Appl. Spectrosc.*, **59**, 769–775.

59 Wu, M., Ray, M., Fung, K.H., Ruckman, M.W., Harder, D., and Sedlacek, A.J. (2000) Stand-off detection of chemicals by UV Raman spectroscopy. *Appl. Spectrosc.*, **54**, 800–806.

60 Carter, J.C., Scaffidi, J., Burnett, S., Vasser, B., Sharma, S.K., and Angel, S.M. (2005) Stand-off Raman detection using dispersive and tunable filter based systems. *Spectrochim. Acta A*, **61**, 2288–2298.

61 Harvey, S.D., Peters, T.J., and Wright, B.W. (2003) Safety considerations for sample analysis using a near-infrared (785nm) raman laser source. *Appl. Spectrosc.*, **57**, 580–587.

62 Stuart, D.A., Biggs, K.B., and Van Duyne, R.P. (2006) Surface-enhanced Raman spectroscopy of half-mustard agent. *Analyst*, **131**, 568–572.

63 Zhang, X., Young, M.A., Lyandres, O., and van Duyne, R.P. (2005) Rapid detection of an anthrax biomarker by surface-enhanced Raman spectroscopy. *J. Am. Chem. Soc.*, **127**, 4484–4489.

64 Patel, C.K.N. (2008) Laser photoacoustic spectroscopy helps fight terrorism: High sensitivity detection of chemical Warfare Agent and explosives. *Eur. Phys. J. - Spec. Top.*, **153**, 1–18.

65 Prasad, R.L., Prasad, R., Bhar, G.C., and Thakur, S.N. (2002) Photoacoustic spectra and modes of vibration of TNT and RDX at CO_2 laser wavelengths. *Spectrochim. Acta A*, **58**, 3093–3102.

66 Webber, M.E., Pushkarsky, M., Patel, C., and Kumar, N. (2005) Optical detection of chemical warfare agents and toxic industrial chemicals: Simulation. *J. Appl. Phys.*, **97**, 113101.

67 Dunayevskiy, I., Tsekoun, A., Prasanna, M., Go, R., and Patel, C.K.N. (2007) High-sensitivity detection of triacetone triperoxide (TATP) and its precursor acetone. *Appl. Opt.*, **46**, 6397–6404.

68 Wolfenstein, R. (1895) Über die einwirkung von wasserstoff superoxyd auf aceton und mesityloxyd. *Chem. Ber.*, **28**, 2265–2269.

69 Willer, U. and Schade, W. (2009) Photonic sensor devices for explosive detection. *Anal. Bioanal. Chem.*, **395**, 275–282.

70 Ramos, C. and Dagdigian, P.J. (2007) Detection of vapors of explosives and

explosive-related compounds by ultraviolet cavity ringdown spectroscopy. *Appl. Opt.*, **46**, 620–627.

71 Dudovich, N., Oron, D., and Silberberg, Y. (2002) Single-pulse coherently controlled nonlinear Raman spectroscopy and microscopy. *Nature*, **418**, 512–514.

72 Oron, D., Dudovich, N., Yelin, D., and Silberberg, Y. (2002) Quantum control of coherent anti-Stokes Raman processes. *Phys. Rev. A*, **65**, 043408.

73 Dudovich, N., Oron, D., and Silberberg, Y. (2002) Single-pulse phase-contrast nonlinear raman spectroscopy. *Phys. Rev. Lett.*, **89**, 273001.

74 Dudovich, N., Oron, D., and Silberberg, Y. (2003) Femtosecond phase-and-polarization control for background-free coherent anti-stokes raman spectroscopy. *Phys. Rev. Lett.*, **90**, 213902.

75 Dudovich, N., Oron, D., and Silberberg, Y. (2003) Single-pulse coherent anti-Stokes Raman spectroscopy in the fingerprint spectral region. *J. Chem. Phys.*, **118**, 9208.

76 Katz, O., Natan, A., Rosenwaks, S., and Silberberg, Y. (2008) Shaped femtosecond pulses for remote chemical detection. *Opt. Photonics News*, **19**, 47.

77 Schippers, W., Gershnabel, E., Burgmeier, J., Katz, O., Willer, U., Averbukh, I.Sh., Silberberg, Y., and Schade, W. (2011) Stimulated Raman rotational photoacoustic spectroscopy using a quartz tuning fork and femtosecond excitation. *Appl. Phys. B*, **105**, 203–211.

78 Daugey, N., Shu, J., Bar, I., and Rosenwaks, S. (1999) Nitrobenzene detection by one-color laser-photolysis/laser-induced fluorescence of NO (v'=0–3). *Appl. Spectrosc.*, **53**, 57–64.

79 Shu, J., Bar, I., and Rosenwaks, S. (1999) Dinitrobenzene detection by use of one-color laser photolysis and laser-induced fluorescence of vibrationally excited NO. *Appl. Opt.*, **38**, 4705–4710.

80 Shu, J., Bar, I., and Rosenwaks, S. (2000) The use of rovibrationally excited NO photofragments as trace nitrocompounds indicators. *Appl. Phys. B*, **70**, 621–625.

81 Shu, J., Bar, I., and Rosenwaks, S. (2000) NO and PO photofragments as trace analyte indicators of nitrocompounds and organophosphonates. *Appl. Phys. B*, **71**, 665–672.

82 Portnov, A., Rosenwaks, S., and Bar, I. (2003) Identification of organic compounds in ambient air via characteristic emission following laser ablation. *J. Lumin.*, **102–103**, 408–413.

83 Portnov, A., Rosenwaks, S., and Bar, I. (2003) Emission following laser-induced breakdown spectroscopy of organic compounds in ambient air. *Appl. Opt.*, **42**, 2835–2842.

84 Shen, Y.C., Lo, T., Taday, P.F., Cole, B.E., Tribe, W.R., and Kemp, M.C. (2005) Detection and identification of explosives using terahertz pulsed spectroscopic imaging. *Appl. Phys. Lett.*, **86**, 241116-241116-3.

85 Federici, J.F., Schulkin, B., Huang, F., Gary, D., Barat, R., Oliveira, F., and Zimdars, D. (2005) THz imaging and sensing for security applications—explosives, weapons and drugs. *Semicond. Sci. Tech.*, **20**, S266–S280

86 Passaro, V.M.N., Dell'Olio, F., and De Leonardis, F. (2007) Ammonia optical sensing by microring resonators. *Sensors*, **7**, 2741–2749.

87 Chao, C.-Y. and Guo, L.J. (2003) Biochemical sensors based on polymer microrings with sharp asymmetrical resonance. *Appl. Phys. Lett.*, **83**, 1527–1529.

88 De Vos, K., Bartolozzi, I., Schacht, E., Bienstman, P., and Baets, R. (2007) Silicon-on-Insulator microring resonator and label-free biosensing. *Opt. Expr.*, **12**, 7610–7615.

89 Kim, G.-D., Son, G.-S., Lee, H.-S., Kim, K.-D., and Lee, S.-S. (2008) Integrated photonic glucose biosensor using a vertically coupled microring resonator in polymers. *Opt. Commun.*, **281**, 4644–4647.

90 Kwon, M.-S. and Steier, W.H. (2008) Microring-resonator-based sensor measuring both the concentration and temperature of a solution. *Opt. Expr.*, **16**, 9372–9377.

91 Luchansky, M.S. and Bailey, R.C. (2010) Silicon photonic microring resonators

for quantitative cytokine detection and T-cell secretion analysis. *Anal. Chem.*, **82**, 1975–1981.

92 Dionne, B.C., Rounbehler, D.P., Achter, E.K., Hobbs, J.R., and Fine, D.H. (1986) Vapor pressure of explosives. *J. Energ. Mater.*, **4**, 447–472.

93 Orghici, R., Lützow, P., Burgmeier, J., Koch, J., Heidrich, H., Schade, W., Welschoff, N., and Waldvogel, S.R. (2010) A microring resonator sensor for sensitive detection of 1,3,5-trinitrotoluene (TNT). *Sensors*, **10**, 6788–6795.

94 Rabus, D.G. (2007) *Integrated Ring Resonators*, Springer-Verlag, Berlin Heidelberg.

95 Schopohl, M.C., Faust, A., Mirk, D., Fröhlich, R., Kataeva, O., and Waldvogel, S.R. (2005) Synthesis of rigid receptors based on triphenylene ketals. *Eur. J. Org. Chem*, **14**, 2987–2999.

96 Orghici, R., Willer, U., Gierszewska, M., Waldvogel, S.R., and Schade, W. (2008) Fiber optic evanescent field sensor for detection of explosives and CO_2 dissolved in water. *Appl. Phys.*, **B90**, 355–360.

97 Rapp, M., Bender, F., Lubert, K.-H., Voigt, A., Bargon, J., Wächter, L., Klesper, G., Klesper, H., and Fusshöller, G. (2005) Vorrichtung zum Aufbringen von Electrospraybeschichtungen auf elektrisch nicht leitfähigen Oberflächen. German Patent No. DE 10,344,135.

98 Lubczyk, D., Siering, C., Lörgen, J., Shifrina, Z.B., Müllen, K., and Waldvogel, S.R. (2010) Simple and sensitive online detection of triacetone triperoxide explosive. *Sensor Actuat. B*, **143**, 561–566.

99 Lützow, P., Pergande, D., and Heidrich, H. (2011) Integrated optical sensor platform for multiparameter bio-chemical analysis. *Opt. Expr.*, **19** (14), 13277–13284.

100 Bauer, C., Willer, U., and Schade, W. (2010) Use of quantum cascade lasers for detection of explosives: progress and challenges. *Opt. Eng.* **49** (11), 1111261–1111267.

101 Orghici, R., Spad, C., Lützow, P., and Schade, W. (accepted for submission 2012) Microring resonators for optical sensing of hazardous materials. *Appl. Phys. B*.

Part Four
High Throughput and Content Screening

Handbook of Biophotonics. Vol.3: Photonics in Pharmaceutics, Bioanalysis and Environmental Research, First Edition.
Edited by Jürgen Popp, Valery V. Tuchin, Arthur Chiou, and Stefan Heinemann

9
High-Throughput and -Content Screening/Screening for New Pharmaceutics

Astrid Tannert and Michael Schaefer

Abbreviations

ADP	adenosine diphosphate
ALPHA	amplified luminescence homogeneous assay
AM	acetoxymethyl ester
ATP	adenosine triphosphate
cAMP	cyclic adenosine monophosphate
BRET	bioluminescence resonance energy transfer
cGMP	cyclic guanosine monophosphate
cpFP	circular permuted FP
CRE	cAMP response element
CREB	CRE binding protein
DELFIA	dissociation-enhanced lanthanide fluorescent immunoassay
DTT	dithio-treitole
ER	endoplasmic reticulum
FCS	fluorescence correlation spectroscopy
FIDA	fluorescence intensity distribution analysis
FLT	fluorescence lifetime technology
FP	fluorescent protein
FRET	fluorescence resonance energy transfer
GPCR	G-protein-coupled receptor
HCS	high-content screening
HTS	high-throughput screening
LSC	laser scanning cytometry
NMR	nuclear magnetic resonance
TR-FRET	time-resolved FRET

Handbook of Biophotonics. Vol.3: Photonics in Pharmaceutics, Bioanalysis and Environmental Research, First Edition.
Edited by Jürgen Popp, Valery V. Tuchin, Arthur Chiou, and Stefan Heinemann

9.1 Introduction

The need for new pharmaceutics is underlined by the fact that about 90% of human diseases are not sufficiently controlled, let alone cured, by currently available drugs.

Screening approaches are consistently among the primary tools for developing new pharmaceutics. In contrast to so-called rational approaches, screening utilizes large collections of chemical compounds (libraries) to be tested on biological targets. Compounds identified in a primary screen to bind to the target or to modify its biological activity are designated as hits. The primary hit list might be long and most of the hits might not be relevant for further development. Therefore, known or non-relevant compounds have to be identified, a procedure called dereplication. Confirmed hits might become lead molecules for further optimization. Starting from the lead molecules, other compounds are developed with better physico-chemical or biological properties, which might eventually lead to potential new drugs in clinical testing. Optimizing lead molecules for drug candidates includes improving their binding properties, specificity and tissue distribution, decreasing their toxicity, and characterization and optimization of their metabolic profile and elimination pathway.

Compound libraries can be tested for their ability to modify a biological target in many different ways. The simplest form to identify potential interacting molecules is to test for binding of compounds to isolated targets, or to prevent binding of an established ligand by compound competition. More relevant information may be obtained by testing compounds for their biological activity in functional assays. Assay formats that detect photons as readout (i.e., most commonly fluorescence or bioluminescence) are currently the most widely applied methods in screening campaigns. Optical assays generally are cheap, simple and fast to perform, and inherently amendable to miniaturization. They are characterized by a high sensitivity and selectivity combined with a large dynamic range. For cell-based studies, using photonics is non-invasive, at least if no exposure to excitation light of high intensity and short wavelength is necessary. Imaging at a nanometer scale provides resolution almost unmatched. Other techniques applied for affinity-based screening approaches include mass spectrometry, NMR and X-ray crystallography, which have a decreased throughput rate and often require more material than optical techniques [1]. Alternative functional assays include automated patch clamp approaches in the field of ion channel studies, which again yield much lower throughput than optical methods. In former screening approaches, the use of radioactive tracers was very popular. Environmental and safety concerns have led to an extensive replacement of radioactive techniques by fluorescent or bioluminescent methods.

To enable high-throughput screening (HTS), an assay should be as easy as possible while being robust in the generation of the required readout. To achieve this, the number of handling steps has to be reduced, for instance by avoiding separation steps in signal generation. Screening approaches should be feasible to miniaturization to reduce consumption of reagents and, thus, to save money and time.

The logical course of events in most screening campaigns is to identify a target related to disease, and then screen for compounds that alter the target's biological activity. Alternatively, one can apply models of certain diseases and screen for compounds that improve the disease-related parameters, a process known as phenotypic screening. In this case, the underlying molecular processes of the compound's action have to be identified thereafter.

HTS is mostly achieved by assay designs that yield one simple readout, such as fluorescence intensity, bioluminescence, or fluorescence polarization. With today's increased computational power, it is now possible to acquire and analyze multiple parameters from a single assay, thus gaining additional information, which can help to further characterize the effects of the tested compound on a biological system. Simple HTS assays are nowadays often substituted or complemented by multi-parametric screening approaches like automated image acquisition both in academic and in industrial screening campaigns. Because of the high content of information deduced from such approaches, they are summarized under the term high-content screening (HCS).

The aim of industrial screening is to develop pharmaceutics that have fewer side effects than current drugs or that target diseases where no adequate treatment is available. Consequently, industrial screening campaigns are influenced by the potential market and have to apply economic considerations for drug development. Even though the number of identified targets is increasing rapidly, the number of new chemical entities developed into approved drugs is not growing in a similar manner. One reason for this is that industrial drug development is focussed on a few highly relevant drug targets. Academic screening can potentially address the gaps in industrial screening by enabling a characterization of new drug targets and present hits, addressing those targets currently under-represented in industrial screening campaigns. Academic screening, therefore, often addresses broader questions, including targets where the market is not big enough for industry to set up screening campaigns [2]. The basic aim in academic screening is to develop new probes for the investigation of biological processes, which might in some cases finally also lead to the development of new drugs.

In this chapter, we briefly summarize the general considerations for screening. These considerations include target classes, compound libraries and data analysis aspects. We will focus on biophotonic assays applied in screening campaigns, starting from simple assays that achieve very high-throughput to complex screening formats that acquire multiple parameters.

9.2 Targets

Screening for new pharmaceutics primarily means testing libraries of molecules for their ability to act on a preselected biological target. A good target in a medical sense is a biological pathway which is associated with disease and whose modification is likely to provide a therapeutic option. In a chemical sense, a good target can be easily addressed by a small molecule, meaning the target is "druggable".

About half of the current pharmaceutics target only a few protein families, namely G-protein-coupled receptors (GPCRs), enzymes (especially protein kinases and phosphodiesterases), nuclear hormone receptors, or voltage-gated and ligand-gated ion channels [3, 4]. Other targets are severely under-represented by current drugs, mainly because they do not provide binding sites for small molecules from available libraries. The need to develop pharmaceutics that address such target classes is underlined by the lack of suitable therapies for many diseases in which interactions of proteins with other intracellular proteins or nucleic acids play essential roles.

Target selection for screening approaches might be influenced by previously successful targets. Receptors or other proteins that belong to an established target class, but currently lack suitable drugs that modify their activity are addressed in this case. The advantage of this approach is that screening assays for these classes are established and specialized compound libraries can be generated by similarity to known modulators of the target family [3, 5].

To develop screening strategies for targets which are less frequently modulated by existing drugs, greater efforts in assay development and library configuration are necessary. By expanding the range of possible drugs from small molecules to nucleic acids, proteins and peptides (so called biologicals), targets that were previously defined as "undruggable" can now be addressed [3].

Whereas industrial screening mainly aims for drug or lead discovery, academic screening rather covers the area of target validation, that is, investigators search for molecules that can be utilized to analyze signaling pathways or protein activities, to more precisely understand physiological and pathophysiological roles of these targets [2]. For these tool-like compounds less chemical criteria have to be fulfilled, that is, oral bioavailability or possible covalent target modification is not a major concern for research targets [6]. Target identification and validation can also be achieved by gene knock down or silencing. The number of published siRNA screens is growing rapidly, and knockout mice have also been applied for target identification, suggesting that about 10% of human genes are disease-associated [7]. Nonetheless, the acute modulation of protein functions by biologically active small molecules will always remain an important source for understanding the role of a protein within its signaling network. Since most drugs act by competing with naturally occurring molecules for binding sites on proteins, new targets might also be identified by metabolic profiling [8].

The use of optical methods is well established in screening for most common drug targets like G-protein-coupled receptors (GPCRs) and many enzymes. They have been less frequently applied in screening of ion channel function, which is classically studied by electrophysiology (though with low-throughput), but fluorescence-based functional screening assays are now also becoming common for this target class.

9.3
Substance Libraries

Classical libraries applied in drug screening consist of small molecules, mostly of less than 500 Da, which share building blocks with basic structural features of existing drugs. Lipinski introduced the so-called "rule of five", which suggests whether a compound is drug-like with respect to hydrogen-bond donors, molecular mass, lipophilicity, and the sum of nitrogen and oxygen atoms [9]. This rule has strongly influenced the selection of compounds included in small molecule libraries. There are a number of commercially available small molecule libraries, ranging from small collections containing about 1000 approved drugs and natural compounds (e.g., LOPAC1280™ or Spectrum Collection™) to huge libraries containing several 100 000 compounds. In addition, libraries exist that are enriched for compounds expected to bind to a specific target class. Industrial companies host their own preferred huge libraries (in the range of one million compounds). The arrangement of these libraries is governed by the company's intellectual property, as resulting hits are more easily developable into patented drugs.

The existing small molecule libraries are frequently not successful in identifying modulators of less common targets. Since a considerable part of natural products and metabolite scaffolds are not included in current libraries, a new trend emerges to incorporate compounds based on structures that are biologically validated but less represented in current libraries (e.g., less hydrophobic compounds and natural products or metabolite-like compounds) [10–12]. Extending the chemical space covered by screening libraries, using combinatorial chemistry or diversity-oriented synthesis, may also lead to drug developments for targets that were difficult to address previously [13].

Fragment-based libraries have been developed recently and provide an alternative lead discovery approach. These libraries are usually relatively small (1000–10 000 compounds), containing considerably smaller molecules of less than 300 Da. These so-called fragments will mostly bind target proteins with lower potency (K_D between 100 μM and 10 mM). However, they possess a high ligand efficiency, meaning that they are extremely potent with regard to their size. Fragment hits have to be further developed to yield leads, for example, by fragment linking, self-assembly or fragment optimization [14]. Starting from fragment-based libraries, drug development might succeed where classical small molecule libraries have failed [15].

Peptides are important modulators of signaling pathways, and can, thus, represent suitable pharmaceutics to address disease-relevant targets, both in basic research and in drug formulations. To identify peptides that modulate biological targets, peptide libraries are available. They can either be biologically encoded, for example, using the phage display technique, or produced by synthetic approaches. Peptide microarrays generated by SPOT-synthesis [16] can be screened by positional scanning using fluorescent readouts. Random or combinatorial peptide libraries can be used to identify biologically active substances, like kinase substrates or inhibitors [17], or GPCR modulators [18]. Moreover, libraries derived from protein-sequences are applied to characterize protein–protein interaction sites by scanning the binding of a target protein to small linearized epitopes of its interaction partner.

To increase throughput and multiplexing in affinity assays, DNA-encoded libraries were developed where each small molecule compound is tagged by a specific DNA sequence, enabling identification of binding molecules by subsequent sequencing of their DNA barcode [19, 20].

For target identification and validation, RNA interference (RNAi) libraries, relying on double-stranded short RNAs or short-hairpin RNA can be used (see [21] for a detailed discussion on strategies in RNAi library design and application).

9.4 Biomarkers and Labels

Nowadays, the most widely applied detection techniques in screening approaches include fluorescence or chemiluminescence. Label-free optical techniques like surface plasmon resonance or Raman scattering are applied in some screening approaches, but often suffer from a decreased sensitivity or limited throughput [1]. Utilizing absorbance limits assay miniaturization for HTS as absorption depends on the optical pathlength [22]. In contrast, assays that are based on photon emission are very sensitive, easily adaptable to biological targets and feasible for automation methods and miniaturization [23]. The fundamental difference between fluorescence and chemiluminescence assays is that the former requires excitation light to induce photon emission from a chromophore, while the latter converts chemical energy to light emission. An enzyme and its substrate must be available to induce a signal in bioluminescence, a special case of chemiluminescence. Emission is usually much brighter in fluorescence than in bioluminescence assays, because the excited state can be induced quickly by a high rate of excitation photons. In contrast, bioluminescence assays generally exhibit much lower background signals than fluorescence assays, since no external light has to be introduced. For the same reason, bioluminescence assays are less influenced by absorbing or fluorescent library compounds. For assays based on the detection of subcellular structures using microscopy, fluorescent probes are preferred, to allow the collection of a sufficient number of photons. The same is true for fast dynamic processes where short detection times are required. For macroscopic measurements where background signal interference is more severe, bioluminescence assays may be preferable over fluorescence [23]. Most bioluminescence assays are based on firefly or *Renilla* luciferases. The former emits light at 560 nm utilizing D-luciferin and ATP, while the latter catalyses the oxidation of coelentrazine, yielding light at 480 nm. Moreover, aequorin, a luminescent sensor which, upon calcium binding, emits blue light, is frequently applied. While fluorescence assays may be more prone to interference from colored compounds, there are a substantial number of compounds in common libraries that inhibit luciferase activity, making counterscreening utilizing a different assay format indispensable [24].

Quantum dots feature increased brightness, uniform emission profiles and resistance to photobleaching while excited at a single wavelength. Their application in screening approaches is still scarce, mainly because of difficulties in associating them with cells, though they have been used as indicators for cell position in cell microarrays [25].

9.4.1 Labels for Cell-Free Assays

A number of bioluminescence assays are applied for the detection of enzyme activity or in affinity assays. Since the intensity of bioluminescence is correlated to the concentration of reaction components, variation of one component can be translated into a signal, while holding the other reagents constant. Bioluminescence assays are well-known for the detection of ATP, but also substrate (using specifically developed "pro-luciferins" that become cleaved during the assay) or the enzyme itself may be the variable assay component. Because of their low background, bioluminescence assays yield a linear range over several orders of magnitude [23].

Fluorescence readouts include labeled antibodies detecting products from enzyme activity, labeled ligands for affinity assays, or labeled substrates for enzyme reactions. Specifically designed fluorophores may be used to sense chemical modifications. A classical example is the detection of phosphorylation by Lewis metal chelates, which is now frequently used in ALPHA assays to replace phosphorylation-specific labeled antibodies. Using fluorescent readouts, to circumvent compound interference, lanthanides may be applied, either for fluorescence intensity measurement or as FRET donors. Because of their extremely long fluorescence lifetime, gate detection can be applied. In this so-called "time-resolved" fluorescence detection method, molecules are excited with a short light pulse and fluorescence detection is delayed by about 100 μs. During this time, fluorescence of most fluorophores that may occur in compound libraries has decayed and only lanthanide fluorescence is detected. When using lanthanides in FRET assays it might be counterproductive to decrease the distance between donor and acceptor below a certain value, since this will also dramatically decrease the donor fluorescence lifetime and, thus, interfere with gated detection [26]. Using classical fluorophores, interference from fluorescent compounds can be reduced by applying red-shifted dyes [27]. Moreover, ideal fluorophores should exhibit a high quantum yield, high photostability, a large Stokes shift, and insensitivity to environmental factors like pH.

9.4.2 Labeling of Cells

In cell-based assays, fluorophores have to be introduced externally or cells have to be modified genetically to express auto-fluorescent proteins or luciferases to yield a photon emission signal. To ensure stability in screening assays, labeling procedures have to be specific, quantitative and feature high reproducibility. Moreover, interference with biological processes, other than the one they have been designed for, has to be minimized [28].

9.4.2.1 Synthetic Fluorophores

A huge number of chemical probes has been designed and is commercially available, partitioning into subcellular structures (like mitochondria, nucleus, Golgi apparatus, plasma membrane). These probes can, therefore, serve as compartment markers.

The visualization of specific proteins within the cell is usually achieved by labeled antibodies, which requires fixation and permeabilization of cells if the epitope is not exposed at the cell surface. Unfortunately, this kind of labeling typically restricts cellular imaging to endpoint measurements [29].

Other probes may indicate cellular parameters like ion concentration. Most widely applied indicators of cellular parameters are calcium indicators and probes for changes in membrane potential.

For toxicity screening a number of fluorescent or fluorogenic indicators are available (reviewed in [30]), indicating either membrane integrity, metabolic cell activity, or cellular respiration.

9.4.2.2 **Genetically-Encoded Marker Proteins**

Modern biology was greatly influenced by the possibility to introduce genes that produce a photonic signal into cells. Genetically encoded fluorescent proteins (FPs) as well as luciferases can be utilized as genetic reporters under the transcriptional control of a target molecule (see Section 9.6.7) [31].

FPs are also frequently applied to monitor the influence of compound libraries on protein activity or signaling pathways. Activation of signaling pathways or protein function can be monitored in three ways using FPs: by altered optical properties (fluorescence intensities or spectra) of the FP, by an altered FRET ratio between two FPs (or one FP and one small fluorophore), or by changes in subcellular localization of the reporter protein within the cell [32]. While the first two mechanisms can be followed in bulk measurements, the latter usually requires automated microscopy (see Section 9.6.11). Since immunostaining of cells is labor- and cost-intensive, FP fusion proteins may be preferred in localization studies over immunoassays. Moreover, once stably transfected, FP-expressing cells can be readily and repetitively monitored using microscopy, enabling the detection of dynamic processes. One limitation may arise from the size of the FP, which can alter the function of the protein it is fused to.

FPs are nowadays available in a variety of colors, which enables the use of the red and far-red spectral range, where phototoxicity of the excitation light is reduced, and less interference from cellular and compound autofluorescence is expected [33].

To probe for protein–protein interaction, protein cleavage, or protein phosphorylation, a variety of FRET probes have been designed. The dynamic range, which is defined as the difference between the minimum and maximum signal, is frequently low in FRET-based sensors, making these probes less suited for a robust HTS assay format. To overcome these drawbacks, either the linker between FPs and the sensor sequence can be altered, or circularly permuted FPs (cpFPs) can be introduced, increasing the dynamic range of the FRET-based sensors [32]. cpFPs are usually more sensitive to their physicochemical environment and are, therefore, highly suitable as sensors, for example, for pH changes. Insertion of a sensor domain of a protein within a cpFP can yield highly sensitive biosensors, for instance for calcium or hydrogen peroxide [34].

Specialized FPs have been engineered, for example, with the property of induced-color switching (Kaede, EosFP, Dendra), or that change color while aging (fluorescent timers). Moreover, FP that release reactive oxygen species upon illumination (KillerRed) can be used to degrade proteins in their proximity [29]. For investigation of chloride-conductive ion channels, halide-sensitive FP variants have been developed, which sense chloride influx by decreased fluorescence intensity [35]. These applications, though not widely used for HTS today, might enable the design of new HTS and HCS assay formats.

Using FPs in screening approaches requires transfection of DNA into cells which can be difficult for some cell types. Moreover, timing and level of expression are difficult to control [36]. If cell lines stably express the protein of interest, however, screening is straightforward since it does not require dye-loading steps.

9.5 Instrumentation

For detection of optical readouts a number of different devices can be applied. Nowadays, the standard format for many HTS applications is a 384-well or a 1536-well plate. Due to limitations that arise from capillary forces, current miniaturization attempts translate into the development of microarrays and microfluidic devices for screening applications. Here, the spatial resolution has to be maintained by patterning the stationary phase, for example, with printing techniques.

Averaged readouts of entire wells are the simplest form to obtain signals from a multi-well plate. They can be collected either in a well-by-well mode or by simultaneous assessing of the signals from all wells. For averaged well-by-well measurements, a number of commercially available plate readers can be used. Typical readouts from plate readers include absorbance, fluorescence intensity, fluorescence polarization, and luminescence. Some plate readers may also be used for time-resolved gated fluorescence measurements, or for fluorescence lifetime analysis. Standard plate readers generally apply photomultiplier tubes to detect light while scanning the plate well by well. The excitation and emission wavelengths can be selected either by the use of monochromators or filters, the former improving flexibility and offering spectrometry, the latter bearing the potential of higher sensitivity and facilitating multi-channel readouts. For simultaneous whole plate readouts, CCD-based fluorometric imaging plate readers have been developed, enabling faster data acquisition and multiplexing.

Spatially resolved readouts from microplates, enabling single cell observation, can be obtained by using laser scanning cytometry or high resolution automated microscopy (see Section 9.6.11).

Flow cytometry allows the detection of cells or particles according to their spectral properties in a homogeneous assay format where different reactions are identified by optical rather than spatial addressing [37]. When using inhomogeneous cell populations, where the target-relevant cell type represents only a minor fraction of all cells, biological signals from these cells are frequently missed in averaged measurements. Especially in these cases, flow cytometry is an important tool that easily allows the analysis and quantification of signals from cell sub-populations by means of gated detection.

The development of microfluidic techniques has made tremendous progress in the past decade. It has the potential to increase throughput dramatically while decreasing reaction size and, thus, required biological material and cost [38]. Microfluidic platforms, commercially available from Caliper, have already been successfully applied in screening for kinase inhibitors [39, 40].

In order to miniaturize cell-based screening approaches, methods to attach live cells to microarrays are being developed [41, 42]. In a non-compartmentalized cell culture approach, a continuous cell layer was exposed to spots containing different substances, using a sophisticated microdelivery system [43]. Another recent approach is the delivery of substances to cells using sandwiched microarrays, one containing cells seeded in microwells onto which a compound-loaded array is sealed [44]. Reverse transfection on microarrays [45] can be used to screen for RNA interference in a miniaturized way [41].

In addition to optical detection systems, the integration of automated liquid handling and compound transfer systems is essential for HTS, but will not be further discussed here.

9.6 Assays

There is a huge variety of assays to screen for new pharmaceutics. Historically, assays that utilized radioactive tracers were very popular. Nowadays, they are more and more replaced by optical methods. Screening assays can be classified into binding studies and functional assays. Simple binding assays, which achieve a very high throughput, detect the binding of compounds to purified target proteins. In contrast, functional assays measure the compound-induced modification of a biological function, either by detecting products from biochemical pathways or by phenotypic alterations. Functional assays may either detect accumulated metabolic products and second messengers or phenotypes in end-point measurements, or monitor compound-induced changes in a kinetic way.

One disadvantage of binding studies using isolated protein targets is that they do not allow conclusions about the kind of modification the binding molecule might induce in the protein of interest. For this reason, one cannot predict if a hit from an affinity assay might prove as a full or partial agonist, neutral or inverse antagonist. Therefore, intensive secondary characterization of primary hits becomes essential. Moreover, allosteric modulators of protein function might be missed in binding studies.

Classical binding assays require a separation step to remove unbound molecules either by filtration, centrifugation or dialysis. To minimize working steps “mix and run assays” that require no washing step are now widely applied [46]. These include non-separation assays (containing a solid and a liquid phase) or homogeneous phase assays. Nowadays, available homogeneous assay formats allow detection of binding of compounds to isolated targets by competition with labeled ligands, or by directly coupling screening compounds to fluorophores. Ligand binding can also be detected in a label-free manner using surface plasmon resonance. In addition to the application of fluorescence or luminescence-based binding assays to detect target-compound interaction, these assay formats can also be used in functional approaches. Here, metabolic products are detected by binding to specific antibodies. The principle of some widely applied competition immunoassays to detect metabolic analytes is illustrated in Figure 9.1.

Functional assays can be classified as cell-free (i.e., measuring the activity of an isolated enzyme) or cell-based assays, assessing the activity of a protein by monitoring the biological response of a cell that expresses the target protein. The latter has the advantage of simultaneously providing information about off-target effects of compounds on living cells, for example, allowing the detection of toxic compounds, but requires extensive cell culture work. Using cell-based assays in signal transduction, one might detect early (e.g., second messenger production) or later (e.g., gene expression) events, regarding modulation of the protein/receptor of interest. Distal readouts, like expression of reporter genes, might increase signal strength due to cellular amplification steps, but will also increase the number of false positive hits due to additional effects of the investigated compounds on downstream processes. Therefore, screening for a gain of function rather than a suppression in downstream signals is considered to yield more robust results with respect to possible off-target or toxic effects of compounds.

In order to increase throughput, multiplexing approaches, that is, the simultaneous acquisition of multiple data sets (each representing one analyte), are becoming more and more popular. One possibility for easy implementation of multiplexing is the application of flow cytometric approaches [47].

Another current trend in screening approaches is the application of multiparametric analysis. The simultaneous measurement of several different characteristics of live cells can be achieved by flow cytometric approaches or by using automated imaging techniques.

9.6.1 Fluorescence Polarization

Fluorescence polarization measurements provide a cheap and easy method to detect a variety of molecular interactions and enzyme activities in a homogeneous assay format. In fluorescence polarization assays, a small molecule of interest is coupled to a fluorophore (e.g., fluorescein) and excited with polarized light. Fluorescence intensity is detected with polarization filters, measuring the polarization of the emission light parallel ($I_{||}$) and perpendicular ($I_{\perp}$) to that of the excitation. Fluorescence polarization is calculated according to:

$$FPol = \frac{I_{||} - I_{\perp}}{I_{||} + I_{\perp}}$$

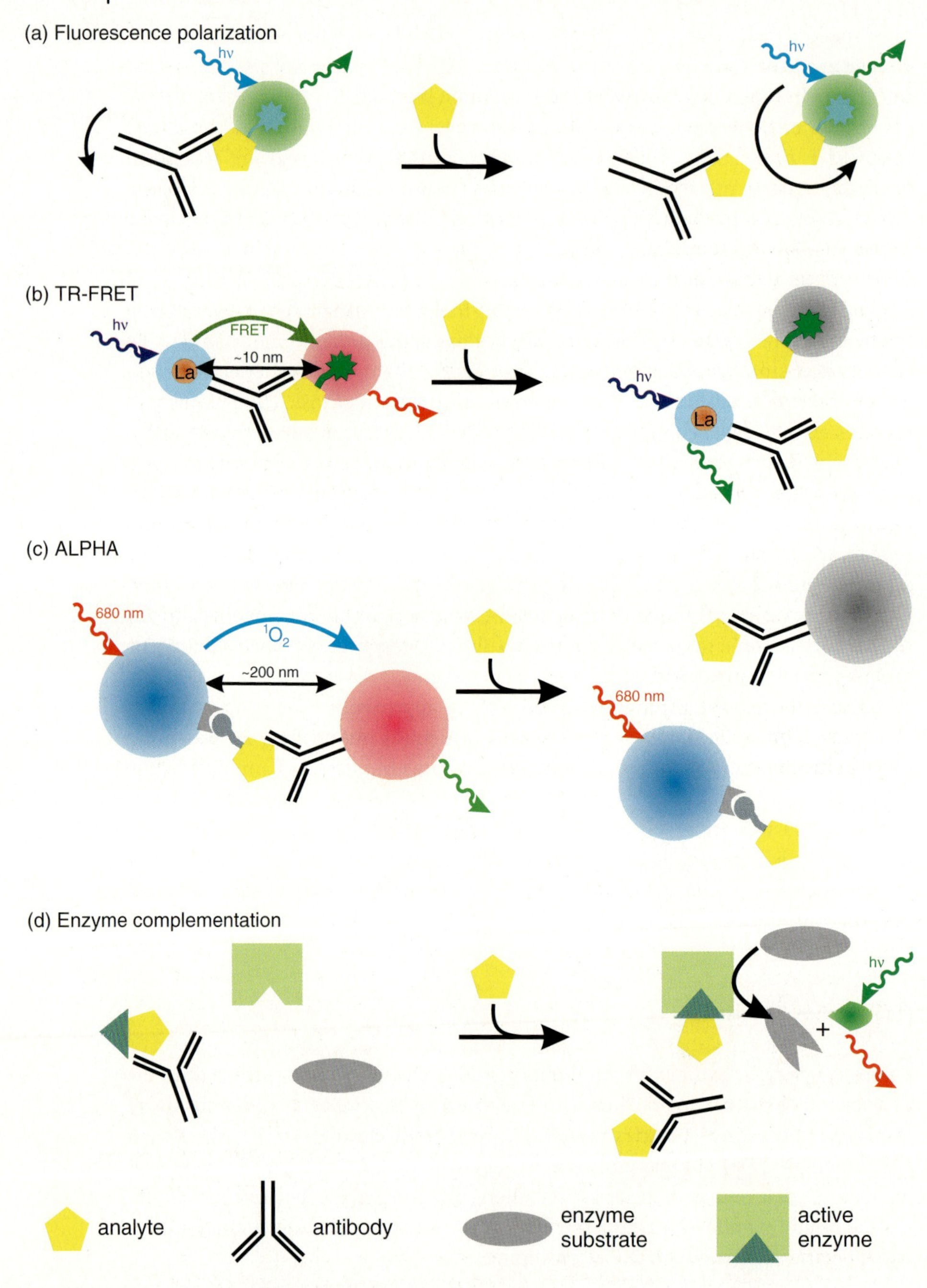

(a) Fluorescence polarization
hν
hν
(b) TR-FRET
hν
FRET
La
~10 nm
hν
La
(c) ALPHA
680 nm
$^1O_2$
~200 nm
680 nm
(d) Enzyme complementation
hν
+
analyte
antibody
enzyme substrate
active enzyme

If rotation of the fluorophore is fast with respect to its fluorescence lifetime, its emission is strongly depolarized, and both detection channels record nearly the same intensity. Binding of a larger structure (e.g., an antibody) to the fluorophore analyte complex slows down the rotation of the fluorophore. The more slowly the fluorophore rotates, the more photons will be detected in the parallel detection channel compared to the perpendicular channel, resulting in an increased fluorescence polarization value. In order to reach a sufficiently small fluorescence polarization value for the unbound analyte fluorophore complex, the analyte size has to be small and the fluorescence lifetime of the fluorophore should be relatively long [48].

Fluorescence polarization differs from other affinity detection methods in that bound and unbound molecules are measured simultaneously [49]. This may lead to low signal-to-noise ratios and only minor total shifts in fluorescence polarization values. Nevertheless, since the fluorescence polarization measurements are very precise and steady, high *Z*-factors (see Section 9.6.1) may be achieved with this technique. Fluorescence polarization assays can be applied in competition assays for antibody-based second messenger detection, to analyze displacement of labeled ligand from their receptors, or for direct assessment of binding of a small labeled ligand to a larger structure. Utilizing labeled substrates, enzyme activity can be assessed by fluorescence polarization upon binding of the converted substrate to specific antibodies, or in the case of phosphorylation to immobilized metal affinity nanoparticles (IMAP™), resulting in an increased fluorescence polarization value. Fluorescence polarization assays are simple and cost-efficient, however fluorescent compounds may interfere with the assay signal. Such cases might be detected by their unusually high total fluorescence intensity, or by spectroscopically detecting an overlapping fluorochrome. A number of commercially available plate readers include the option to measure fluorescence polarization.

Figure 9.1 Schematic representation of some competition affinity assays. (a) A labeled analyte competes with the free analyte for binding to a bulky antibody. If the labeled analyte is displaced from the binding site, it rotates faster, leading to a decrease in detected fluorescence polarization. (b) Molecules labeled with a resonance energy transfer acceptor for lanthanide fluorescence compete with unlabeled molecules of interest for binding sites at a lanthanide-coupled antibody. The FRET signal is measured in a time-gated fashion, to minimize unspecific signal. Displacement of labeled molecules leads to a decreased FRET signal. (c) Biotinylated analyte binds to donor beads coated with streptavidine. Free analyte competes for antibody coupled to acceptor beads. Upon excitation at 680 nm, singlet oxygen is produced which diffuses about 200 nm before decaying. If singlet oxygen reaches a closely positioned acceptor bead, it produces a fluorescence at about 520–600 nm, this signal decreases with the amount of competing non-biotinylated molecules. (d) Analytes of interest are bound to a peptide, which will complement a mutated enzyme to restore its activity. Antibodies against the molecule of interest prevent enzyme complementation, free analytes displace peptide-bound molecules from the antibody leading to a restoration of enzyme activity and the production of fluorescent or luminescent products from nonfluorescent substrates.

9.6.2 Time-Resolved Fluorescence

To circumvent interferences from fluorescent compounds, gated detection can be applied (see Section 9.4.1). This is most frequently done in time-resolved FRET (TR-FRET) assays like LANCE®, homogeneous time-resolved fluorescence (HTRF®) or LanthaScreen™. In TR-FRET, lanthanides that exhibit a very long fluorescence lifetime act as donor molecules. Acceptor molecules are classically red fluorescent, but can also be green fluorescent molecules when utilizing a terbium instead of a europium donor, which has an additional emission band in the green spectrum. This extends the spectrum of possible acceptor fluorophores to genetically encoded autofluorescent proteins like GFP, which simplifies assay development and allows, for example, a kinase substrate to be expressed in a cell [50]. Rare earth ions alone are not absorbing and, therefore, require light-collection assisted by chelates or cryptates to be excited. Cryptates are more stable under certain chemical conditions [51]. The assay principle is similar for various applications. An antibody recognizing the analyte under investigation is coupled to the lanthanide donor. The acceptor is either coupled to the molecules that will be detected directly or binds the molecule of interest via streptavidin-biotin or immunological interaction. When the acceptor comes close to the donor (usually in the range of 10 nm), energy transfer occurs and acceptor fluorescence is detectable. TR-FRET assays generally detect fluorescence at both the donor and the acceptor emission wavelength, allowing for some compensation of compound interference. Additionally, fluorescence is detected after a time-window of about 50–100 μs after excitation, when interfering fluorescence from compounds has already decayed. In kinase assays, the donor is a lanthanide-coupled antibody recognizing the phosphorylated substrate. The substrate is labeled with the acceptor, leading to increased acceptor fluorescence upon kinase activity. In contrast, in competition assays (i.e., to detect second messengers) acceptor-labeled molecules compete with cellular produced molecules for antibody binding, leading to a decrease in acceptor fluorescence with increasing second messenger concentrations.

Ready-to-use TR-FRET assay systems are available to detect a variety of second messengers like cAMP [52] or inositol 1,4,5-trisphosphate (IP_3), for monitoring kinase activity [53] by utilizing labeled kinase substrates or by detecting ADP generation, for a variety of other enzyme reactions, or for the analysis of protein–protein interactions.

In contrast to TR-FRET assays, dissociation-enhanced lanthanide fluorescent immunoassay (DELFIA®) is a heterogeneous time-resolved assay based on fluorescence enhancement of lanthanides after dissociation. Its principle is similar to that of an ELISA assay. In competition assays, antibodies specific for the analyte of interest are immobilized at the plate surface, for example, via biotin–streptavidin interaction. The plates are incubated with a mixture of lanthanide-labeled and competing analytes. Unbound molecules are washed away, the lanthanides are dissociated into solution, and incubated with an enhancement solution, leading to the formation of highly fluorescent chelates. Competing molecules are detected by a decreased fluorescence, since less labeled molecules remain bound to the surface. For enzyme assays the substrate is immobilized and incubated with the enzyme. After washing, the plates are incubated with lanthanide-coupled antibodies specifically recognizing the modulated substrate. Unbound antibodies are washed away, and detection occurs as in competition assays. Here, a high fluorescence intensity indicates high enzyme activity. As in TR-FRET, emission is detected delayed after excitation, allowing interfering compounds to decay within the delay window.

DELFIA® can be multiplexed by using lanthanide probes with different emission characteristics. It shows superior sensitivity and dynamic range in kinase assays and for detection of cAMP [52], which is achieved by the enhancement step. However, due to the heterogeneous nature of the assay, it requires a number of washing steps, making it less suited for HTS and placing its application rather in secondary hit confirmation.

A number of state-of-the-art plate readers are compatible with time-resolved fluorescence measurements.

9.6.3 Proximity Assay

Amplified luminescent proximity homogeneous assay (ALPHAScreen®) is a homogeneous immunoassay that utilizes donor and acceptor beads of about 200 nm size. Donor beads contain phtalocyanine that, upon excitation at 680 nm, produces singlet oxygen (about 200 molecules per excited donor bead). Singlet oxygen has a half-life of about 4 μs in which it can diffuse about 200 nm. Acceptor beads are coupled to three dyes (thioxene, anthracene and rubrene), of which thioxene, when reacting with singlet oxygen, produces a UV luminescence, which is transferred by energy transfer to the other dyes, eventually leading to emission at 520–620 nm. Emission occurs only when donor and acceptor beads are at a distance of less than 200 nm. This distance is about one order of magnitude longer than that feasible for resonance energy transfer. Therefore, larger binding partners, like large proteins or even phages, which cannot be analyzed by the TR-FRET method, can be studied using ALPHA technology. The surface of the beads is coated with latex-based hydrogels and reactive aldehydes, reducing unspecific binding and providing a surface to covalently attach various binding partners. Often, the donor bead is coated with streptavidin to bind a biotinylated small molecule, protein or an antibody against the molecule of interest, while the acceptor bead is coupled to an antibody detecting a different binding site of the analyte [54].

AlphaLISA™ is a variation of the ALPHAScreen® technology, utilizing europium as acceptor fluorophore, which provides a high intensity emission at 615 nm, enabling a narrow detection band. AlphaLISA™ can also be more easily adapted to smaller sample volumes required for higher throughput.

The ALPHA technology can be used to detect second messengers like cAMP after cell lysis [52], to monitor enzyme activity from both purified enzymes and cell lysates, and to analyze protein–protein interaction. The method is characterized by a high sensitivity with low background signal but is also prone to temperature-induced signal variabilities, sensitive to light and interference of compounds with antioxidant properties. Owing to the longer excitation compared to the emission wavelength, ALPHAScreen®-based technologies require specialized plate readers.

9.6.4 Protein Complementation Assays

In protein complementation assays (PCA) a reporter protein is split into two fragments that cannot fold when separated. Each protein fragment is fused to one of two interacting partners. Once interaction occurs, the two fragments are able to form a functional protein [55]. Since protein complementation relies on protein folding rather than assembly of folded subunits, it cannot be applied for measurements of fast kinetics. Moreover, some split proteins, including FP variants, form quite stable proteins after interaction, making them unsuitable for the investigation of dynamic processes. Irreversible association might also lead to higher background signals due to spontaneous interaction, especially if the expression level is high. A luciferase-based assay has overcome these problems and was shown to efficiently report the disruption of protein–protein interaction [56].

DiscoveRx offers a huge variety of assay systems based on complementation of β-galactosidase fragments to generate a fluorescence or luminescence signal. Besides cell-based complementation assays that report on protein–protein interaction [57], there is a panel of assays to detect the production of signaling molecules or enzyme activity compatible with standard plate readers. The principle is similar for a variety of target assays: a β-galactosidase missing certain amino acids for functionality (enzyme acceptor) is complemented with a small fragment peptide that binds to the acceptor with high affinity, enabling substrate hydrolysis, and thus generating the signal. This complementation can only occur when the fragment binding site for the acceptor is not covered by a bulky structure. The signal can, therefore, be produced by cleavage processes or by competition of fragment-bound small molecules with molecules of interest for antibody binding sites, enabling the detection of a variety of processes [58]. Available assays include detection of cAMP, activation of nuclear hormone receptors, protein kinase assays, and protease assays that are either cell-based or biochemical.

9.6.5 Resonance Energy Transfer

Fluorescence or Förster resonance energy transfer (FRET) can be used to detect binding of a ligand to its receptor in a homogeneous assay format. FRET occurs when the emission spectrum of the donor overlaps significantly with the excitation spectrum of a closely positioned acceptor fluorophore. FRET efficiency declines with the 6th power of the distance between donor and acceptor, giving efficient energy transfer when the distance between donor and acceptor is in the lower nanometer range for most fluorophores. One disadvantage of FRET is that it requires labeling of ligand and receptor. Since spectral separation of FRET-based indicators may be low, even weak background fluorescence and absorbance of screening compounds might considerably interfere with the assay. By using lanthanides with long excited state lifetimes as donor molecules, and measuring FRET in a time-resolved manner, the latter problem might be solved [46].

A number of FRET sensors based on genetically encoded FPs have been designed. The most commonly applied FRET systems apply variants of cyan FP as donor and yellow FP or its improved variants as acceptor molecules. When donor and acceptor are fused to different proteins, interaction of these proteins is sensed by an increased FRET ratio. In other FRET sensors, donor and acceptor chromophores are intramolecularly fused to two different positions within a target molecule, detecting conformational changes by an altered FRET ratio. Such intelligent designed FRET sensors are applied for the detection of a huge variety of cellular signaling events, including protease, kinase, and GTPase activity. The susceptibility to interfering background signals might be overcome in cell-based assays by a higher expression rate of fluorescent proteins. Bioluminescence resonance energy transfer (BRET) utilizes a luminescent donor and a fluorescent acceptor protein, meaning that no external excitation is required and reducing the influence of background interference.

In cell-based resonance energy transfer assays, a spatially resolved readout, applying imaging techniques, is often used. This allows analysis of subcellular localization of the signaling molecules and reduces the influence of background signals. Technically, FRET can be detected by analyzing the changes in donor fluorescence lifetime, by using spectral readouts like sensitized acceptor emission, or by detecting donor quenching/unquenching.

9.6.6 Fluorescence Fluctuation Approaches

In fluorescence fluctuation techniques, samples diffuse through a diffraction-limited focus of a laser beam, and variations in fluorescence intensity are monitored. Fluorescence correlation spectroscopy (FCS) is based on autocorrelation curves, from which the diffusion time of molecules can be calculated. Since small fluorescent molecules diffuse significantly faster than molecules attached to a larger structure, this technique can be applied to the detection of binding reactions. The concentration of labeled molecules has to be relatively low to ensure single molecule detection. As a consequence, longer measuring times are required for robust statistics, making this technique less suitable for HTS [59]. A complementary approach is fluorescence intensity distribution analysis (FIDA) based on the statistics of fluorescence fluctuations, where slightly higher concentrations of labeled molecules and much shorter detection times are applied. In FIDA, the molecular species in the sample are quantified according to their molecular brightness, using frequency histograms of signal amplitudes. Compared to fluorescence polarization measurements, FIDA is less sensitive to compound autofluorescence and light scattering [60]. In 2D-FIDA, signals from two detection channels (either based on polarization or emission bands) are analyzed, resulting in improved performance [61].

9.6.7
Reporter Gene Expression

The activation of signaling cascades can be monitored by observing the expression of reporter genes, whose transcription is controlled by specific promoters. These promoters contain binding sites for transcription factors produced downstream of the investigated receptor or regulatory protein [62]. A common example for the detection of cAMP production is the CRE sequence, which is activated by CREB. Widely applied reporter gene products include fluorescent proteins, luciferases, β-galactosidase and β-lactamase. A variety of detection methods exists for β-galactosidase, enabling cheap assays with fluorescent, luminescent or colorimetric readouts. FPs have the advantage that they are autofluorescent and do not require co-factors, so they can be measured in live cells. FPs and β-galactosidase possess relatively long lifetimes, which might lead to higher background signals. To overcome this limitation, genetically modified FPs with decreased stability have been constructed. In contrast, luciferase and β-lactamase have shorter half-lives per se. For firefly luciferase detection, various reagents are commercially available. For the detection of β-lactamase activity, a cell permeable FRET sensor (CCF2/AM) has been developed, which is cleaved by the enzyme, resulting in reduced FRET efficiency. This non-toxic sensor enables ratiometric measurements in live cells [63].

Since protein synthesis has to occur for the signal to develop, these assays require longer incubation times. Moreover, since the expression level may depend on certain factors not influenced by the test compound, results deviate with uneven cell seeding or metabolic integrity. Expression of a second reporter gene, which is constitutively active, enables correction for these fluctuations but requires either another colored variant or sequential reading of both products, which may prolong assay time considerably. A typical reporter gene assay, correlating reporter gene expression to constitutive expression is shown schematically in Figure 9.2. Dual reporter gene assays may also be used to simultaneously assess the activation of two different signaling pathways by monitoring the expression of two different reporter genes [64]. Luciferase isoforms emitting at different wavelengths but using the same substrate have been developed, enabling parallel measurement of different gene products [65].

9.6.8
Measurement of Intracellular Calcium

Changes in the intracellular calcium concentration can arise from the activation of GPCRs coupled to $G\alpha_q$, which, via the activation of PLC β and production of IP_3, leads to a calcium release from the ER. With several hundred members in the human genome, non-olfactory GPCRs form a large druggable family. Note that activation of other GPCR family members that couple to $G\alpha_s$ or $G\alpha_i$ leads to changes in cellular cAMP rather than calcium due to the activation or inhibition of adenylyl cyclase, respectively. However, these pathways can be switched to result in

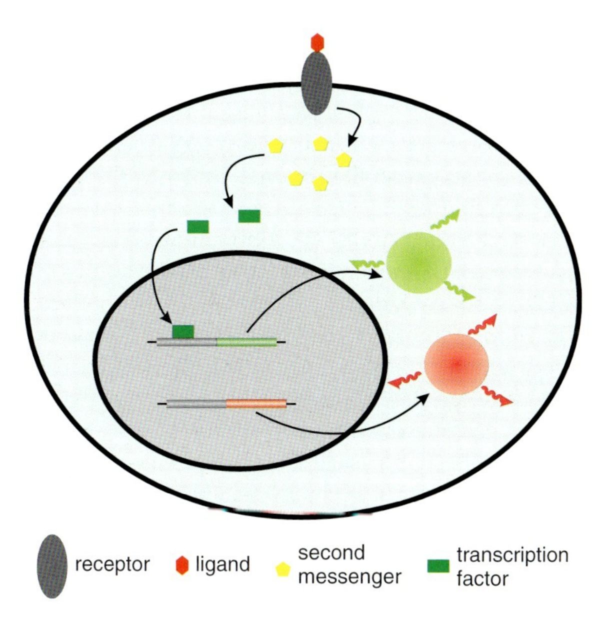

Figure 9.2 Schematic drawing of a dual reporter gene expression assay: After ligand binding to a receptor, second messengers are produced, which then control the activation of transcription factors. Binding of these transcription factors to the promoter of the reporter gene induces protein expression, and a detectable fluorescence or luminescence signal is established. A protein of different spectral properties, which is constitutively expressed, serves as an internal control for differences in cell viability, metabolic fidelity, or uneven seeding.

a calcium increase by using chimeric or promiscuous Gα subunits [66, 67]. Coexpression of these subunits is frequently applied, since calcium measurements are most easily incorporated in many HTS platforms. In addition, this technique is widely used for de-orphaning GPCRs with as yet unknown physiological activators. Intracellular cAMP or cGMP production can also be detected indirectly by coexpressing variants of cyclic nucleotide-gated cation channels (CNGs) that translate cyclic nucleotide accumulation into a calcium signal [68, 69], enabling high-throughput kinetic measurements of these messengers [70–72]. Other important pharmacological targets that causes changes in intracellular calcium levels, are calcium-permeable ion channels, like voltage-, ligand-, or second messenger-gated calcium channels.

There is a huge variety of small chemical calcium indicators available (for a comprehensive review of their properties see [73]). For easy loading of cells, these dyes can be used in the acetoxymethyl ester (AM) form, which facilitates membrane crossing and, when hydrolyzed by cellular esterases, the anionic dyes remain trapped within the cells. In addition to various calcium affinities, chemical calcium sensors can be classified according to their spectral properties as single wavelength indicators and ratiometric dyes. The former change their fluorescence intensity upon calcium binding, while the latter undergo a shift in either their excitation or emission spectra.

While single wavelength dyes like Calcium Green-1, Fluo-3, Fluo-4, and Oregon Green 488 BAPTA have a low spectral bandwidth and, thus, can be easily combined with other fluorophores, they are more prone to photobleaching, uneven dye loading or interferences from compound fluorescence. The most widely applied ratiometric calcium sensor is Fura-2, which emits above 510 nm and shifts its absorption maximum from 380 nm in the calcium-free to 340 nm in the calcium-bound state. The ratiometric approach enables calibration as well as corrections for certain assay interferences due to the added compound. However, it requires a device with dual UV channel excitation.

When choosing the right calcium sensor for a screening approach, the spectral properties, as well as the calcium affinity of the dyes must be considered. Generally, the K_D of the dye should be in the same range as the measured calcium concentration. While high affinity sensors have a higher sensitivity, they will also buffer a greater amount of calcium and might, thus, affect physiological signaling. Another consideration for choosing the sensor for HTS might be the minimization of working steps. Loading cells with the classical calcium sensors requires removal of the dye that has not been internalized. When using cell suspensions, this can easily be achieved by loading and washing the cells before dispensing into a microplate. For adherent cells that are cultured within a multiwell plate, the newly developed no wash dyes [74], which use nonmembrane-permeable quenchers to suppress extracellular reporter fluorescence, might be advantageous.

In addition, genetically encodable calcium sensors, like variants of the luminescent aequorin [75], and fluorescent proteins that either change their fluorescence intensity (like camgaroos or pericams) or FRET-ratio (cameleon) may be used [76]. These proteins have the advantage of being targetable to specific subcellular environments or fusable to proteins of interest. Their disadvantages include a lower sensitivity and slower response rates, and they are more difficult to introduce into some cells or tissues [76].

Due to the easy application of calcium measurements, they are widely applied in HTS, especially when screening for GPCR or calcium channel modulators. Since the increase in intracellular calcium is rapid and transient in most cases, calcium measurements do not represent equilibrium conditions for receptor–ligand interaction, and pharmacological parameters obtained under these conditions are indirect. This may cause different conclusions about agonist/antagonist potency, or even reversal in agonist potency orders when different assay systems are used (see [77] for a detailed discussion).

Calcium measurements can be easily performed using plate imagers, plate readers, or microchip approaches. When subcellular localization or intracellular spreading of the calcium signals is of interest, automated microscopy may be applied. Most physiological calcium signals are rapid and transient (i.e., in the range of seconds after agonist addition). Thus, systems are required that measure with a high temporal resolution [78]. The fastest kinetic acquisitions are achieved with imaging plate readers, enabling parallel measurement of an entire multiwell plate. State-of-the-art plate readers nowadays achieve a temporal resolution that is sufficient for most applications in calcium signaling by measuring all wells of one plate before proceeding to the next time point of the kinetics. However, when cellular imaging is required, a kinetic measurement has to be completed within one well before moving to the next, decreasing the throughput considerably.

9.6.9
Indicators for Ion Channel Activity

Ion channel activity of calcium-permeable channels can be easily detected by measuring changes in the intracellular Ca^{2+} (see above). Other ion channels have been a difficult target class for drug development, partly because some functional assays do not fulfill the demands of HTS [79]. Functional screening for ion channel modulators can be performed using ion flux measurements but these suffer from poor detection specificity. Recently, platforms for automated patch clamp measurements (like IonWorks® Quattro platform, QPatch™ 16 or PatchLiner® 8) have been introduced, however, the throughput of these devices is still considerably slower than that of optical methods. They are, therefore, rather suited for confirmation of primary hits acquired from a fluorescence-based screening assay [80], and are applied in testing drug candidates for unwanted effects (e.g., long-QT syndrome) to meet safety regulations.

Methods based on fluorescent indicators remain the only assays that allow real high-throughput. Indicators for ion channel activity can be classified into those sensing changes in the membrane potential and dyes detecting ions that pass through the channel.

Sensors for ions other than calcium are not that well established for HTS. A Tl^{+} uptake assay to monitor opening of K^{+} channels has been developed. Loading cells with Tl^{+}-detecting fluorophores leads to an increase in fluorescence intensity upon uptake of Tl^{+} through open K^{+} channels [81, 82]. Other available ion indicators include sodium green for the detection of sodium, and several indicators that detect halide ions via diffusion-limited collisional quenching. Chloride or iodide fluxes can be also detected by variants of FPs, which have been applied in screening for cystic fibrosis transmembrane conductance regulator modulators [35, 83].

Voltage-sensing probes are organic fluorophores that bind to the plasma membrane. According to their mechanism of action, these dyes can be subdivided into probes that, upon changes in membrane potential, rearrange their transmembrane distribution, and probes that change their electronic structure, resulting in altered fluorescence properties. The latter exhibit a fast response to voltage changes, which is especially required for fast transient processes. However, the potential-dependent changes in fluorescence signals are tiny, resulting in low signal-to-noise ratios, which make these dyes unfavorable for HTS. In contrast, dyes that cross the membrane, like cationic rhodamines or anionic oxonols, have been successfully applied in HTS. Redistribution of dyes across the membrane is rather slow (equilibrium is reached in seconds to minutes) [84]. Negatively charged oxonols are probably the most commonly applied voltage sensors for screening. Upon hyperpolarization, they leave the cells and are quenched in the extracellular environment, leading to a decrease in fluorescence signal, whereas depolarization results in increased fluorescence intensity [79]. Fluorescence changes can be increased by adding quenching dyes to the extracellular compartment and no-wash kits with improved fluorescence properties and kinetics are commercially available [85, 86].

Moreover, ratiometric approaches have been developed to sense membrane potential changes, relying on FRET, for example, between a coumarin-linked phospholipid at the outer leaflet of the plasma membrane as donor and a negatively charged oxonol as acceptor. Upon depolarization, the acceptor accumulates at the inner plasma membrane leaflet, leading to a decrease in FRET efficiency [80, 87]. As explained above, ratiometric approaches are less prone to artefacts from unequal dye loading and cell density as well as autofluorescence of compounds, but require dual excitation or emission and occupy a broader spectral range.

Since, in most cases, the number of ion channels that have to open to induce changes in membrane potential is small compared to the total number of channels of the investigated type, assays that rely on membrane potential changes may yield low EC_{50} values in agonist screening but higher IC_{50} values and, thus, reduced sensitivity in antagonist screenings [87].

Potentiometric indicators and ion-sensitive probes are compatible with imaging plate readers [88].

9.6.10
Flow Cytometry

Flow cytometry, originally developed to characterize and sort cells based on their fluorometric properties, can also be extended to the analysis of other fluorescent particles of similar size, like synthetic microspheres. Microspheres may serve as solid supports for antibodies in immunoassays or molecular assemblies of drug targets. Flow cytometric applications in drug screening include competition assays where compounds are identified that block the binding of fluorescent ligands to their receptor expressed on cells. Also, protease activity can be monitored using a fluorescent substrate attached to a microsphere [89]. Flow cytometry is capable of detecting bound molecules in the presence of about 100 nM of free label in a homogeneous assay format.

Modern flow cytometers are capable of acquiring multiple parameters, ranging from fluorescence intensities at different wavelengths to light scattering which reports on the size of the analyzed particle. Multiparametric analysis or multiplexing by color coding different microspheres is thus supported [37]. In the Luminex® approach, a reporter molecule is labeled with one fluorophore and the interaction with the microsphere surface containing possible binding partners is analyzed. These microspheres contain two fluorophores of varying concentration allowing multiplexing of up to 100 different binding partners by spectral addressing of their microspheres [47].

Fluorescent cell barcoding with different fluorophores in varying concentrations is another possibility to achieve multiplexing using live cell samples [90].

Flow cytometers adapted for the demands of HTS applying microfluidic dimensions are now commercially available.

9.6.11
Automated Microscopy

Biochemical assays and measurements of bulk cell responses provide the basis for high-throughput screening approaches. Although information on target-specific compound activity can be gained rapidly by these assays, certain important information on phenotypic changes on a cellular level might be missed. The development of automated microscopic workstations has greatly facilitated the application of image-based assays in screening approaches. Because of the high amount of information gained using imaging of cells and subcellular structures, the term high-content screening was coined. Image-based screening has been primarily applied for secondary screening or target identification and validation, as evident by the huge number of published high-content siRNA screens. With the sophisticated hardware and software solutions for HCS now available, there is a trend in academia, as well as in industry, to apply HCS also in primary compound screening campaigns. The advantage of using multiplexed imaging approaches in primary screening is gaining the additional information on solubility, permeability and stability of compounds in a cellular context, which has to be obtained in secondary screening for hits when using classical biochemical approaches [91].

Automated microscopy, when applied at low resolution, can serve to analyze phenotypic changes of larger cell populations at a single cell level, while high resolution microscopy enables the detection of subcellular effects [92]. HCS permits the use of assays that do not rely on changes in the overall fluorescence or luminescence intensity of a cell mixture. Information on single cells and subcellular structures can be gained, for example, changes in subcellular location of fluorescent biomarkers can be assessed [33].

For HCS, a number of specialized instrumental solutions are required (see [93] for a detailed review). To achieve reasonable throughput, all steps in HCS have to be automated, starting from sample preparation via image acquisition and analysis to data storage and handling [94]. The amount of data generated by HCS experiments is enormous compared to classical biochemical assays, requiring a high degree of bioinformatic processing.

Since HCS is a young and upcoming technique, we will briefly discuss some aspects of image acquisition in this section. When choosing an automated microscopy platform, speed and flexibility, sample positioning, available excitation sources and detection methods (confocal versus wide-field), as well as the possibility to add substances before or during measurement have to be considered [93, 95]. Microscopes should provide different objectives to allow imaging at various magnifications. As a rule of thumb, the objective with the lowest magnification that is still capable of resolving the structures of interest should be used to increase the size of the field of view. By this scheme, more cells can be measured per image acquisition, decreasing the required number of fields to be imaged per well [93]. When live cell imaging is intended, exposure times with excitation light have to be minimized to reduce phototoxicity. Consequently, a very sensitive detection system is required, and the objectives should feature a high numerical aperture [96]. Moreover, automated microscopes may be equipped with an incubation chamber which allows control of the temperature, humidity and CO_2 level to enable prolonged live cell imaging.

Choosing a confocal detection system will usually provide sharper images with improved resolution and lower background, since out-of-focus light is eliminated. Image acquisition in confocal systems is, however, restricted to certain excitation laser lines and, generally, much slower than wide-field detection since scanning is required. Faster throughput is achieved with confocal systems by applying line scanning instead of point scanning or Nipkow spinning disc systems. By contrast, wide-field microscopes provide fast image acquisition, and their excitation source is usually a cheaper metal halide lamp or long-life light-emitting diodes. When using metal halide lamps, excitation wavelengths are selected using appropriate filters or a monochromator device. When LEDs are used as the excitation source, several LEDs of different colors are usually collimated and then combined by dichroic mirrors to offer various excitation wavelengths, with fast interchange and almost unlimited lifetime of the light source.

Automated microscopy requires focus control, which can be achieved by two different methods. For software-based autofocus an image stack in the z-direction is rapidly acquired, and online image analysis identifies the correct focal plane. Hardware-based autofocus implementations use optical methods (mostly an additional laser in the IR spectral range) to locate the bottom of the substrate and acquire images at a user-defined offset. While software-based methods are generally slow and might induce photobleaching or phototoxicity due to the additional excitation, they provide superior focus quality for culture plates with irregularities at the bottom or variations in cell size [94]. To minimize photo-induced damage, digital focus control using DIC images has been developed [97, 98]. Hardware-based autofocus may operate faster and still achieve good quality images with standard 384-well culture plates. Several imaging systems also use combinations of hardware and software-based autofocus, identifying the coarse focus optically and then using software autofocus for fine adjustment.

There are a number of commercially available fully automated imaging platforms. Laser scanning cytometers (LSC) have been developed, in analogy to flow cytometers, and are now offered by a couple of companies. They usually scan a whole multi-well plate with low magnification, enabling the identification of individual cells. In contrast to the flow cytometer, LSC uses spatial information of cell location and can thus repetitively scan the same cells, thereby enabling kinetic measurements. LSC use certain integrated laser lines as the excitation source. Higher resolution and often broader excitation spectra can be achieved with automated microscopes.

Microscopy-based screening approaches can either be adopted for endpoint measurements, or to analyze the dynamic behavior of cells. For endpoint measurements, cells are usually fixed using automated preparation protocols. Subcellular localization of proteins or compartments can be visualized by labeling with small molecules, expression of genetically encoded FPs, or immunostaining. Endpoint measurements feature high reproducibility and relatively high throughput, while not requiring specialized incubation procedures during imaging. However, fixation might induce artefacts. In the case of immunofluorescence, labeling protocols might be labor-intensive, require expensive reagents, and only static information can be gained [95]. In contrast, dynamic processes can be followed in live cells by time-lapse microscopy, yielding even more information. Consequently, for live cell kinetic measurements, data evaluation is considerably more complex [94]. Most HCS assays apply cell lines, often stably expressing fluorescent marker proteins. Primary cells, though better representing the physiological conditions of the target in the human body, are often difficult to obtain in sufficient quantity and with reproducible quality. Embryonic, adult, or reprogrammed stem cells may be used as an alternative [92].

The big advantage of HCS over classical biochemical assays is that library compounds are not only tested for their influence on a specific target, but also for their effects on other related or unrelated processes. This includes information on cytotoxicity and cellular morphology. Multiplexing can principally be achieved in two different ways. One is to use a single staining and to analyze various cellular objects by supervised machine learning to classify multiple phenotypes. This approach requires considerable input from the field of bioinformatics. The other is to employ different fluorescent reporters for distinct signaling processes. If spectra of different fluorophores overlap, spectral unmixing may be applied prior to a morphological image analysis. The number of channels that can be simultaneously acquired is limited due to the spectral properties of the available fluorophores [94].

Currently, time resolution of automated microscopy seriously lowers throughput, especially when live cell kinetic measurements are performed. Since many signal transduction pathways give signals that last for seconds to minutes only, parallelization of image acquisition from all wells of one plate is difficult. Mostly, image acquisition has to be completed in one well before proceeding to the next when fast dynamic processes are investigated.

HCS has great potential, but also requires very careful assay design and optimization, especially when no commercially pre-optimized assays are being used [92].

9.6.11.1 Image Analysis

With modern HCS platforms, images are now captured much faster than is analyzable by human observation. Visual inspection and manual scoring, though still frequently used, are slow and prone to biasing by the experimenter. Fortunately, in recent years, the progress in developing computer-assisted analysis methods has been significant. The simplest goal of automated image analysis is, therefore, to reduce time-consuming labor, requiring human researchers to visually inspect single images [99]. In providing objective and quantitative measurements, automated image analysis extracts information that can be fed into statistical analysis algorithms. In some cases, computer-assisted analysis may even detect phenotypic changes (e.g., by analyzing changes in intensity distribution) that are difficult to annotate by visual inspection. Modern software can be trained by supervised machine learning, for example, to classify subcellular structures. However, there are still applications, where computer vision performs much worse than human examination [100]. It is, therefore, important to develop simple assay readouts that fit to a subsequent computer-assisted automated image analysis.

Image processing steps in automated analysis include correction for uneven illumination, background correction, segmentation, and identification of subcellular compartments. Simple outputs from automated analysis can be count, size, shape, and mean intensity of objects. Moreover, information about texture and localization can be extracted and statistically analyzed [99]. Image analysis on a single cell level generally provides a number of advantages compared to averaging over cells. These include subpopulation analysis, cellular statistics and cell filtering [91].

HCS is characterized by the acquisition of multiple parameters. According to the values of these parameters, cells are classified into distinct categories. For robust statistical analysis, one should choose the minimal parameter set that enables discrimination between different categories [100].

A number of software solutions are available for HCS image analysis. Specialized automated imaging platforms usually come with their own data analysis software which is tightly adjusted to the assay format provided with this system. Online image analysis during the screening procedure is often available with these commercial systems. Some third-party commercial software packages offer processing and statistical analysis of image data from screens. In addition, there are a number of open source software packages, comprising flexible analysis tools that can be adapted for specific tasks by a skilled user. These open source packages include CellProfiler [101, 102] and CellClassifier [103, 104], CellCognition [105], packages of the Open Microscopy Environment (OME), and CellHTS, which is part of the Bioconductor/R project.

9.7 Data Mining and Quality Control

Screening campaigns produce large data sets that must be statistically analyzed and evaluated to identify those compounds that possess a desirable biological activity. In this process, quality control and identification of possible errors is essential to reduce the number of false-positive hits or false-negative results. False-positive hits will increase the secondary screens work burden, false-negatives may cause hits to be missed or falsely dropped in secondary screens. In this section we briefly summarize some methods to inspect the quality of screening data and to identify active compounds. For a more detailed discussion on this topic, the reader is referred to some excellent recent reviews [106–108]. Appropriate statistical data analysis procedures are provided by commercially available software packages or by open source software like web cellHTS2 [109].

9.7.1 Quality Control

To increase the quality of a screening campaign, certain measures have to be applied before starting the actual screen. These assay validation procedures include pilot tests to identify the appropriate assay format, to maximize the signal window, to minimize variabilities and screening with a subset of the compound library. From this test screen, statistical variability and suitable parameters for hit identification (see Section 9.7.2) can be deduced. Moreover, each screen should contain an appropriate amount of control samples where either no compounds are added or (if available) compounds with known biological activity are included.

Typical parameters to evaluate data quality are signal-to-noise or signal-to-background ratios. Variability of the observed parameter and signal strength together determine the sensitivity and robustness of an assay. To obtain a quantitative measure of assay reliability, the dimensionless Z factor [110] has been introduced to assess the quality of screening data:

$$Z = 1 - \frac{3\ \text{SD of sample} + 3\ \text{SD of control}}{|\text{mean of sample-mean of control}|}$$

where the control in an activation assay is a positive control and in an inhibition assay it is a negative control. Alternatively for assay validation, Z' can be calculated, including only positive and negative control samples to test the suitability of a chosen assay for screening. In the case of an ideal assay, Z would reach its maximal value, which is one. Generally, in good screening assays Z is higher than 0.5. When Z becomes negative, the separation of samples showing activity from inactive samples is impossible.

Several sources of error might interfere with the assay quality. These include random and systematic errors, as well as interferences from the tested compounds with the assay signal.

9.7.1.1 **Random and Systematic Errors**

Random errors, often referred to as noise, arise from technical failures like pipetting errors, robotic failures, and unequal compound concentration due to solvent evaporation or solubility issues. Realtime quality control procedures during the screen may help to identify and deal with these problems promptly. Moreover, the influence of random errors is easily recognized, and strongly diminished, if screens are run in duplicate.

Systematic errors may arise from temporal and spatial effects using multi-well plates, where inaccuracies are caused by location on the plate (e.g., edge effects) and time of measurement [106]. There are a number of statistical methods attempting to deal with these systematic errors. Most of them assume random distribution of compounds on the plates and identify segments of the plate that deviate significantly from the random activity. These algorithms have to be used with care when structurally similar compounds are places in adjacent wells.

Spatial systematic errors may impact significantly on assay quality if hit identification is based on activity in control samples (see Section 9.6.2), which are often placed in border columns, where edge effects can severely alter readout values [107]. Therefore, when possible, control samples should be placed randomly across a multi-well plate.

9.7.1.2 **Compound Interferences**

In HTS assays, it may be challenging to discriminate between hits that represent compound activity against the investigated biological target and off-target assay interferences [111]. A very common assay interference is due to the spectroscopic properties of the tested compounds, especially when the concentration of the read-out substance is low and compounds are tested in the micromolar range. Assay interference is more common in gain-of-function assays (i.e., expecting an increase in the fluorescence signal) where it is caused by compound fluorescence or light scattering. However, interference may also occur in loss-of-function assays due to quenching and absorbance [22]. Pretesting of the compound library before assay initiation is one possibility to identify these false positive hits. Fluorescence properties of certain libraries are also available from the literature [112]. Since the number of molecules that are fluorescent using UV or blue excitation is generally much higher than by exciting with longer wavelengths, the use of red-shifted dyes as assay readout can reduce interference [27]. Ratiometric approaches are generally less prone to interferences, however, shifts in fluorophore quantum yield due to compound binding or formation of scattering aggregates might constrict such assays, necessitating the definition of appropriate cut-off criteria [22]. Applying gated detection using lanthanides with extremely long fluorescence lifetimes generally reduces interference from fluorescent compounds considerably. Another possibility is to apply fluorescence lifetime technology (FLT) to discriminate between interfering compounds and the detection signal [113]. Since fluorescence lifetimes are largely dependent on the physicochemical environment of a fluorophore, they can also be utilized as readout for physiological processes, for example, the detection of phosphorylation events in an antibody-independent manner. FLT requires longer measurement times per well and specialized plate-readers which are now becoming commercially available.

Certain compounds tend to aggregate under specific conditions leading to particles of 50–400 nm in size that may interfere with the assay by sequestering enzymes on the particle surface. In cell-free assays, using low concentrations of nonionic detergents, like Triton X-100, effectively prevents enzyme inhibition by this process [114]. Further assay interferences may arise from redox-active compounds if DTT is used in the assay, which is commonly done to keep enzymes in a reduced state. In this case counterscreens to identify these compounds, for example, by the detection of the produced H_2O_2, are needed [115]. Again, for some substance libraries, redox cycling compounds are already identified and published [116]. In the case of target oxidation by the compound, the addition of DTT to the assay solution may prevent this effect. Furthermore, counterscreens for cytotoxicity using standard assays should be performed.

A number of assays utilize enzymes like firefly luciferase, phosphatase or peroxidase activities, which might be inhibited by the screened compound, producing false-positive results. In such cases, orthogonal assays utilizing another reporter are necessary to identify the hits corresponding to the biological target during secondary screening.

9.7.2
Identification of Hits

There are numbers of approaches to the identification of hits from screening campaigns. The best suited available method for a given assay must be identified in advance, preferably by conducting pretests [117].

Most hit annotation approaches use some kind of data normalization. In principle, sample values can be compared with values of negative control wells, or with the mean of all sample values, assuming that most compounds have no biological activity. If samples are compared to controls, the number of control wells must be sufficiently high to enable robust statistics. Using mean sample values for normalization may be error-prone when a higher number of compounds per plate induces increased signal activity.

For hit identification one can define signal thresholds above which compounds are regarded to be active. Sometimes a certain percentage of compounds, showing the highest activity is chosen. The Z-score is a simple measure for rescaling screening values based on within-plate variations:

$$Z = \frac{X_i - \bar{X}}{S_x}$$

where X_i is the signal produced by compound i, $\bar{X}$ is the mean value of all signals per plate and S_x is the mean standard deviation of all signals. To correct for systematic errors associated with the plate layout, the B-score has been introduced [118], which requires more sophisticated statistical analysis. Since mean and standard deviation are strongly influenced by statistical outliers, more resistant scale estimations, such as the median absolute deviation (MAD), can be used.

For further developmental steps, compounds have to be scored according to their potential to have biological activity and to be developable into drugs. Sometimes this scoring is not only based on the activity of the tested compound, but also on the activity that chemically related compounds have shown in the screen. The rationale behind this process, also referred to as "hit-clustering", is that certain scaffolds produce biological activity and a compound is more likely to be truly active when compounds sharing the same or similar scaffolds also show some kind of activity [106].

9.8
Conclusions

Photonics approaches are, and will presumably remain, the most widely applied techniques in industrial and academic screening campaigns for drug development. The trend to miniaturize screening formats is compatible with photonic detection methods. Especially, the ability to excite sub-picoliter volumes with a diffraction-limited light focus and to detect fluorescence at the single molecule level renders this technique most relevant today and in future.

Owing to the increasing complexity and availability of compound libraries, fueled by a variety of chemical approaches and an increasing number of biological targets that are identified by genetic screens, the need for fast and reliable screening methods with ultra-high throughput increases. Further developments in optical assays and miniaturization approaches applying new microfluidic and chip technologies will certainly help to meet these demands.

In spite of the request for methods to achieve even higher throughput, we notice a trend to deduce more information from the primary screen, which is achieved by multiparametric and high content rather than simple assay readouts. The reason for this development is that unsuitable compounds are pinpointed early in the screen, saving material and reducing the load of follow-up studies. Multiparametric assays require careful statistical analysis and data mining procedures, which represents another rapidly developing field.

Acknowledgment

We thank Christian Hellwig, Anke Klein, and Sebastian Tannert for critically reading the manuscript.

References

1 Zhu, Z. and Cuozzo, J. (2009) Review article: high-throughput affinity-based technologies for small-molecule drug discovery. *J. Biomol. Screen.*, **14** (10), 1157–1164.

2 Baker, M. (2010) Academic screening goes high-throughput. *Nat. Methods*, **7**, 787–792.

3 Betz, U.A.K. (2005) How many genomics targets can a portfolio afford? *Drug Discov. Today*, **10** (15), 1057–1063.

4 Overington, J.P., Al-Lazikani, B., and Hopkins, A.L. (2006) How many drug targets are there? *Nat. Rev. Drug Discov.*, **5** (12), 993–996.

5 Heilker, R., Wolff, M., Tautermann, C.S., and Bieler, M. (2009) G-protein-coupled receptor-focused drug discovery using a target class platform approach. *Drug Discov. Today*, **14** (5–6), 231–240.

6 Lipinski, C. and Hopkins, A. (2004) Navigating chemical space for biology and medicine. *Nature*, **432** (7019), 855–861.

7 Walke, D.W., Han, C., Shaw, J., Wann, E., Zambrowicz, B., and Sands, A. (2001) In vivo drug target discovery: identifying the best targets from the genome. *Curr. Opin. Biotechnol.*, **12** (6), 626–631.

8 Hopkins, A.L. and Groom, C.R. (2002) The druggable genome. *Nat. Rev. Drug Discov.*, **1** (9), 727–730.

9 Lipinski, C.A., Lombardo, F., Dominy, B.W., and Feeney, P.J. (2001) Experimental and computational approaches to estimate solubility and permeability in drug discovery and development settings. *Adv. Drug Deliv. Rev.*, **46** (1–3), 3–26.

10 Dobson, P.D., Patel, Y., and Kell, D.B. (2009) "Metabolite-likeness" as a criterion in the design and selection of pharmaceutical drug libraries. *Drug Discov. Today*, **14** (1–2), 31–40.

11 Bauer, R.A., Wurst, J.M., and Tan, D.S. (2010) Expanding the range of "druggable" targets with natural product-based libraries: an academic perspective. *Curr. Opin. Chem. Biol.*, **14** (3), 308–314.

12 Drewry, D.H. and Macarron, R. (2010) Enhancements of screening collections to address areas of unmet medical need:

an industry perspective. *Curr. Opin. Chem. Biol.*, **14** (3), 289–298.

13 Spandl, R.J., Bender, A., and Spring, D.R. (2008) Diversity-oriented synthesis; a spectrum of approaches and results. *Org. Biomol. Chem*, **6** (7), 1149–1158.

14 Rees, D.C., Congreve, M., Murray, C.W., and Carr, R. (2004) Fragment-based lead discovery. *Nat. Rev. Drug Discov.*, **3** (8), 660–672.

15 Carr, R.A.E., Congreve, M., Murray, C.W., and Rees, D.C. (2005) Fragment-based lead discovery: leads by design. *Drug Discov. Today*, **10** (14), 987–992.

16 Volkmer, R. (2009) Synthesis and application of peptide arrays: quo vadis SPOT technology. *Chembiochem*, **10** (9), 1431–1442.

17 Kim, M., Shin, D.S., Kim, J., and Lee, Y.S (2010) Substrate screening of protein kinases: detection methods and combinatorial peptide libraries. *Biopolymers*, **94** (6), 753–762.

18 Gruber, C.W., Muttenthaler, M., and Freissmuth, M. (2010) Ligand-based peptide design and combinatorial peptide libraries to target G protein-coupled receptors. *Curr. Pharm. Des.*, **16** (28), 3071–3088.

19 Scheuermann, J. and Neri, D. (2010) DNA-encoded chemical libraries: a tool for drug discovery and for chemical biology. *Chembiochem*, **11** (7), 931–937.

20 Buller, F., Mannocci, L., Scheuermann, J., and Neri, D. (2010) Drug discovery with DNA-encoded chemical libraries. *Bioconjug. Chem.*, **21** (9), 1571–1580.

21 Fuchs, F. and Boutros, M. (2006) Cellular phenotyping by RNAi. *Brief Funct. Genomic Proteomic*, **5** (1), 52–56.

22 Gribbon, P. and Sewing, A. (2003) Fluorescence readouts in HTS: no gain without pain? *Drug Discov. Today*, **8** (22), 1035–1043.

23 Fan, F. and Wood, K.V. (2007) Bioluminescent assays for high-throughput screening. *Assay Drug. Dev. Technol.*, **5** (1), 127–136.

24 Thorne, N., Inglese, J., and Auld, D.S. (2010) Illuminating insights into firefly luciferase and other bioluminescent reporters used in chemical biology. *Chem. Biol.*, **17** (6), 646–657.

25 Walling, M.A., Wang, S., Shi, H., and Shepard, J.R.E. (2010) Quantum dots for positional registration in live cell-based arrays. *Anal. Bioanal. Chem.*, **398** (3), 1263–1271.

26 Vogel, K.W. and Vedvik, K.L. (2006) Improving lanthanide-based resonance energy transfer detection by increasing donor-acceptor distances. *J. Biomol. Screen.*, **11** (4), 439–443.

27 Vedvik, K.L., Eliason, H.C., Hoffman, R.L., Gibson, J.R., Kupcho, K.R., Somberg, R.L., and Vogel, K.W. (2004) Overcoming compound interference in fluorescence polarization-based kinase assays using far-red tracers. *Assay Drug. Dev. Technol.*, **2** (2), 193–203.

28 Pepperkok, R. and Ellenberg, J. (2006) High-throughput fluorescence microscopy for systems biology. *Nat. Rev. Mol. Cell Biol.*, **7** (9), 690–696.

29 Wolff, M., Kredel, S., Wiedenmann, J., Nienhaus, G.U., and Heilker, R. (2008) Cell-based assays in practice: cell markers from autofluorescent proteins of the GFP-family. *Comb. Chem. High Throughput Screen.*, **11** (8), 602–609.

30 Fritzsche, M. and Mandenius, C.F. (2010) Fluorescent cell-based sensing approaches for toxicity testing. *Anal. Bioanal. Chem.*, **398** (1), 181–191.

31 Roda, A., Pasini, P., Mirasoli, M., Michelini, E., and Guardigli, M. (2004) Biotechnological applications of bioluminescence and chemiluminescence. *Trends Biotechnol.*, **22** (6), 295–303.

32 VanEngelenburg, S.B. and Palmer, A.E. (2008) Fluorescent biosensors of protein function. *Curr. Opin. Chem. Biol.*, **12** (1), 60–65.

33 Wolff, M., Wiedenmann, J., Nienhaus, G.U., Valler, M., and Heilker, R. (2006) Novel fluorescent proteins for high-content screening. *Drug Discov. Today*, **11** (23–24), 1054–1060.

34 Chudakov, D.M., Matz, M.V., Lukyanov, S., and Lukyanov, K.A. (2010) Fluorescent proteins and their applications in imaging living cells and tissues. *Physiol. Rev.*, **90** (3), 1103–1163.

35 Verkman, A.S., Lukacs, G.L., and Galietta, L.J.V. (2006) CFTR chloride channel drug discovery-inhibitors as

antidiarrheals and activators for therapy of cystic fibrosis. *Curr. Pharm. Des.*, **12** (18), 2235–2247.
36 Morris, M.C. (2010) Fluorescent biosensors of intracellular targets from genetically encoded reporters to modular polypeptide probes. *Cell Biochem. Biophys.*, **56** (1), 19–37.
37 Sklar, L.A., Carter, M.B., and Edwards, B.S. (2007) Flow cytometry for drug discovery, receptor pharmacology and high-throughput screening. *Curr. Opin. Pharmacol.*, **7** (5), 527–534.
38 Kintses, B., van Vliet, L.D., Devenish, S.R.A., and Hollfelder, F. (2010) Microfluidic droplets: new integrated workflows for biological experiments. *Curr. Opin. Chem. Biol.*, **14** (5), 548–555.
39 Perrin, D., Frémaux, C., and Scheer, A. (2006) Assay development and screening of a serine/threonine kinase in an on-chip mode using caliper nanofluidics technology. *J. Biomol. Screen.*, **11** (4), 359–368.
40 Blackwell, L.J., Birkos, S., Hallam, R., Carr, G.V.D., Arroway, J., Suto, C.M., and Janzen, W.P. (2009) High-throughput screening of the cyclic AMP-dependent protein kinase (PKA) using the caliper microfluidic platform. *Methods Mol. Biol.*, **565**, 225–237.
41 Castel, D., Pitaval, A., Debily, M.A., and Gidrol, X. (2006) Cell microarrays in drug discovery. *Drug Discov. Today*, **11** (13–14), 616–622.
42 Zawko, S.A. and Schmidt, C.E. (2010) Simple benchtop patterning of hydrogel grids for living cell microarrays. *Lab Chip*, **10** (3), 379–383.
43 Upadhyaya, S. and Selvaganapathy, P.R. (2010) Microfluidic devices for cell based high throughput screening. *Lab Chip*, **10** (3), 341–348.
44 Wu, J., Wheeldon, I., Guo, Y., Lu, T., Du, Y., Wang, B., He, J., Hu, Y., and Khademhosseini, A. (2011) A sandwiched microarray platform for benchtop cell-based high throughput screening. *Biomaterials*, **32** (3), 841–848.
45 Ziauddin, J. and Sabatini, D.M. (2001) Microarrays of cells expressing defined cDNAs. *Nature*, **411** (6833), 107–110.
46 de Jong, L.A.A., Uges, D.R.A., Franke, J.P., and Bischoff, R. (2005) Receptor-ligand binding assays: technologies and applications. *J. Chromatogr. B Analyt. Technol. Biomed. Life Sci.*, **829** (1–2), 1–25.
47 Krishhan, V.V., Khan, I.H., and Luciw, P.A. (2009) Multiplexed microbead immunoassays by flow cytometry for molecular profiling: Basic concepts and proteomics applications. *Crit. Rev. Biotechnol.*, **29** (1), 29–43.
48 Burke, T.J., Loniello, K.R., Beebe, J.A., and Ervin, K.M. (2003) Development and application of fluorescence polarization assays in drug discovery. *Comb. Chem. High Throughput Screen.*, **6** (3), 183–194.
49 Banks, P. and Harvey, M. (2002) Considerations for using fluorescence polarization in the screening of G protein-coupled receptors. *J. Biomol. Screen.*, **7** (2), 111–117.
50 Vogel, K.W., Zhong, Z., Bi, K., and Pollok, B.A. (2008) Developing assays for kinase drug discovery - where have the advances come from? *Exp. Opin. Drug Discov.*, **3** (1), 115–129.
51 Degorce, F., Card, A., Soh, S., Trinquet, E., Knapik, G.P., and Xie, B. (2009) HTRF: A technology tailored for drug discovery - a review of theoretical aspects and recent applications. *Curr. Chem. Genomics*, **3**, 22–32.
52 Gabriel, D., Vernier, M., Pfeifer, M.J., Dasen, B., Tenaillon, L., and Bouhelal, R. (2003) High throughput screening technologies for direct cyclic AMP measurement. *Assay Drug. Dev. Technol.*, **1** (2), 291–303.
53 Jia, Y. (2008) Current status of HTRF technology in kinase assays. *Exp. Opin. Drug Discov.*, **3** (12), 1461–1474.
54 Eglen, R.M., Reisine, T., Roby, P., Rouleau, N., Illy, C., Bossé, R., and Bielefeld, M. (2008) The use of AlphaScreen technology in HTS: current status. *Curr. Chem. Genomics*, **1**, 2–10.
55 Michnick, S.W., Ear, P.H., Manderson, E.N., Remy, I., and Stefan, E. (2007) Universal strategies in research and drug discovery based on protein-fragment complementation assays. *Nat. Rev. Drug Discov.*, **6** (7), 569–582.
56 Remy, I. and Michnick, S.W. (2006) A highly sensitive protein-protein interaction assay based on gaussia luciferase. *Nat. Methods*, **3** (12), 977–979.

57 Olson, K.R. and Eglen, R.M. (2007) Beta galactosidase complementation: a cell-based luminescent assay platform for drug discovery. *Assay Drug. Dev. Technol.*, **5** (1), 137–144.

58 Eglen, R.M. (2002) Enzyme fragment complementation: a flexible high throughput screening assay technology. *Assay Drug. Dev. Technol.*, **1** (1 Pt 1), 97–104.

59 Eggeling, C., Brand, L., Ullmann, D., and Jäger, S. (2003) Highly sensitive fluorescence detection technology currently available for HTS. *Drug Discov. Today*, **8** (14), 632–641.

60 Rüdiger, M., Haupts, U., Moore, K.J., and Pope, A.J. (2001) Single-molecule detection technologies in miniaturized high throughput screening: binding assays for g protein-coupled receptors using fluorescence intensity distribution analysis and fluorescence anisotropy. *J. Biomol. Screen.*, **6** (1), 29–37.

61 Kask, P., Palo, K., Fay, N., Brand, L., Mets, U., Ullmann, D., Jungmann, J., Pschorr, J., and Gall, K. (2000) Two-dimensional fluorescence intensity distribution analysis: theory and applications. *Biophys. J.*, **78** (4), 1703–1713.

62 Michelini, E., Cevenini, L., Mezzanotte, L., Coppa, A., and Roda, A. (2010) Cell-based assays: fuelling drug discovery. *Anal. Bioanal. Chem.*, **398** (1), 227–238.

63 Williams, C. (2004) cAMP detection methods in HTS: selecting the best from the rest. *Nat. Rev. Drug Discov.*, **3** (2), 125–135.

64 Hill, S.J., Baker, J.G., and Rees, S. (2001) Reporter-gene systems for the study of G-protein-coupled receptors. *Curr. Opin. Pharmacol.*, **1** (5), 526–532.

65 Nakajima, Y., Kimura, T., Sugata, K., Enomoto, T., Asakawa, A., Kubota, H., Ikeda, M., and Ohmiya, Y. (2005) Multicolor luciferase assay system: one-step monitoring of multiple gene expressions with a single substrate. *Biotechniques*, **38** (6), 891–894.

66 Conklin, B.R., Farfel, Z., Lustig, K.D., Julius, D., and Bourne, H.R. (1993) Substitution of three amino acids switches receptor specificity of Gq alpha to that of Gi alpha. *Nature*, **363** (6426), 274–276.

67 Kostenis, E., Waelbroeck, M., and Milligan, G. (2005) Techniques: promiscuous Galpha proteins in basic research and drug discovery. *Trends Pharmacol. Sci.*, **26** (11), 595–602.

68 Rich, T.C., Tse, T.E., Rohan, J.G., Schaack, J., and Karpen, J.W. (2001) In vivo assessment of local phosphodiesterase activity using tailored cyclic nucleotide-gated channels as cAMP sensors. *J. Gen. Physiol.*, **118** (1), 63–78.

69 Wunder, F., Stasch, J.P., Hütter, J., Alonso-Alija, C., Hüser, J., and Lohrmann, E. (2005) A cell-based cGMP assay useful for ultra-high-throughput screening and identification of modulators of the nitric oxide/cGMP pathway. *Anal. Biochem.*, **339** (1), 104–112.

70 Reinscheid, R.K., Kim, J., Zeng, J., and Civelli, O. (2003) High-throughput real-time monitoring of Gs-coupled receptor activation in intact cells using cyclic nucleotide-gated channels. *Eur. J. Pharmacol.*, **478** (1), 27–34.

71 Wunder, F., Kalthof, B., Müller, T., and Hüser, J. (2008) Functional cell-based assays in microliter volumes for ultra-high throughput screening. *Comb. Chem. High Throughput Screen.*, **11** (7), 495–504.

72 Titus, S., Neumann, S., Zheng, W., Southall, N., Michael, S., Klumpp, C., Yasgar, A., Shinn, P., Thomas, C.J., Inglese, J., Gershengorn, M.C., and Austin, C.P. (2008) Quantitative high-throughput screening using a live-cell cAMP assay identifies small-molecule agonists of the TSH receptor. *J. Biomol. Screen.*, **13** (2), 120–127.

73 Paredes, R.M., Etzler, J.C., Watts, L.T., Zheng, W., and Lechleiter, J.D. (2008) Chemical calcium indicators. *Methods*, **46** (3), 143–151.

74 Zhang, Y., Kowal, D., Kramer, A., and Dunlop, J. (2003) Evaluation of FLIPR calcium 3 assay kit-a new no-wash fluorescence calcium indicator reagent. *J. Biomol. Screen.*, **8** (5), 571–577.

75 Brini, M. (2008) Calcium-sensitive photoproteins. *Methods*, **46** (3), 160–166.

76 McCombs, J.E. and Palmer, A.E. (2008) Measuring calcium dynamics in living cells with genetically encodable calcium indicators. *Methods*, **46** (3), 152–159.

77 Charlton, S.J. and Vauquelin, G. (2010) Elusive equilibrium: the challenge of interpreting receptor pharmacology using calcium assays. *Br. J. Pharmacol.*, **161** (6), 1250–1265.

78 Liu, K., Southall, N., Titus, S.A., Inglese, J., Eskay, R.L., Shinn, P., Austin, C.P., Heilig, M.A., and Zheng, W. (2010) A multiplex calcium assay for identification of GPCR agonists and antagonists. *Assay Drug. Dev. Technol.*, **8** (3), 367–379.

79 Lü, Q. and An, W.F. (2008) Impact of novel screening technologies on ion channel drug discovery. *Comb. Chem. High Throughput Screen.*, **11** (3), 185–194.

80 Zheng, W., Spencer, R.H., and Kiss, L. (2004) High throughput assay technologies for ion channel drug discovery. *Assay Drug. Dev. Technol.*, **2** (5), 543–552.

81 Weaver, C.D., Harden, D., Dworetzky, S.I., Robertson, B., and Knox, R.J. (2004) A thallium-sensitive, fluorescence-based assay for detecting and characterizing potassium channel modulators in mammalian cells. *J. Biomol. Screen.*, **9** (8), 671–677.

82 Beacham, D.W., Blackmer, T., Grady, M.O., and Hanson, G.T. (2010) Cell-based potassium ion channel screening using the fluxOR™ assay. *J. Biomol. Screen.*, **15** (4), 441–446.

83 Sui, J., Cotard, S., Andersen, J., Zhu, P., Staunton, J., Lee, M., and Lin, S. (2010) Optimization of a Yellow fluorescent protein-based iodide influx high-throughput screening assay for cystic fibrosis transmembrane conductance regulator (CFTR) modulators. *Assay Drug. Dev. Technol.*, **8** (6), 656–668.

84 Przybylo, M., Borowik, T., and Langner, M. (2010) Fluorescence techniques for determination of the membrane potentials in high throughput screening. *J. Fluoresc.*, **20** (6), 1139–1157.

85 Molokanova, E. and Savchenko, A. (2008) Bright future of optical assays for ion channel drug discovery. *Drug Discov. Today*, **13** (1–2), 14–22.

86 Baxter, D.F., Kirk, M., Garcia, A.F., Raimondi, A., Holmqvist, M.H., Flint, K.K., Bojanic, D., Distefano, P.S., Curtis, R., and Xie, Y. (2002) A novel membrane potential-sensitive fluorescent dye improves cell-based assays for ion channels. *J. Biomol. Screen.*, **7** (1), 79–85.

87 González, J.E. and Maher, M.P. (2002) Cellular fluorescent indicators and voltage/ion probe reader (VIPR™) tools for ion channel and receptor drug discovery. *Receptors Channels*, **8** (5–6), 283–295.

88 Dabrowski, M.A., Dekermendjian, K., Lund, P.E., Krupp, J.J., Sinclair, J., and Larsson, O. (2008) Ion channel screening technology. *CNS Neurol. Disord. Drug Targets*, **7** (2), 122–128.

89 Edwards, B.S., Young, S.M., Saunders, M.J., Bologa, C., Oprea, T.I., Ye, R.D., Prossnitz, E.R., Graves, S.W., and Sklar, L.A. (2007) High-throughput flow cytometry for drug discovery. *Exp. Opin. Drug Discov.*, **2** (5), 685–696.

90 Krutzik, P.O. and Nolan, G.P. (2006) Fluorescent cell barcoding in flow cytometry allows high-throughput drug screening and signaling profiling. *Nat. Methods*, **3** (5), 361–368.

91 Zanella, F., Lorens, J.B., and Link, W. (2010) High content screening: seeing is believing. *Trends Biotechnol.*, **28** (5), 237–245.

92 Lang, P., Yeow, K., Nichols, A., and Scheer, A. (2006) Cellular imaging in drug discovery. *Nat. Rev. Drug Discov.*, **5** (4), 343–356.

93 Vaisberg, E.A., Lenzi, D., Hansen, R.L., Keon, B.H., and Finer, J.T. (2006) An infrastructure for high-throughput microscopy: instrumentation, informatics, and integration. *Methods Enzymol.*, **414**, 484–512.

94 Starkuviene, V. and Pepperkok, R. (2007) The potential of high-content high-throughput microscopy in drug discovery. *Br. J. Pharmacol.*, **152** (1), 62–71.

95 Lee, S. and Howell, B.J. (2006) High-content screening: emerging hardware and software technologies. *Methods Enzymol.*, **414**, 468–483.

96 Bickle, M. (2010) The beautiful cell: high-content screening in drug discovery. *Anal. Bioanal. Chem.*, **398** (1), 219–226.

97 Shen, F., Hodgson, L., and Hahn, K. (2006) Digital autofocus methods for

automated microscopy. *Methods Enzymol.*, **414**, 620–632.

98 Shen, F., Hodgson, L., Price, J.H., and Hahn, K.M. (2008) Digital differential interference contrast autofocus for high-resolution oil-immersion microscopy. *Cytometry A*, **73** (7), 658–666.

99 Ljosa, V. and Carpenter, A.E. (2009) Introduction to the quantitative analysis of two-dimensional fluorescence microscopy images for cell-based screening. *PLoS Comput. Biol.*, **5** (12), e1000603.

100 Wollman, R. and Stuurman, N. (2007) High throughput microscopy: from raw images to discoveries. *J. Cell Sci.*, **120**Pt (21), 3715–3722.

101 Carpenter, A., Jones, T., Lamprecht, M., Clarke, C., Kang, I., Friman, O., Guertin, D., Chang, J., Lindquist, R., Moffat, J., Golland, P., and Sabatini, D. (2006) CellProfiler: image analysis software for identifying and quantifying cell phenotypes. *Genome Biol.*, **7** (10).

102 Jones, T., Kang, I., Wheeler, D., Lindquist, R., Papallo, A., Sabatini, D., Golland, P., and Carpenter, A. (2008) CellProfiler Analyst: data exploration and analysis software for complex image-based screens. *BMC Bioinformatics*, **9**, 482.

103 Rämö, P., Sacher, R., Snijder, B., Begemann, B., and Pelkmans, L. (2009) CellClassifier: supervised learning of cellular phenotypes. *Bioinformatics*, **25** (22), 3028–3030.

104 Misselwitz, B., Strittmatter, G., Periaswamy, B., Schlumberger, M.C., Rout, S., Horvath, P., Kozak, K., and Hardt, W.D. (2010) Enhanced CellClassifier: a multi-class classification tool for microscopy images. *BMC Bioinformatics*, **11**, 30.

105 Held, M., Schmitz, M.H.A., Fischer, B., Walter, T., Neumann, B., Olma, M.H., Peter, M., Ellenberg, J., and Gerlich, D.W. (2010) CellCognition: time-resolved phenotype annotation in high-throughput live cell imaging. *Nat. Methods*, **7** (9), 747–754.

106 Harper, G. and Pickett, S.D. (2006) Methods for mining HTS data. *Drug Discov. Today*, **11** (15–16), 694–699.

107 Malo, N., Hanley, J.A., Cerquozzi, S., Pelletier, J., and Nadon, R. (2006) Statistical practice in high-throughput screening data analysis. *Nat. Biotechnol.*, **24** (2), 167–175.

108 Birmingham, A., Selfors, L.M., Forster, T., Wrobel, D., Kennedy, C.J., Shanks, E., Santoyo-Lopez, J., Dunican, D.J., Long, A., Kelleher, D., Smith, Q., Beijersbergen, R.L., Ghazal, P., and Shamu, C.E. (2009) Statistical methods for analysis of high-throughput RNA interference screens. *Nat. Methods*, **6** (8), 569–575.

109 Pelz, O., Gilsdorf, M., and Boutros, M. (2010) web cellHTS2: a web-application for the analysis of high-throughput screening data. *BMC Bioinformatics*, **11**, 185.

110 Zhang, J., Chung, T., and Oldenburg, K. (1999) A simple statistical parameter for use in evaluation and validation of high throughput screening assays. *J. Biomol. Screen.*, **4** (2), 67–73.

111 Thorne, N., Auld, D.S., and Inglese, J. (2010) Apparent activity in high-throughput screening: origins of compound-dependent assay interference. *Curr. Opin. Chem. Biol.*, **14** (3), 315–324.

112 Simeonov, A., Jadhav, A., Thomas, C.J., Wang, Y., Huang, R., Southall, N.T., Shinn, P., Smith, J., Austin, C.P., Auld, D.S., and Inglese, J. (2008) Fluorescence spectroscopic profiling of compound libraries. *J. Med. Chem.*, **51** (8), 2363–2371.

113 Gakamsky, D.M., Dennis, R.B., and Smith, S.D. (2011) Use of fluorescence lifetime technology to provide efficient protection from false hits in screening applications. *Anal. Biochem.*, **409** (1), 89–97.

114 McGovern, S.L., Helfand, B.T., Feng, B., and Shoichet, B.K. (2003) A specific mechanism of nonspecific inhibition. *J. Med. Chem.*, **46** (20), 4265–4272.

115 Johnston, P.A., Soares, K.M., Shinde, S.N., Foster, C.A., Shun, T.Y., Takyi, H.K., Wipf, P., and Lazo, J.S. (2008) Development of a 384-well colorimetric assay to quantify hydrogen peroxide generated by the redox cycling of compounds in the presence of reducing agents. *Assay Drug. Dev. Technol.*, **6** (4), 505–518.

116 Soares, K.M., Blackmon, N., Shun, T.Y., Shinde, S.N., Takyi, H.K., Wipf, P., Lazo,

J.S., and Johnston, P.A. (2010) Profiling the NIH small molecule repository for compounds that generate H_2O_2 by redox cycling in reducing environments. *Assay Drug. Dev. Technol.*, 8 (2), 152–174.

117 Shun, T.Y., Lazo, J.S., Sharlow, E.R., and Johnston, P.A. (2011) Identifying actives from HTS data sets: practical approaches for the selection of an appropriate HTS data-processing method and quality control review. *J. Biomol. Screen.*, **16** (1), 1–14.

118 Brideau, C., Gunter, B., Pikounis, B., and Liaw, A. (2003) Improved statistical methods for hit selection in high-throughput screening. *J. Biomol. Screen.*, 8 (6), 634–647.

10
Optical Measurements for the Rational Screening of Protein Crystallization Conditions

Christoph Janzen and Kurt Hoffmann

10.1
Introduction

The determination of protein structures by X-ray diffraction is a complex process because optimal solution parameters must be found for the growth of single crystals, and these are usually defined empirically. In this chapter we describe a rational crystallization process, combining automated sample preparation with the objective evaluation of crystallization trials based on optical measurements. The computerized analysis of such measurements allows the optimum crystallization conditions to be deduced rationally. We developed three new optical measurement techniques for the characterization of small protein droplets. First, static laser light scattering (SLS) allows thermodynamic interaction parameters (e.g., osmotic virial coefficients) to be determined, indicating the attractive and repulsive forces between solute particles. Second, dynamic laser light scattering (DLS) allows the determination of particle radius distributions and the detection of aggregation. Finally, quantitative polarization light microscopy generates a set of three images for transmission, the amplitude of birefringence, and the orientation of the axis of the sample's indicatrix, allowing the detection and quantitative analysis of birefringent structures (e.g., protein crystals or crystalline precipitates). By combining SLS, DLS and quantitative polarization microscopy, it is possible to carry out a rational optical analysis of the pre-crystallization, crystallization and post-crystallization phases. All three methods were developed in combination with a novel crystallization protocol that can be applied to droplets with a volume $<1\,\mu l$. A demonstrator was built that contains an optical measurement unit for light scattering and microscopy, and a liquid-handling unit for sample preparation. This demonstrator provides a unique and novel platform for the development of rational crystallization strategies.

Handbook of Biophotonics. Vol.3: Photonics in Pharmaceutics, Bioanalysis and Environmental Research, First Edition.
Edited by Jürgen Popp, Valery V. Tuchin, Arthur Chiou, and Stefan Heinemann

10.2
State of the Art of Protein Crystallization Techniques

The three-dimensional structures of macromolecules such as proteins provide useful reference data and facilitate downstream biotechnology applications, such as drug development, because they allow drug–target interactions to be modeled accurately. The three-dimensional structures of biological macromolecules are usually determined by X-ray diffraction analysis, which requires the preparation of macromolecules and their complexes as individual monocrystals.

Protein crystals are usually obtained from highly-purified protein solutions in the presence of salts or polymers (such as polyethylene glycol) that act as precipitation reagents. Water is extracted slowly and in a controlled manner so that the protein solution becomes oversaturated. Local fluctuations in protein concentration can then lead to the formation of aggregates that nucleate to seed crystals. Although nucleation is a prerequisite for the formation of crystals, the more nucleation seeds that are formed simultaneously, the more likely that polycrystalline precipitates will form, and these cannot be used for X-ray diffraction. Optimal crystallization conditions are achieved when very few nucleation seeds are formed, and these grow into large single crystals. The protein solution is said to exist in a metastable state when the solid phase of the protein is thermodynamically favored over the solute form, yet the appearance of seed crystals is hindered due to the large free surface enthalpy that must be overcome. All crystallization experiments try to achieve this metastable state because this favors individual nucleation events that can grow to form large single crystals. The crystallization conditions are usually tested empirically by selecting different protein concentrations, different salts at varying ionic strengths, and different solution parameters (e.g., pH, buffers, polymers, organic solvents and temperatures). However, this results in a vast matrix of distinct multi-parameter conditions [1].

There are many crystallization methods. In a batch process, the target protein is added to an aqueous solution containing buffer, salts and other crystallizing reagents (the so-called crystallization solution) or vice versa [2]. The two solutions are combined in a closed sample compartment, such as a microtiter plate with a closure head or layer of paraffin oil to inhibit evaporation. In these closed systems, there is no enrichment of the precipitation reagents or proteins over a prolonged period and all concentrations can be adjusted directly by applying the protein and crystallization solutions in different relative proportions (Figure 10.1). However, the batch process does not support the slow and continuous change of conditions that allow entry into the metastable region of the protein phase diagram, so the great merit of classical crystallization processes is lost (kinetic and dynamic properties of the end point, see Figure 10.2).

A modified batch process method replaces the pure paraffin oil with paraffin oil containing dimethylsiloxane (DMS), which allows the continuous diffusion (evaporation) of water from the covered solution. This gradually increases the protein and salt concentrations, and the metastable region can thus be reached (Figure 10.2). However, since this is an open system, an end point cannot be set. The main

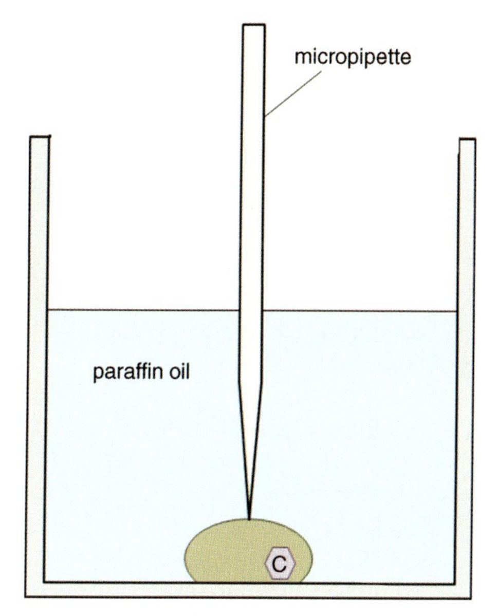

Figure 10.1 Schematic illustration of a micro-batch experiment. A drop is covered with immersion oil, and liquids can be added under the oil with a micropipette. If pure paraffin oil is used, no water loss is expected (batch experiment), but where an open diffusion oil is used (such as paraffin with dimethylsiloxane) the water can evaporate and the protein concentration increases.

disadvantage is that the protein solution will dry out completely over a period time and this approach is therefore only of limited use (see "diffusion oil" path, Figure 10.2).

The traditional and most widely used protein crystallization techniques are hanging and sitting droplets. Here a drop of protein solution (mixed with reservoir solution) is incubated in a closed vessel with a reservoir of aqueous solution containing higher concentrations of salt (Figure 10.3). Over time, water evaporates from the droplet and is transported to the reservoir solution so that the concentration of the protein and the precipitant increases continuously. An end point is reached when the droplet is in equilibrium with the reservoir (identical chemical potential) (see "h/s-drop" path, Figure 10.2). The salt concentration in the reservoir solution determines the end point of the diffusion process.

Crystals suitable for structure determination by X-ray diffraction are very difficult to prepare using the above conventional methods, and it is often not possible to grow crystals of a suitable size and quality. Crystallization often fails for unknown reasons. The interplay between physical and chemical processes during the formation of crystals is complex and still not completely understood.

Over the last few years, we have witnessed rapid progress in the automated determination of gene and protein sequences, and it would be advantageous to apply the same high-throughput paradigm to protein structure determination. This has become possible, in theory, because of recent advances in X-ray crystallography, such

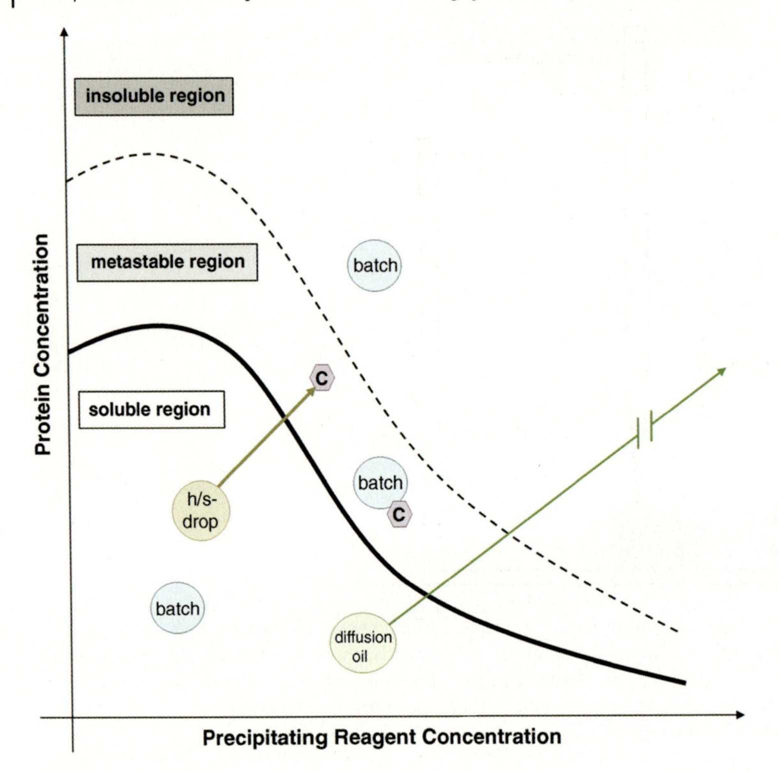

Figure 10.2 Simplified phase diagram summarizing typical crystallization techniques. The soluble, metastable and insoluble regions for the protein are shown. The encircled "batch" indicates crystallization trials performed under paraffin oil, where the initial concentrations do not change, whereas "diffusion oil" indicates a modified batch experiment with oil that allows diffusion covering the protein solution. No endpoint for water loss is determined. Finally, "h/s-drop" shows the classical approach to crystallization (hanging/sitting drop) with a defined end point for the diffusion process.

as the availability of high-brilliance X-ray sources (synchrotrons), improved hardware, and software that allows rapid data interpretation and structure calculations. Unfortunately, conventional methods for protein crystallization have not evolved in the same manner and, despite there being numerous automated, high-throughput approaches, no process exists that easily yields protein crystals of sufficient size and quality for X-ray diffraction [3]. The achievement of optimal conditions for the formation of single crystals is still a labor-intensive empirical process that depends mainly on the intuition of the crystallographer. Protein crystallography is therefore the bottleneck in the structural determination of biological macromolecules and there is a great demand for novel protein crystallization methods that can be automated and evaluated objectively, thus enabling the systematic production of crystals.

Automated diffusion experiments based on the hanging-drop or sitting-drop methods are performed in systems that are closed with siliconized glass lids or

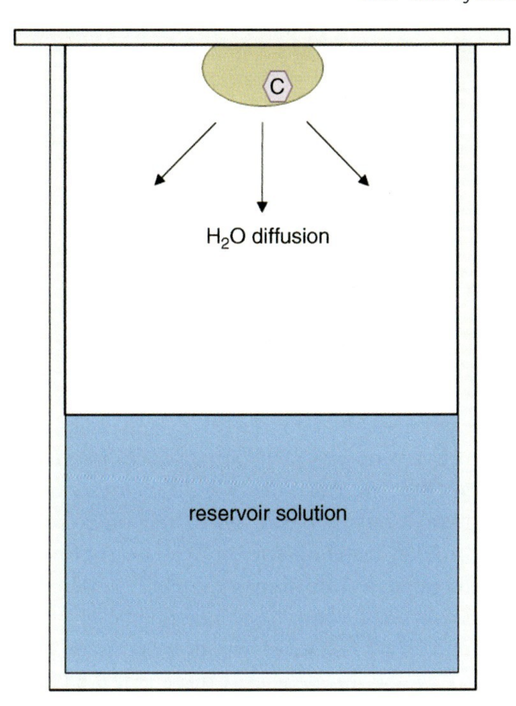

Figure 10.3 Schematic illustration of a hanging drop crystallization experiment. The liquid in the hanging drop is in diffusion contact with the reservoir solution, which has a higher salt concentration. Water evaporates from the drop and is deposited in the reservoir, the protein concentration in the drop rises. When the concentrations are equilibrated, the net transport process stops (defined end point).

self-adhesive transparent films. The automation of liquid handling systems requires many complex steps (e.g., closing the lid) and current systems tend to be expensive, slow, irreproducible and incompatible with the smallest sample volumes (in the microliter to nanoliter range). Protection against evaporation is required for these small sample volumes as long as the systems are not closed by the application of self-adhesive foils or glass lids, because water losses can vary considerably and thus affect the delicately balanced concentrations of proteins and precipitation reagents.

In automated batch processes, robots are used to deposit protein and crystallization solutions directly under paraffin oil to prevent evaporation. Such automated methods are associated with the usual disadvantages of batch processes (i.e., no change in concentration under pure oil, and no end to the diffusion process under open diffusion oil). Nevertheless, most automated high-throughput crystallization systems in use today are based on batch processes. Large numbers of crystallization trials are necessary to test different combinations of conditions empirically, with automated camera systems used to identify amorphous precipitates, microcrystalline precipitates or large single crystals. The parameter field under investigation is based upon the experience and intuition of the crystallographer. With the exception of the post-crystallization phase, no optical analysis of these experiments is possible, therefore the outcome depends on trial and error and thousands of experiments may be required to produce adequate crystals, if this is possible at all.

In order to improve the process of crystallization significantly, we have developed a combination of novel rational methods for protein crystallization based on optical measurement techniques, and these are discussed in more detail below.

10.3
A New Crystallization Method that Enables the Use of Optical Measurement Technologies

We have developed a new crystallization method to integrate light scattering measurements within a screening process using only small amounts of protein [4]. The method uses two oils and combines the advantages of vapor diffusion methods with the automation capability of micro-batch methods. As in the micro-batch method, the total system is protected from evaporation, for example, by paraffin oil (first phase). However, the protein/precipitant mixture (fourth phase) can equilibrate with a reservoir solution (third phase) via a second layer of oil (second phase) that is placed under the paraffin oil. The second oil is immiscible with the paraffin oil, and does not interact with or influence the properties of phases three and four. The system is summarized in Figure 10.4.

The system can be set up in several ways, using different plate geometries. In one approach, the protein/precipitant and the reservoir can be applied under the paraffin oil, resulting in a classic micro-batch experiment. When the second oil is added, water begins to diffuse from the protein droplet to the reservoir solution. The removal of phase II allows dynamic control of water diffusion and, because the system is protected from evaporation by the paraffin oil, small volumes of protein solution can

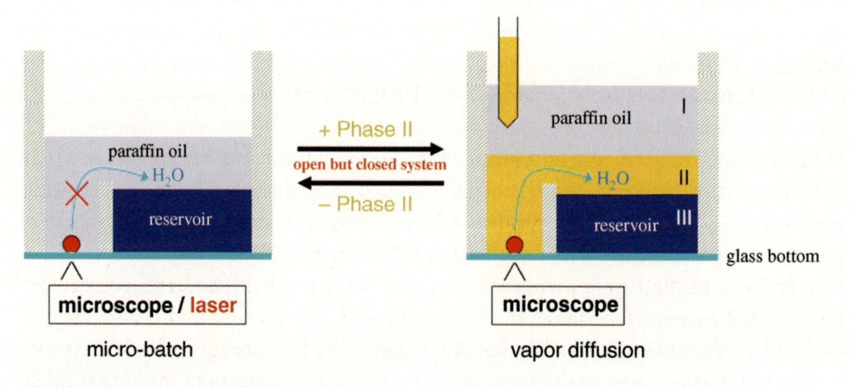

Figure 10.4 A new crystallization method based on a specially-designed oil phase that allows controlled water diffusion from the protein sample to the reservoir phase. The paraffin oil protects the system from evaporation and the glass bottoms of the plates enable optical measurements for the characterization of the protein samples.

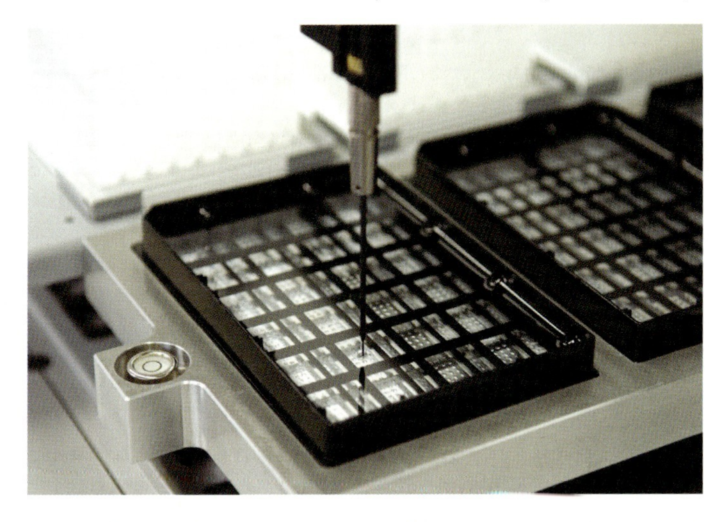

Figure 10.5 A novel microplate was developed for the new crystallization system. This figure shows how the protein solution is applied under paraffin oil. A series of six droplets of various protein concentration are applied for the determination of the second virial coefficient. 96 virial coefficients can be determined within a single microplate.

be applied using a conventional liquid-handling system. The microtiter plates are equipped with a glass bottom that allows optical analysis of the protein sample without distortions from curved meniscus surfaces (Figure 10.5). The droplets of protein solution sit on the glass surface and adopt a hemispherical form. This can be supported by structuring the glass surface with a polar and nonpolar surface pattern. The light scattering optics can then be placed under the microtiter plate and the laser beam does not need to pass through curved surfaces on its way to the sample, so the laser beam can be focused precisely inside the sample even in small sample droplets.

10.4 Optical Measurements for a Rational Crystallization Process

Optical measurements can be used to characterize all three phases of the crystallization process, which avoids the need for trial and error strategies to find optimal solution conditions. Before nucleation takes place (pre-nucleation phase) the solute protein molecules interact with each other and with the solvent molecules, and this interaction potential can be measured by static light scattering. The formation of seed crystals begins with the aggregation of proteins (nucleation phase) which can be detected by dynamic light scattering. After the nucleation and growth phase (post-nucleation phase), the resulting single crystals, microcrystalline, or amorphous precipitates can be detected and characterized by quantitative polarization microscopy, which is better suited than standard microscopy for the detection and evaluation of crystalline features.

10.4.1 Static Light Scattering for the Analysis of the Pre-Nucleation Phase

Crystals form when attractive interactions between protein molecules are sufficient to drive them from the solute state to a thermodynamically more stable ordered solid state. If protein molecules repel each other, crystals do not form and the protein remains in solution. However, if the attractive interactions are too strong, massive precipitation may occur, yielding polycrystalline or even amorphous precipitates, because in this case even a disordered solid state is thermodynamically favored over the solute state. Therefore, conditions must be selected to favor slightly attractive interactions that achieve the necessary delicate balance between the absence of nucleation processes and the presence of too many nucleation processes. Only under these conditions is it possible to grow large single crystals.

In the pre-nucleation phase, static light scattering can be used to analyze the interaction potential between dissolved protein molecules. Static light scattering is a technique that measures the absolute intensity of the scattered light as a function of the scattering angle or the sample concentration. For static light scattering experiments, a high intensity monochromatic laser is focused on a protein solution, and one or more detectors are used to measure the scattering intensity at one or more angles. The average molecular weight M_w of a macromolecule, such as a protein, can be calculated from the angular dependence. The scattering intensity of different samples with identical crystallization conditions (salt concentrations, pH-value etc.) but different protein concentrations allows the second osmotic virial coefficient A_2 to be calculated. This empirical parameter can be used to describe the behavior of the osmotic pressure in a real sample, in contrast to an ideal sample. The deviation in osmotic pressure from the ideal behavior is expressed using A_2. To describe the osmotic pressure Π of a solution containing proteins of concentration c_P and molecular weight M_W, the osmotic virial expansion is set up as follows, where R is the universal gas constant and T is temperature:

$$\Pi = RTc_P\left(\frac{1}{M_W} + A_2c_P + \ldots\right)$$

The virial expansion is usually aborted after the second term [5].

If $A_2 = 0$, the virial expansion describes ideal behavior (compare with the corresponding formula for ideal gases). This means that the interaction between protein molecules in solution is neither attractive nor repulsive and they interact with each other as they interact with the solvent. If $A_2 > 0$, the osmotic pressure rises above ideal behavior, indicating repulsion between protein molecules. For negative A_2 values, the osmotic pressure is lower than under ideal conditions, and in this case the protein molecules attract each other more than they attract the solvent molecules. From a series of static light scattering measurements taken from solutions containing different protein concentrations c_P, the osmotic second virial coefficient A_2 and the molecular weight M_W can be deduced from the slope and the intercept of the so-called

Zimm plot based on the following equation:

$$\frac{Kc_P}{R} = \frac{1}{M_W} + 2A_2c_P$$

Here, K is an optical constant, representing the laser wavelength, the refractive index of the solvent and the refractive index increment of the sample. R is the so-called Rayleigh ratio, representing the normalized scattering intensity of the protein sample divided by the scattering intensity of toluene (the reference scattering compound). With the help of the reference substance, instrument-specific parameters (e.g., light collection efficiency, detector sensitivity) are regarded. The necessary calculations are explained in more detail by Wyatt [6].

The usefulness of the osmotic virial coefficient for the prediction of protein crystallization parameters was first described in 1994 by George and Wilson [7]. They showed for the first time that protein crystals often form when the precipitant induces a slight attraction between protein molecules in solution, which was a sound perception. Several other studies have demonstrated the importance of the second osmotic virial coefficient for the crystallization behavior of various proteins [8–10]. The point at which the virial coefficient becomes slightly negative, indicating slight but not excessive attraction between dissolved protein molecules, was therefore named the "*crystallization window*". The second virial coefficient can be determined by static light scattering (SLS), but alternative methods have been developed [11]. However, these are not easily combined within a screening process. Figure 10.6 shows several Zimm plots for the protein lysozyme with different salt concentrations.

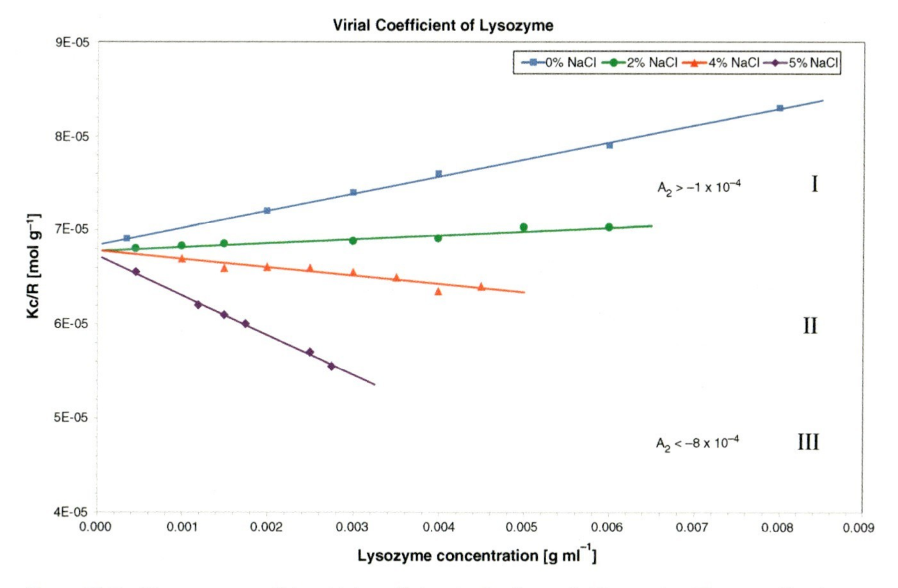

Figure 10.6 Measurement of the virial coefficient in the form of a Zimm plot. The crystallization window (area II) is highlighted in gray. This figure is based on the review by George and Wilson [7].

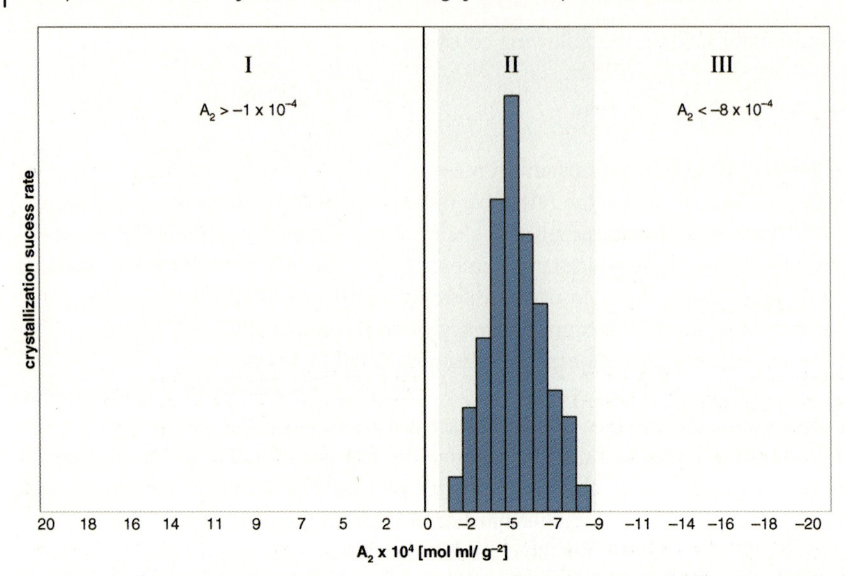

Figure 10.7 Success rate for different crystallization trials as a function of A_2. This figure is based on the review by George and Wilson [7].

Higher NaCl concentrations induce attractive forces between the protein molecules (negative slope in the Zimm plot). Crystals are only formed in area II (highlighted in gray), which has moderately negative A_2 values.

In order to verify the general utility of the crystallization window concept, many different proteins have been analyzed and the crystallization success rate was found to correlate with the second virial coefficient. Figure 10.7 shows that successful crystallization trials are concentrated in the crystallization window.

10.4.2
Dynamic Light Scattering for the Analysis of the Nucleation Phase

Quasielastic light scattering (QLS), also known as dynamic light scattering (DLS), is an optical method well-suited to determination of the diffusion coefficients of particles undergoing Brownian motion in solution. Diffusion coefficients are determined by particle size, shape and flexibility, as well as by inter-particle interactions. The dependence between the diffusion constant D_0 and the hydrodynamic radius R_h is described by the Stokes–Einstein equation:

$$D_0 = \frac{k_B T}{6\pi\eta R_h}$$

Here k_B is the Boltzmann constant, T the temperature and η the viscosity. Once the diffusion constant is known, it is possible to calculate particle sizes and to evaluate the distribution of heterogeneous particle mixtures.

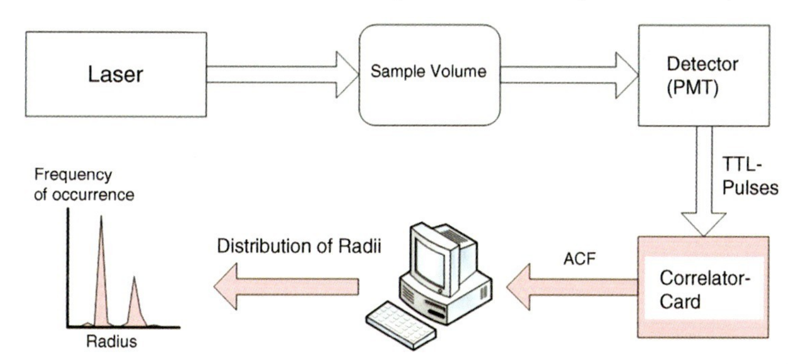

Figure 10.8 Concept of the dynamic light scattering process. The detector is a photon counting photomultiplier, and single photons are converted to electrical TTL-pulses. A correlator card calculates the autocorrelation function (ACF). With a computer, the distribution function of the radii can be calculated from the ACF.

Although the experimental set-ups used to measure DLS and SLS can be comparable, data processing and interpretation differ significantly (see Figure 10.8). A collimated or partially-focused laser beam is directed through the sample and scattered light reaching the detector at a defined collection angle is recorded as a function of time. Because the incident light is coherent, the scattered light forms an interference pattern. The pattern changes with the movement of the sample (e.g., Brownian motion), leading to intensity fluctuations that are related directly to the diffusion of the scattering particles.

Autocorrelation analysis can then be used to extract the diffusion coefficient from the temporal fluctuations in scattering intensity, and by inserting the diffusion coefficient into the Stokes–Einstein equation, it becomes possible to calculate the hydrodynamic radius of the moving particle. The autocorrelation function does not usually fit to a single particle size but to a distribution of hydrodynamic radii. The calculation of the size distribution is carried out with iterative fit-analysis containing inverse Laplace transformations. Specialized software packages such as CONTIN [12] are frequently used for this purpose following DLS experiments.

Figures 10.9 and 10.10 show the autocorrelation function and the calculated radius distribution for a lysozyme sample with a concentration of $50\,\mathrm{mg\,ml^{-1}}$. The measurements were collected using a commercial light scattering instrument [13].

The polydispersity of the proteins in the size distribution is closely correlated to the success of protein crystallization. The presence of non-specific aggregates, various oligiomeric states and impurities widens the size distribution and, at the same time, reduces the likelihood that high-quality crystals will be generated. In this manner, DLS can be used to characterize the quality of a protein sample. With time-resolved measurements, it is also possible to monitor the process of aggregation. Aggregates can be seen as a pre-form to seed crystals, so the initial phase of nucleation can be directly observed with dynamic light scattering [14–16].

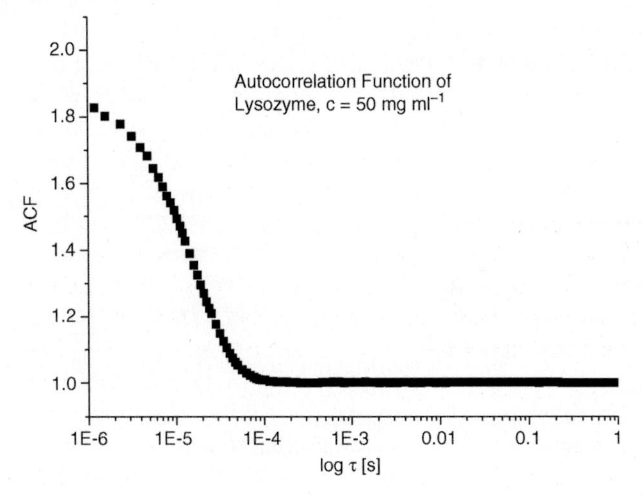

Figure 10.9 Autocorrelation function for lysozyme, recorded with a commercial light scattering instrument.

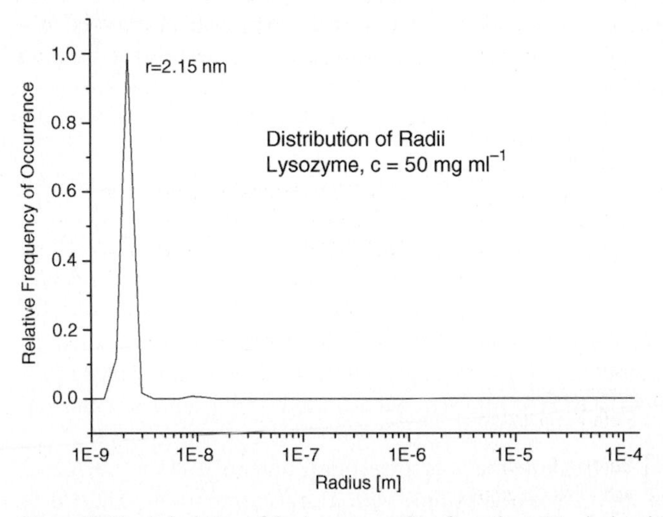

Figure 10.10 Calculation of the lysozyme radius from the ACF calculated in Figure 10.9.

10.4.3
Quantitative Polarization Microscopy for the Analysis of the Post-Nucleation Phase

Quantitative polarization microscopy is a robust and nondestructive method to analyze anisotropic properties of materials. Optical anisotropy of crystals leads to birefringence. The variation of the refractive index is dependent on the orientation of the incident light with respect to the crystal. It can be visualized by a mathematical representation the so-called indicatrix. Light passing such anisotropic material undergoes birefringence corresponding to the ellipse axes length difference (Δn) of the indicatrix cross-section ellipse orthogonal to the propagation direction of the light. In the case of crystals, the shape of the indicatrix is defined by the symmetry of the crystal. In this respect, cubic crystal systems present a spherical indicatrix and shown birefringence. In order to detect birefringence, light with defined polarization states is used to probe the sample and the resulting properties of the light after transmittance through the sample are analyzed.

The technique we use to analyze anisotropic properties of protein crystals is based on the so-called rotating polarizer method [17], as implemented in the Taorad MP-1 microscope. This method allows the birefringence of precipitates and crystals to be determined quantitatively [18], and the diffraction quality of crystals to be predicted accurately [19]. Knowledge of the optical properties of a protein crystal allows the nondestructive determination of crystal diffraction quality before any X-ray data collection is attempted. This enables the pre-selection of high-quality crystals and provides a ranking system for crystallization experiments producing many crystals.

The optical train of a quantitative polarization microscope typically consists of a light source, a bandpass-filter, a (motorized) linear polarizer, a sample/micro-plate holder, a microscope objective, a circular analyzer and a camera (see Figure 10.11). The light intensity, which is recorded by each pixel (i,j) of the camera is described by:

$$I_{i,j} = I_{0_{ij}}[1-\sin 2(\alpha-\varphi)\sin\delta]$$

where $I_{0_{ij}}$ is the absorption affected incident light intensity (transmittance), α describes the relative orientation of the transmission axis of the rotating polarizer, $|\sin\delta|$ the relative optical anisotropy and φ the optical orientation of the indicatrix.

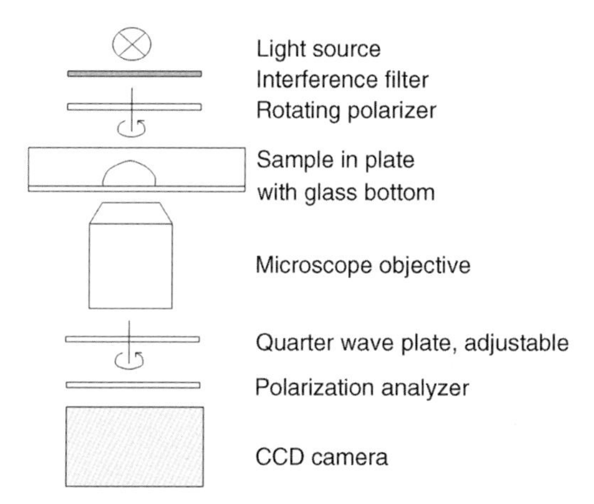

Figure 10.11 Concept of quantitative polarization microscopy. A series of images is recorded using different angular positions (α) of the rotating analyzer.

δ is the phase shift of the light:

$$\delta = \frac{2\pi}{\lambda}\Delta n L$$

where λ is the wavelength of the light, L the thickness of the sample, and $\Delta n = n_1 - n_2$; ΔnL represents the optical retardation.

In order to determine $I_{0_{ij}}$, $|\sin\delta|$ and φ, several images at different linear polarizer orientations (α) are recorded, the data processed pixelwise and represented as grayscale image ($I_{0_{ij}}$) or as false-colored images ($|\sin\delta|$ and φ). $|\sin\delta|$ is related to the thickness of the crystals and φ represents the orientation of the observed slow axis of the elliptical cross-section through the indicatrix on each image pixel of the crystal and can be correlated with the internal order of the crystal.

Quantitative polarization microscopy is not only suitable for analysis of the properties of larger monocrystals, but is also extremely useful to detect anisotropic properties of precipitates (Figure 10.12), and to discriminate between amorphous and microcrystalline precipitates.

Echalier *et al.* showed [18], that the application of quantitative polarization microscopy allowed a much earlier detection of microcrystals as compared to classical microscopic observations (see Figure 10.13).

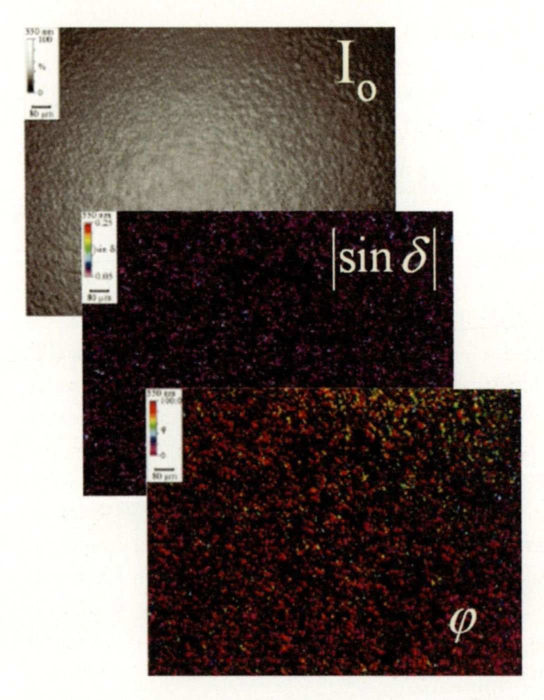

Figure 10.12 Three images for transmission (I_0), birefringence amplitude $|\sin\delta|$ and orientation φ of a microcrystalline sample. This figure is taken from the review by Echalier *et al.* [18].

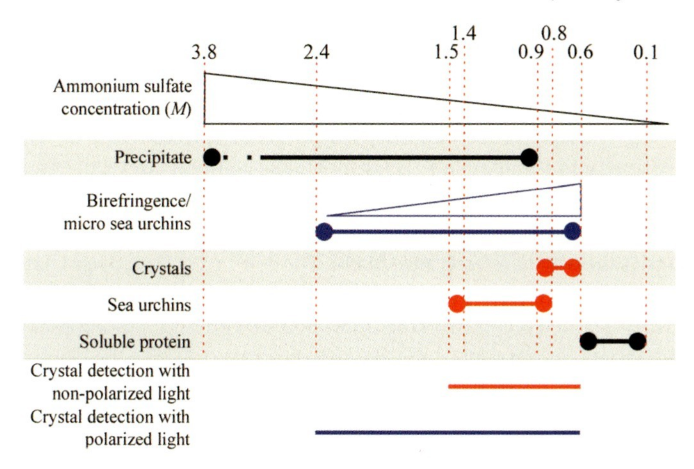

Figure 10.13 Early detection of crystals with the help of quantitative polarization microscopy. Crystallization of glucose isomerase using ammonium sulfate as precipitant was detectable at concentrations between 1.5 and 0.6 M using classical microscopic inspection and from 2.4 to 0.6 M using quantitative polarization microscopy. The crystallization space is more efficiently screened and the risk of missing productive crystallization conditions reduced. This figure is sourced from the review by Echalier *et al.* [18].

10.5 Development of a New Optical Instrument

10.5.1 Measurement of Static and Dynamic Light Scattering in Small Volumes

The new crystallization process is based on the automated optical measurement of crystallization solutions. Because target proteins for crystallization experiments are often expensive and strictly limited in availability, crystallization experiments are usually set up using small sample volumes. Ideally, measurements should be possible in droplets <1 μl in volume. Commercial light scattering instruments for the measurement of virial coefficients use expensive glass cuvettes with volumes typically in the10 μl to 1 ml range. Static measurements in microtiter plates are not possible using commercial instruments because of the suboptimal measurement geometry. Automated sample preparation with liquid-handling robotics can only be achieved using microtiter plates, whereas single glass cuvettes must be cleaned and refilled manually. Plastic cuvettes cannot be used because the optical quality is insufficient.

Our novel crystallization concept, therefore, required the design and development of a new optical instrument allowing precise static and dynamic light scattering measurements to be obtained from small droplets less than 1 μl in volume in microtiter plates [20]. The measurement of dynamic light scattering is based on the evaluation of signal fluctuations caused by particle movement, therefore a background caused by reflections from interfaces that is constant over time does not significantly influence the readings. Static light scattering measurements are more demanding because the absolute scattered light intensity must be measured in a small droplet volume in close proximity to several interfaces (glass/oil, oil/water). The curved surfaces of the droplet create reflections of the incident laser beam that are difficult to control, which is why precise measurements of static light scattering inside small droplets have appeared impossible to achieve. Unlike confocal laser scanning microscopy the static light scattering experiment does not include a frequency shift resulting from the fluorescence phenomenon that can be used to effectively suppress the excitation light with the help of interference filters. The detected scattered light has more or less the same frequency as the excitation laser, so only geometrical approaches can be used to separate scattered light from the unwanted background. The suppression of undesirable reflections and scattering from interfaces is, therefore, the key for successful static light scattering measurements in small volumes. The closer the interfaces are, the more difficult effective suppression becomes.

The solution to this technical challenge is to use an optical device that combines confocal microscopy optics with dark-field illumination. The confocal concept allows three-dimensional limitation of the probed sample volume, but unlike conventional confocal microscopy the illumination has a separate optical path to that used for light collection. An annular parabolic mirror is used to focus a laser beam with the same profile into the sample, and data are collected from the middle of this mirror by the light-detection optics. The system is summarized in Figure 10.14 and the following numbers refer to those in the figure. The laser beam (single mode fiber-coupled diode laser, 658 nm) (1) passes through a beam-shaping optical system (2) to form an

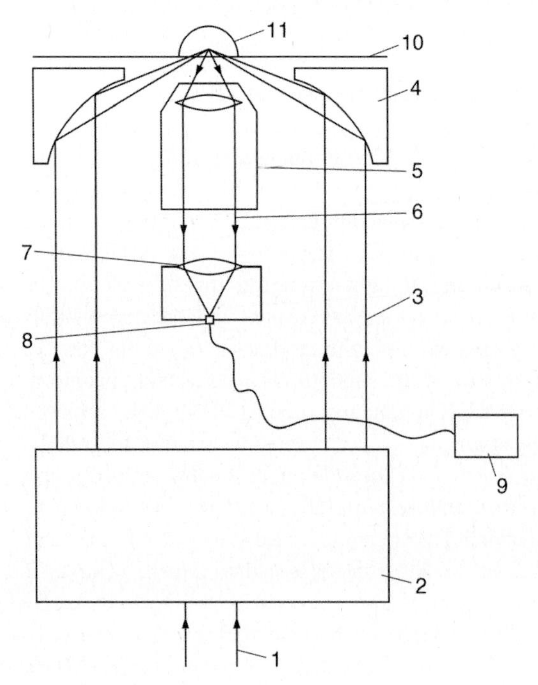

Figure 10.14 Schematic drawing of the confocal dark-field optics used for static and dynamic light scattering measurements in small sample droplets (see text for full explanation).

annular beam (3). The easiest way to generate an annular beam is to enlarge the beam diameter with a telescope and then to use an annular aperture that removes unwanted parts of the beam. The overall transmission of this concept is poor, and if maximum laser intensity is needed for the measurement, special ring-shaping optics can be used with more than 80% transmission. This can be achieved with two glass axicons or with two metallic mirrors in the form of inner and outer cones. The annular beam is then focused by the parabolic mirror (4) into the sample droplet (11), which is placed on a microtiter plate with a glass bottom (10). To prevent the small droplet from evaporating, it is protected under oil. This dark-field illumination strongly suppresses direct reflections from the air/glass and glass/water interfaces because the acceptance angle of the detection optics (microscope objective 5) is much smaller than the illumination angle. From the laser focus inside the sample droplet, light is scattered in all directions and reaches the detection optics (5). Because the focal planes of the parabolic mirror and microscope objective are precisely aligned, light from the focus forms a collimated beam (6). Reflections coming from the curved surface of the sample droplet or from scratches and impurities on the glass surface can also reach the microscope objective, but will form converging or diverging beams as their origin lies outside the objective focal plane. To form a confocal microscope, the scattered light is focused with an achromatic lens (7) onto a small confocal aperture, the front surface of a single mode fiber (8). All unwanted unfocused light collected by the microscope objective is strongly suppressed here and cannot enter the waveguide fiber, which directs the detected photons to a sensitive single-photon counting detector, for example, a photomultiplier (9). Here the integral intensity and the temporal fluctuations of the scattered light are detected and evaluated for static and dynamic measurements. We used a Hamamatsu R2949 photomultiplier with photon-counting ability for this purpose.

The twofold suppression of light reflected from the interfaces (confocal principle and dark-field illumination) generates an excellent signal-to-background ratio and enables static and dynamic light scattering measurements to be collected from

Figure 10.15 Optical apparatus for the measurement of static and dynamic light scattering in small droplets. The dark-field microscope objective with a parabolic mirror can be seen on the right, and the beam-shaping optics for the formation of the annular excitation beam is inside the housing on the left. Two single-mode fibers connect the optics to the excitation laser and detector.

proteins in droplets with a volume as low as 200 nl. The concept of this new optics was patented by the authors [20].

The compact optical apparatus used to achieve the above measurements is shown in Figure 10.15. Both the parabolic mirror and the single-mode fiber-coupling optics must be aligned accurately. Tilting the parabolic mirror with respect to the axis of the annular excitation beam distorts the focus, and maladjustment of the fiber-coupling optics results in a sharp drop-off in signal intensity because the focused scattered light must be aligned precisely onto the single-mode fiber, which has a diameter of just a few micrometers.

10.5.2
Quantitative Polarization Microscopy

The quantitative polarization microscopy module was designed similarly to the system already described by Glazer and colleagues [17], that is, an inverted, infinity-corrected microscope with a motorized rotating polarizer. A fixed offset was introduced between the optical modules for light scattering and polarization microscopy along the x-axis. Both optics are used by the automation module of the demonstrator (see below) to measure laser-light scattering and microscopic data in the same droplets (see Figure 10.16).

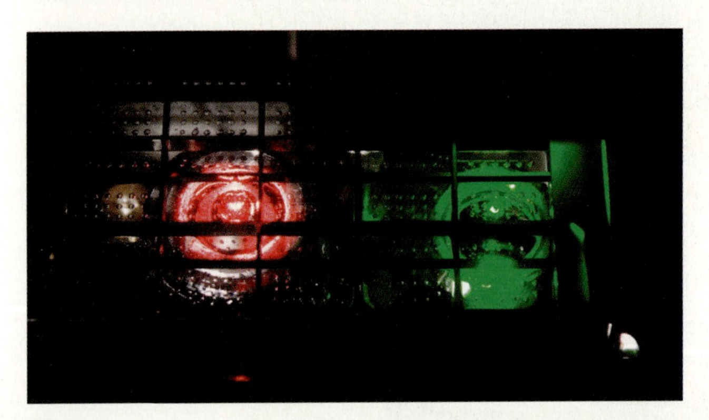

Figure 10.16 The two optical modules for the light scattering (left) and quantitative polarization microscopy (right) are shown within the demonstrator. A crystallization plate containing small droplets of protein solution is placed on top of the modules as shown.

10.5.3
System Integration, Automation and Data Processing

To evaluate the new crystallization procedure, we built a demonstrator containing a liquid-handling system for sample preparation, a light scattering and microscopy module, an x/y translational stage to position microtiter plates, a z translational stage to focus the light scattering and microscopy module, and a rotational stage for the

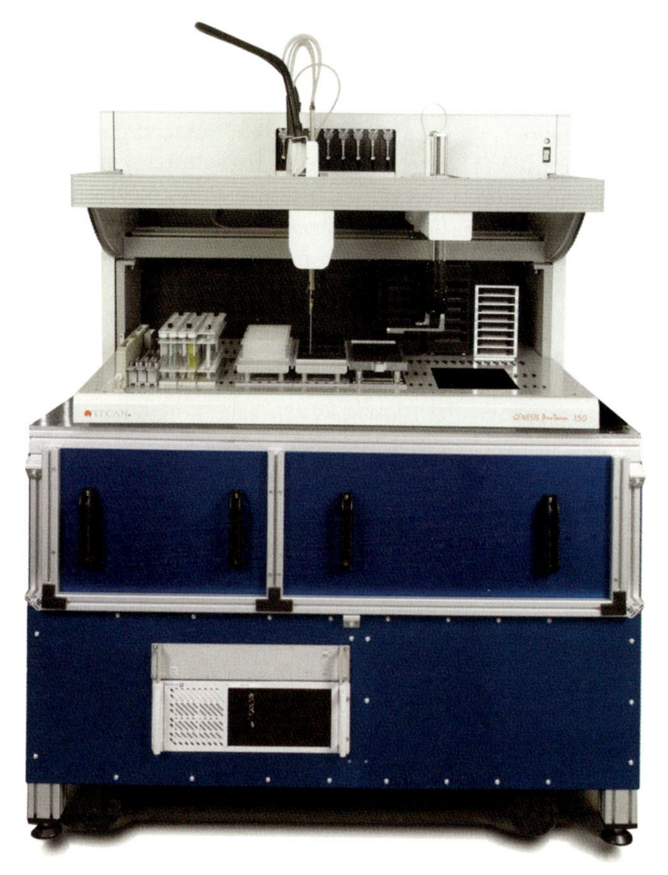

Figure 10.17 Demonstrator for the new crystallization system. A liquid-handling system for sample preparation is positioned above the optical modules inside the housing.

polarization microscopy (Figure 10.17). The system was fully automated and connected to a database application to manage the acquisition, archiving, retrieval and analysis of experimental data.

10.6 Optical Measurements

The demonstrator was used to obtain measurements from small droplets (typically 1 μl volume) in microtiter plates with a glass base. The demonstrator was set up in a temperature-controlled laboratory with 1 °C temperature stability over 24 h. In this section we report the first performance measurements for the light scattering optics, then virial coefficient determinations for different proteins with different solvent conditions. Examples are also provided for measurements recorded by dynamic light scattering and quantitative polarization microscopy.

10.6.1 Static Light Scattering

10.6.1.1 Instrument Performance

The performance of the light scattering optics was tested using solutions containing reference particles, that is, stable latex beads with a uniform diameter of 20 nm. The concentration of the particles was adjusted so that the light scattering signal from the highest bead concentration was comparable to the scattering intensity of a 50 mg ml^{-1} solution of lysozyme.

The most important feature of the light scattering optics is the signal-to-background ratio. The background comprises all the signal components from interfaces in the sample and from the instrument itself. We measured this background using a droplet of filtered water, assuming that the water itself gives no significant scattering signal. The water droplet was then replaced with droplets containing latex bead suspensions of different concentrations. The recorded scattering intensity is shown in Figure 10.18, presented as photon counts per second as a function of the relative bead concentration. The minimal signal generated by the pure water droplet confirms the effective suppression of unwanted reflections and scattering from the interfaces achieved by combining confocal optics and dark-field illumination. The signal is also near linear, allowing the calculation of a regression coefficient of $R = 0.99974$.

The measurement volume of the light scattering optics is much smaller than the diameter of the sample droplet. The size of 1 μl droplets is typically 700 μm, the focus of the detection optics is less than 5 μm in diameter. In order to determine the influence of the droplet walls on the scattering intensity, we scanned the microtiter plate moving the laser focus, and recorded the signal along a trace through the

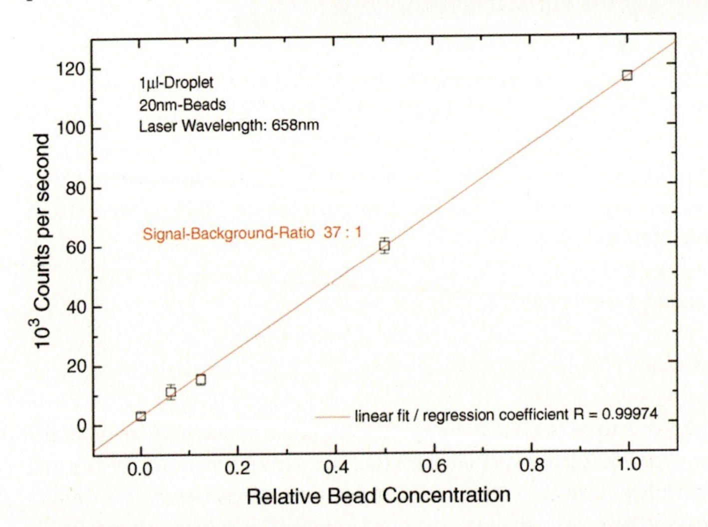

Figure 10.18 Signal-to-background ratio for measurements taken from 1-μl droplets containing latex beads (20 nm diameter). Only minor components of the signal originate from the interfaces, generating an excellent signal-to-background ratio of 37 for the highest bead concentration. The signal also remains linear over a range of relative bead concentrations.

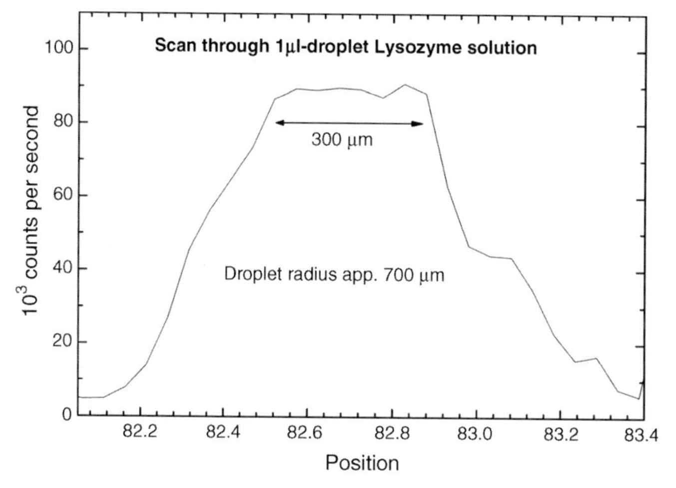

Figure 10.19 Scan through a droplet containing 1 µl of a lysozyme solution. The middle of the droplet is characterized by a constant scattering intensity which decreases as the laser focus moves closer to the wall of the droplet.

droplet. As shown in Figure 10.19 a plateau of relatively stable signal intensity forms in the center of the droplet, and the intensity only decreases when the focus approaches the droplet walls, due to shadowing effects. This behavior confirms that unwanted reflections and scattering signals from the interfaces are strongly and efficiently suppressed by our new optics.

10.6.1.2 **Measuring the Virial Coefficient of Proteins**

Two model proteins were used to test our new system, as this allowed our data to be compared to virial coefficients that have already been published. Lysozyme is a relatively small, compact protein whose crystallization behavior is well known, and it has been extensively studied with various light scattering methods. The second model, glucose isomerase, is a much larger protein than lysozyme, and belongs to a different protein family. In the case of lysozyme, we added NaCl to modify the interaction forces between dissolved molecules, changing them from repulsive to weakly attractive, as previously described [21].

Figure 10.20 shows the scattering intensity measured for the lysozyme solution, and the insert shows a three-dimensional model of lysozyme. Figure 10.21 shows the Zimm plot for the corresponding measurements. Only relative virial coefficients are plotted because no toluene reference measurements were collected. The optical constant K is defined to be one and the protein concentration is divided by the recorded scattering intensity (measured in counts per second, cps) minus the background intensity (the signal that is recorded with crystallization solution without protein, cps[0]). For automated crystallography, it is crucial to observe trends in the relative virial coefficient that correspond to changes in solution conditions. Positive and negative values can be clearly distinguished, and relative virial coefficients differ only by a multiplication factor from absolute virial coefficients.

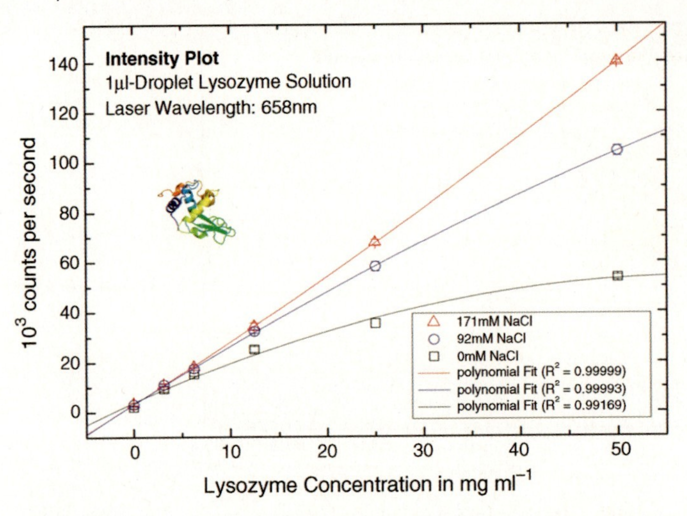

Figure 10.20 Scattering intensities observed in lysozyme solutions with three different sodium chloride concentrations. Measurements were recorded using 1-μl droplets.

The Zimm plot clearly shows the change in the interaction potential between protein molecules. The higher salt concentrations strongly reduce intermolecular repulsive forces and promote intermolecular attraction. As the solutions are clear, microscopic analysis does not yield any information about the status of the crystallization experiment.

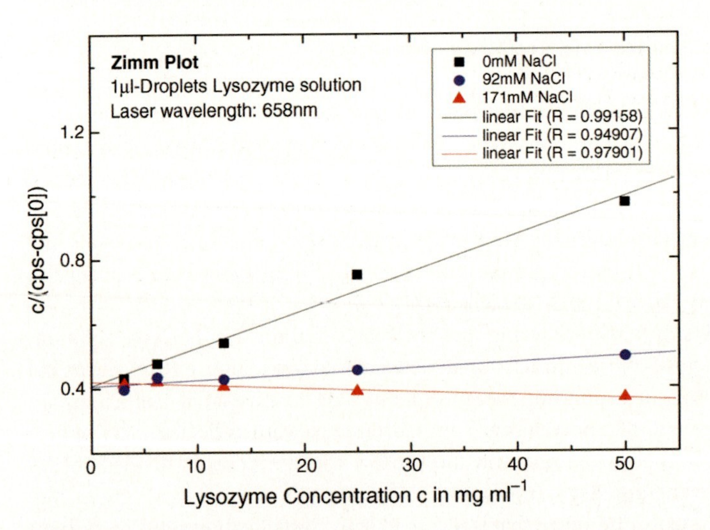

Figure 10.21 Zimm plot for lysozyme solutions under different conditions. The gradient indicates the molecular interaction potential. Positive gradients indicate repulsive forces between molecules, and negative gradients indicate attractive forces. Proteins that interact to the same degree with protein and solvent molecules (ideal solution) generate a horizontal line in the Zimm plot.

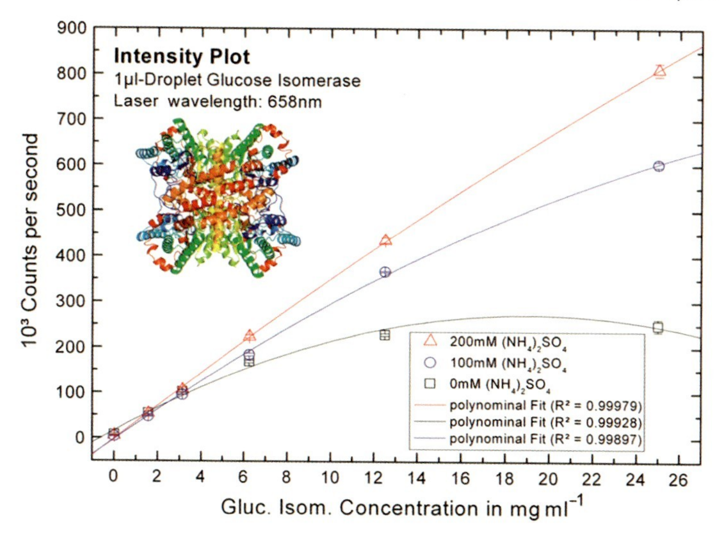

Figure 10.22 Scattering intensities observed in glucose isomerase solutions with three different ammonium sulfate concentrations. Measurements were recorded using 1-μl droplets.

Figure 10.22 shows the scattering intensity measured for the glucose isomerase solution and the corresponding Zimm plot is shown in Figure 10.23. These reveal that interactions between protein molecules are influenced by the addition of ammonium sulfate, changing repulsive forces in water to weakly attractive forces

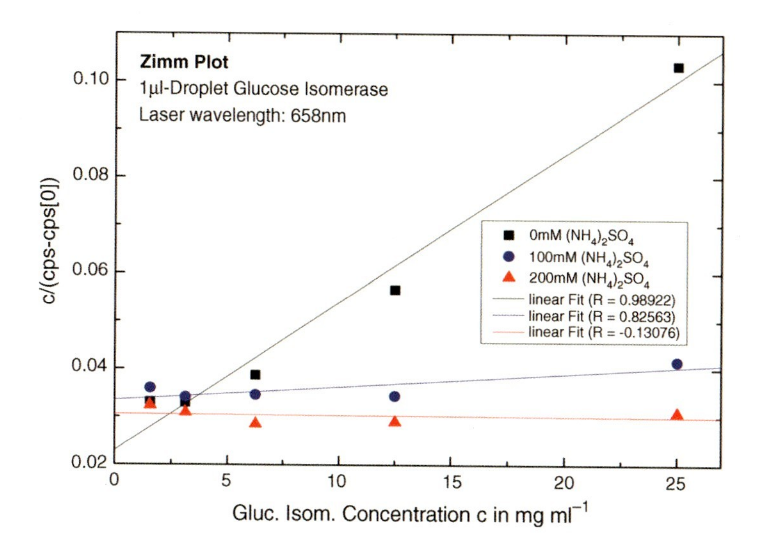

Figure 10.23 Zimm plot for glucose isomerase under different conditions. The gradient indicates the molecular interaction potential. Positive gradients indicate repulsive forces between molecules, and negative gradients indicate attractive forces. Proteins that interact to the same degree with protein and solvent molecules (ideal solution) generate a horizontal line in the Zimm plot.

in 200 mM ammonium sulfate. Glucose isomerase is larger than lysozyme, so the incident light is scattered much more strongly and the scattered light intensities are correspondingly higher.

10.6.1.3 Development of the Virial Coefficient with Different Solvent Parameters

We next set out to determine the robustness of the crystallization window as a requirement to obtain protein crystals (not to be confused with assuming the crystallization window in any way guarantees the formation of protein crystals). For this purpose, we used *Tritirachium album* proteinase K as a model because it forms crystals in the presence of many different precipitant formulations, hence the sparse matrix reagent kits that can be purchased from different providers. The virial coefficients resulting from such productive crystallization conditions should indicate small negative values for the second virial coefficient. The outcome of this experiment clearly validated the hypothesis, since all crystallization conditions yielded a negative virial coefficient even though some precipitants with negative virial coefficients were unable to generate protein crystals (Figure 10.24).

In order to test the validity of our approach in a screening process, we constructed a two-dimensional grid screen and used glucose isomerase as a reference protein. We screened the pH of the solution and the ammonium sulfate concentration, producing

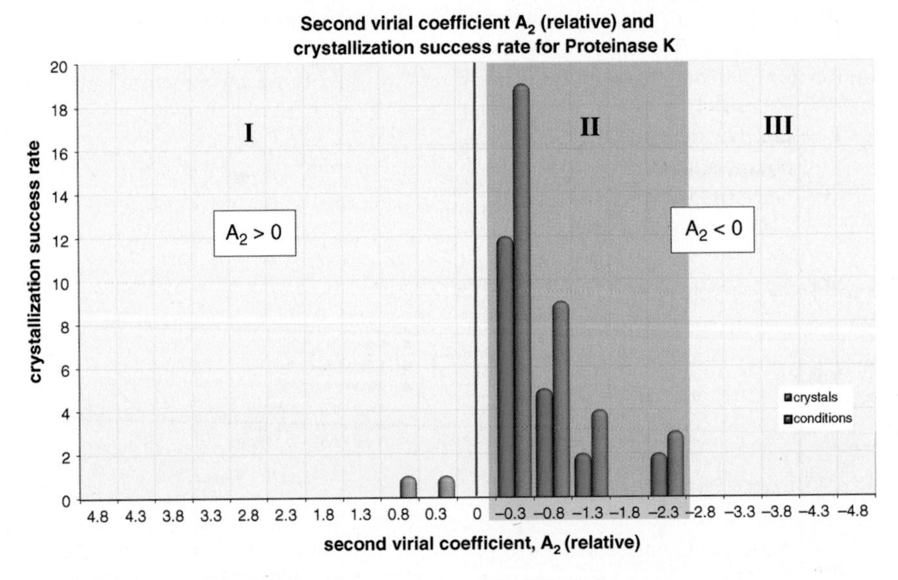

Figure 10.24 Crystallization success rate for *Tritirachium album* proteinase K as a function of the relative second virial coefficient (A_2). We tested 42 different crystallization conditions (Hampton Research, Crystal Screen I). In 37 cases, A_2 was determined via SLS. Each of the 21 experiments that produced crystals had a negative A_2 value. Furthermore, 14 experiments yielded a negative virial coefficient without forming any crystals or precipitates. In 8 of these 14 experiments, crystals were obtained by increasing the precipitant concentration. Two conditions yielded a positive virial coefficient and neither crystals nor a precipitate were formed.

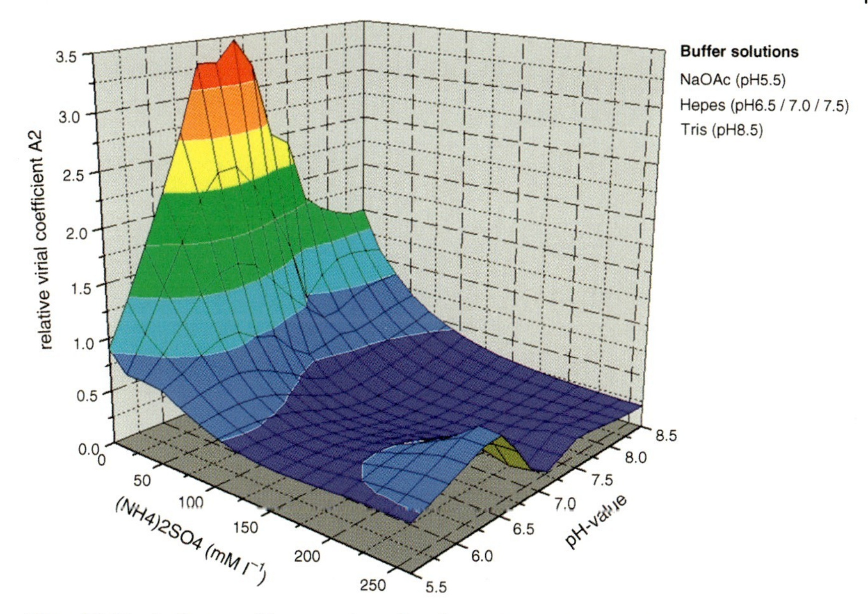

Figure 10.25 Evaluation of the second virial coefficient A_2 as a function of the ammonium sulfate concentration and pH of the precipitant solution.

a three-dimensional representation of the virial coefficients as a function of both parameters. On this surface, we observed a clear tendency towards a productive crystallization solution, as represented by the valley toward a negative virial coefficient (neutral pH at higher ammonium sulfate concentrations) (Figure 10.25). This indicates that a systematic characterization of solution parameters using the second virial coefficient to optimize molecular interactions can help to identify the most productive crystallization conditions objectively. Furthermore, this method can also be used to optimize solution parameters that favor repulsion between protein molecules, for example, in applications such as the formulation of biopharmaceuticals to prevent aggregation/precipitation and increase shelf-life.

10.6.2
Dynamic Light Scattering

We used the demonstrator to carry out a proof-of-concept analysis on the size distributions of bead particles, proteins and protein aggregates. To cover the typical size range for proteins, we measured the light scattered from a small protein (lysozyme) and from 20-nm latex beads. Figures 10.26 and 10.27 show the autocorrelation function and the size distribution for beads with a nominal diameter of 20 nm, and these match the expected values. Figures 10.28 and 10.29 show the autocorrelation function and the corresponding size distribution for lysozyme, and this indicates a hydrodynamic radius of 1.3 nm. A commercial instrument using a

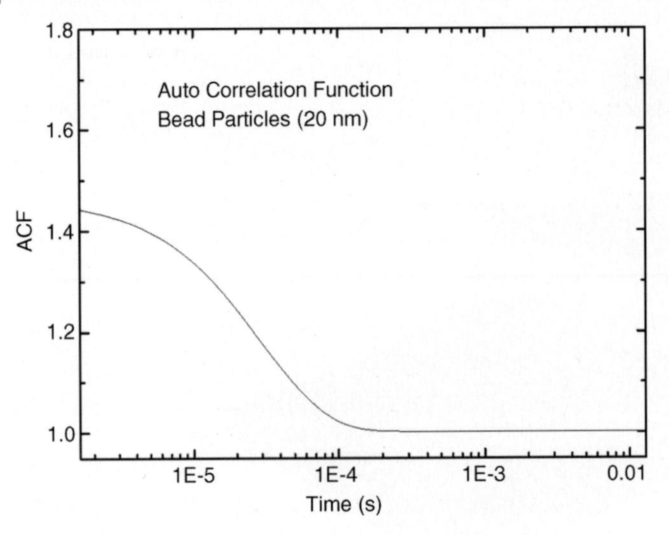

Figure 10.26 Autocorrelation function for bead particles 20 nm in diameter, measured in a 1-μl droplet with the new confocal dark-field optics.

fixed back-scattering angle of 173° indicates a hydrodynamic radius of approximately 1.89 nm [22]. The deviation between these measurements reflects the different light collection geometries of the instruments.

In crystallization experiments, it is particularly important to detect changes in particle size distributions. We used bovine serum albumin (BSA) to verify the ability

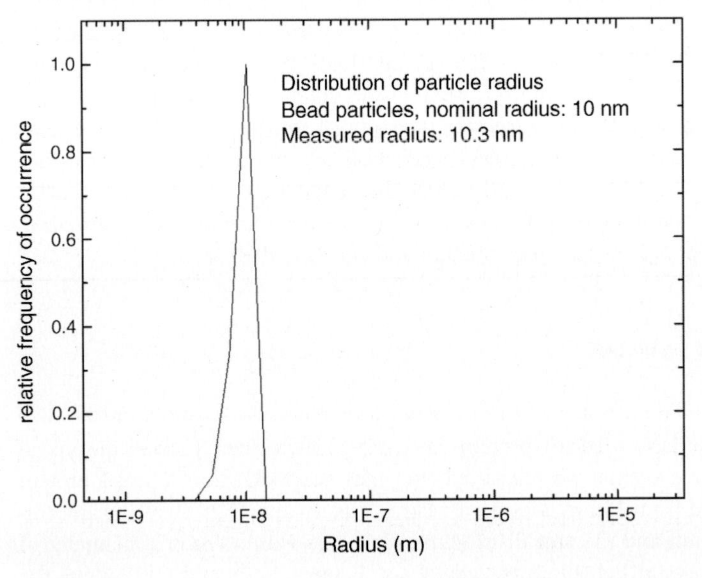

Figure 10.27 Size distribution for bead particles 20 nm in diameter calculated from the autocorrelation function in Figure 10.26, indicating a good agreement between nominal and measured size.

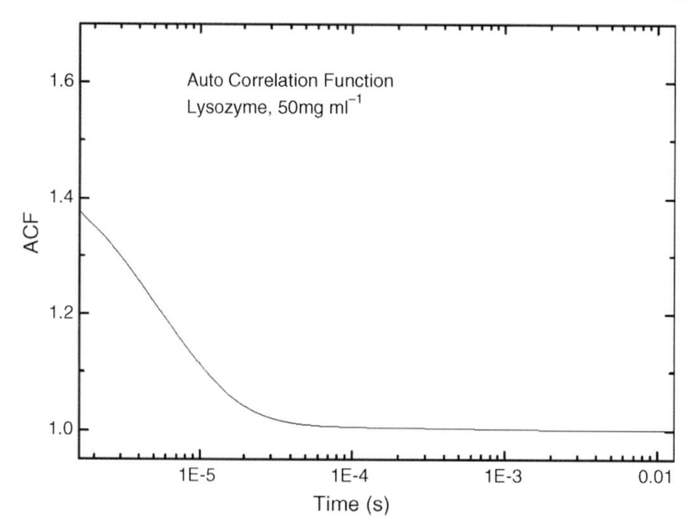

Figure 10.28 Autocorrelation function for lysozyme, measured in a 1-μl droplet with the new confocal dark-field optics.

of our new system to detect such changes because this protein undergoes a conformational change when heated to 70 °C. Heat-denatured BSA molecules aggregate to form larger units, and can therefore be used as a model for aggregation processes in protein solutions. Dynamic light scattering measurements were therefore carried out using a BSA sample before and after heating to 70 °C (Figure 10.30).

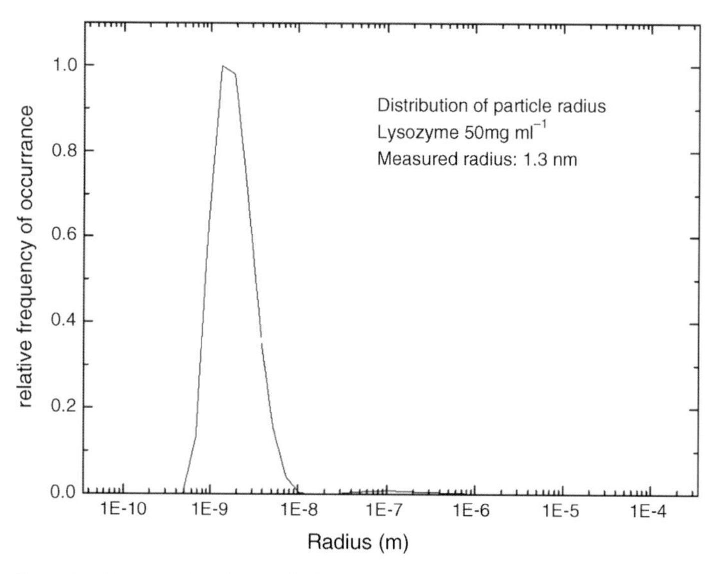

Figure 10.29 Size distribution for lysozyme calculated from the autocorrelation function in Figure 10.28, indicating a good agreement between nominal and measured size.

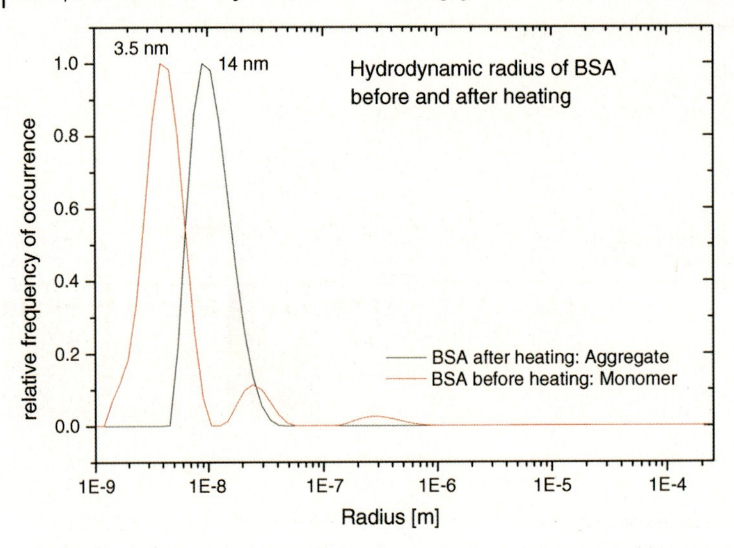

Figure 10.30 Size distribution of a BSA sample before and after heating to 70 °C. The shift in size distribution due to aggregation can be detected clearly.

The aggregation process can be followed clearly using DLS measurements and our results agree well with those already published [23].

10.6.3
Quantitative Polarization Microscopy

The optical properties of the novel crystallization plate provide the means to analyze crystallization experiments by quantitative polarization microscopy. As shown by Echalier *et al.* [18], microcrystals can be identified earlier under more productive crystallization conditions, allowing the efficient screening of crystallization environments to identify productive conditions rapidly. Furthermore, the ability of quantitative polarization microscopy to discriminate between amorphous and microcrystalline precipitates can, for the first time, be introduced into an automated crystallization screening platform.

Quantitative polarization microscopy data are presented as a series of three images, representing the birefringence properties of the precipitates or crystals within the crystallization droplet, namely the transmittance I_0, and the amplitude $|\sin \delta|$ and the orientation of the birefringence φ (Figure 10.31). The orientation of the birefringence, in particular, allows the identification of microcrystals in situations where all known classical methods would fail and even experienced crystallographers would not be able to identify suitable crystals.

Figure 10.32 shows a variety of possible outcomes often observed in classic protein crystallization experiments. The first series of images shows situations that can be difficult to interpret. Phase separation might be suspected in Figure 10.32a because of the droplet-like appearance of a second phase within the crystallization droplet, as seen in the transmittance image. However, the orientation image clearly indicates

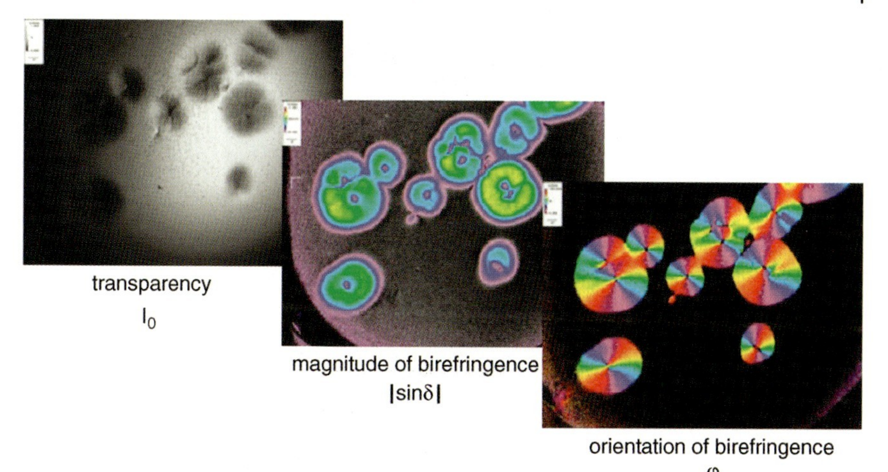

Figure 10.31 Analysis of a classical crystallization experiment, yielding so-called "sea-urchins" (thin needles clustered around a single nucleation site). The results are presented as a calculated transmission image (I_0), an amplitude image ($|\sin\delta|$) of the birefringence, and an orientation image (φ) of the birefringence. Sea-urchins are obtained when a central nucleation event is followed by the growth of a multitude of crystal-needles in all directions in space, as deduced by the centrosymmetric orientation of the microcrystals in the orientation image (φ) of the birefringence.

the presence of microcrystals, even identifiable as "sea-urchins", as already shown in Figure 10.31.

Quantitative polarization microscopy can also be used to identify well-ordered crystals, and to evaluate the impact of potential heterogeneity on crystal diffraction qualities. Figure 10.33 shows how the orientation image (φ) can be used to identify disordered regions of larger monocrystals that might interfere with diffraction analysis. In both panels, the transmission image highlights a potential region of disorder. In panel (a) the disordered region is also revealed in the orientation image, and several crystal lattices are thus shown in the diffraction image. In contrast, the potentially disordered region in panel (b) is not shown in the orientation image (φ). This indicates there is no disturbance of the anisotropic properties of the crystal and the diffraction image clearly shows the presence of a unique crystal lattice.

10.7 Outlook – An Iterative Optimization Process Based on Optical Measurements

To rationally determine crystallization conditions in the future, we developed an integrated and automated crystallization environment, based on a new crystallization method, a new crystallization plate, and new detection systems, allowing for the first time the application of objective measurements like static and dynamic light

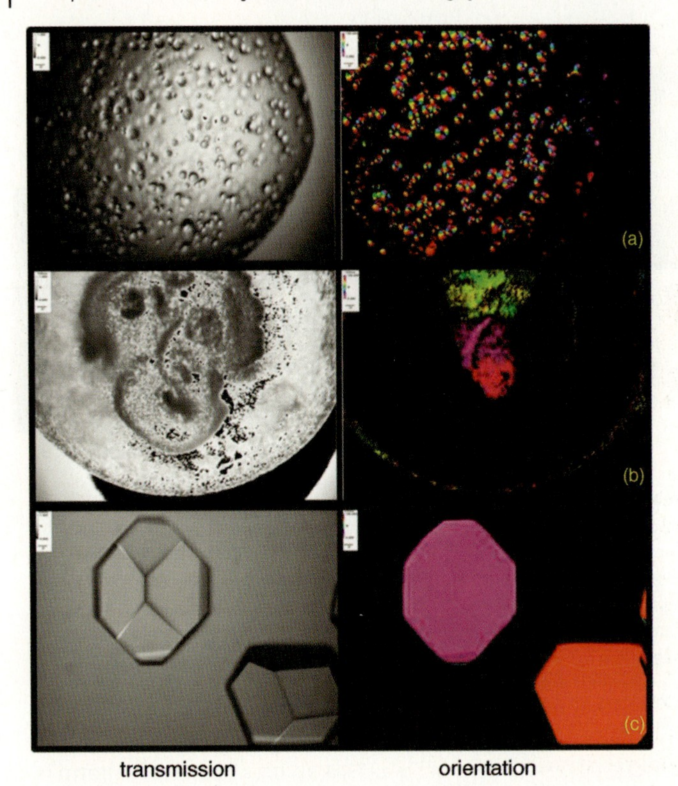

Figure 10.32 Quantitative polarization microscopy data from different crystallization experiments. Each row shows a different case (a, b, c), left side the transmittance image (I_0), right side the orientation image (φ):
(a) discriminating between sea-urchins and phase separation becomes possible, since phase separation would show no birefringence; (b) discriminating between amorphous precipitate (dark region in orientation image) and microcrystals (colored regions in orientation image); here both within the same drop, resulting from fast and slow mixing zones; (c) large and well-ordered monocrystals.

scattering, as well as quantitative polarization microscopy within each single crystallization droplet. The optical properties of our crystallization plate and the triple-phase system will allow us to further integrate valuable techniques, like the autofluorescence-based discrimination between protein and salt crystals using UV-light [24].

The implementation of sound, well-documented and unbiased information derived from thermodynamic solution parameters, such as the second virial coefficient, the detection of aggregates and nucleation sites using dynamic light scattering, the early and objective identification of microcrystalline precipitates, and the qualitative analysis of protein crystals using quantitative polarization microscopy, together with the potential to dynamically intervene within the crystallization process

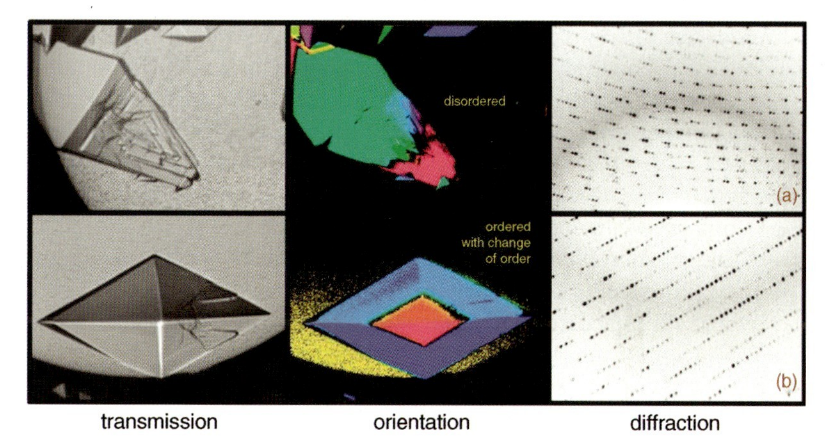

Figure 10.33 (a) transmission and orientation image of a disordered crystal, as predicted by the orientation image and confirmed by the diffraction data on the right side;
(b) transmission and orientation image of an ordered crystal as predicted by the orientation image (despite apparent defects in the transmission image) and confirmed by the diffraction data. The two colors in the orientation image reflect a change in order due to the thickness of the crystal, resulting in a change of $\pi/2$ in the value of the orientation.

using classic liquid handling systems will cause a paradigm shift in the way protein crystallization is approached.

In the future, a fully automated crystallization system might become reality, which has the following unique features:

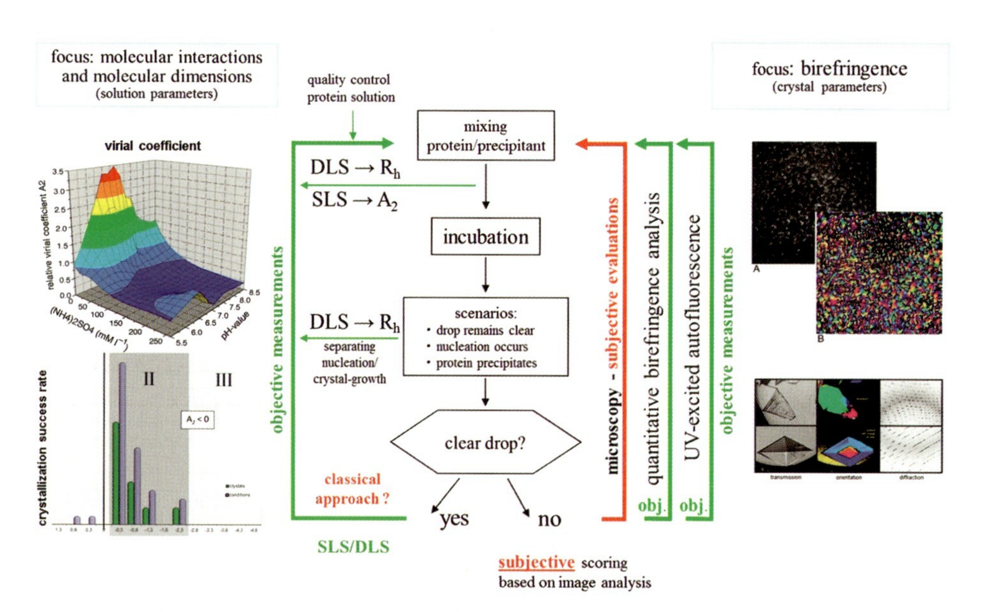

Figure 10.34 The concept for an iterative approach to the optimization of crystallization conditions based upon objective optical measurements.

- An automated iterative process based on cycles of static light scattering experiments to iteratively and automatically prepare precipitant conditions, placing all crystallization experiments within the necessary "*crystallization window*", avoiding non-productive crystallization conditions with positive virial coefficients.
- Quantitative polarization microscopy is used for early detection of microcrystals from unproductive amorphous precipitates.
- Dynamic laser light scattering is used to detect polydisperse protein solutions in order to optimize their monodispersity.
- Nucleation is uncoupled from crystal growth to produce larger crystals [14] by the control of vapor diffusion within the triple-phase system.
- Large single crystals are directly analyzed regarding their diffraction quality using quantitative polarization microscopy.

Figure 10.34 summarizes this new rational approach for a crystallization procedure based on objective optical measurements. The newly developed optical light scattering and polarization microscopy techniques described in this chapter form the basis of such an approach.

The research presented here was sponsored by the German Ministry of Economic Affairs and Employment.

References

1 McPherson, A. (1999) *Crystallization of Biological Macromolecules,* Cold Spring Harbor Press, New York.

2 Chayen, N.E. (1999) Crystallization with oils: a new dimension in macromolecular crystal growth. *J. Cryst. Growth,* **196** (2–4), 434–441.

3 Chayen, N.E. and Saridakis, E. (2002) Protein crystallization for genomics: towards high-throughput optimization techniques. *Acta Crystallogr. D,* **58** (2), 921–927.

4 Hoffmann, K. (2002) Protein crystallization method, patents WO2004012841, AU2003266957, CA2493335, EP2003747867, JP2005506071, KR1020057001501, US10522690.

5 Neal, B.L., Asthagiri, D., Velev, O.D., Lenho, A.M., and Kaler, E.W. (1999) Why is the osmotic second virial coefficient related to protein crystallization? *J. Cryst. Growth,* **196** (2–4), 377–387.

6 Wyatt, P.J. (1993) Light scattering and the absolute characterization of macromolecules. *Anal. Chim. Acta,* **272** (1), 1–40.

7 George, A. and Wilson, W.W. (1994) Predicting protein crystallization from a dilute-solution property. *Acta Crystallogr. D,* **50** (4), 361–365.

8 Neal, B.L., Asthagiri, O.D., Velev, O.D., Lenhoff, A.M., and Kaler, E.W. (1999) Why is the osmotic second virial coefficient related to protein crystallization? *J. Cryst. Growth,* **196** (2–4), 377–387.

9 Liu, T. and Chu, B. (2002) Light scattering by proteins. in *Encyclopedia of Surface and Colloid Science,* (ed. Arthur Hubbard, Marcel Dekker, New York), 3023–3043.

10 Ruppert, S., Sandler, S.I., and Lenhoff, A.M. (2001) Correlation of the osmotic second virial coefficient and the solubility of proteins. *Biotechnol. Prog.,* **17** (1), 182–187.

11 Tessier, P.M. and Lenhoff, A.M. (2003) Measurements of protein self-association as a guide to crystallization. *Curr. Opin. Biotech.,* **14** (5), 512–516.

12 CONTIN is a general-purpose constrained regularization method for continuous

distributions, such as inverse Laplace transforms in relaxation studies and in dynamic light scattering. The program can be downloaded from this site http://s-provencher.com/index.shtml.
13 Spectroscatter 201, RiNA GmbH.
14 Saridakis, E. and Chayen, N.E. (2000) Improving protein crystal quality by decoupling nucleation and growth in vapor diffusion. *Protein Sci.*, **9** (4), 755–757.
15 Chayen, N., Dieckmann, M., Dierks, K., and Fromme, P. (2004) Size and Shape determination of Proteins in Solution by a noninvasive depolarized dynamic light scattering instrument. *Ann. N.Y. Acad. Sci.*, **1027**, 20–27.
16 Lomakin, A., Teplow, D.B., and Benedek, G.B. (2005) Quasielastic light scattering for protein assembly studies, methods in molecular biology, in *Amyloid Proteins: Methods and Protocols*, vol. **299** (ed. E.M. Sigurdsson), Humana Press Inc., Totowa, NJ, pp. 153–174.
17 Geday, M.A., Kaminsky, W., Lewis, J.G., and Glazer, A.M. (2000) I mages of absolute retardance L. Deltan, using the rotating polariser method. *J. Microsc.*, **198** (1), 1–9.
18 Echalier, A., Glazer, R.L., Fulop, V., and Geday, M.A. (2004) Assessing crystallization droplets using birefringence. *Acta. Crystallogr.D*, **60** (4), 696–702.
19 Owen, R.L. and Garman, E. (2005) A new method for predetermining the diffraction quality of protein crystals: using SOAP as a selection tool. *Acta Crystallogr.D*, **61** (2), 130–140.
20 Hoffmann, K., Janzen, C., Noll, R., and Uhl, W. (2007) Device and method for measuring static and dynamic scattered light in small volumes. patents WO2009003714, EP 2008784613, JP2010513791, US12667601, DE 102007031244B3.
21 Muschol, M. and Rosenberger, F. (1995) Interactions in undersaturated and supersaturated lysozyme solutions: Static and dynamic light scattering results. *J. Chem. Phys.*, **103** (24), 10424–10432.
22 Parmar, A.S. and Muschol, M. (2009) Hydration and hydrodynamic interactions of lysozyme: effects of chaotropic versus kosmotropic ions. *Biophys. J.*, **97** (2), 590–598.
23 Bulone, D., Martonaraa, V., and San Biagio, P.L. (2001) Effects of intermediates on aggregation of native bovine serum albumine. *Biophys. Chem.*, **91** (1), 61–69.
24 Judge, R.A., Swift, K., and Gonzalez, C. (2005) An ultraviolet fluorescence-based method for identifying and distinguishing protein crystals. *Acta Crystallogr.D*, **61** (Pt 1), 60–66.

Index

Handbook of Biophotonics. Vol.3: Photonics in Pharmaceutics, Bioanalysis and Environmental Research, First Edition.
Edited by Jürgen Popp, Valery V. Tuchin, Arthur Chiou, and Stefan Heinemann

d

m

q

r

t